Lecture Notes in Nanoscale Science and Technology

More information about this series at http://www.springer.com/series/7544

Peng Yu • Zhiming M. Wang
Editors

Quantum Dot Optoelectronic Devices

Springer

Editors
Peng Yu
University of Electronic Science
and Technology of China
Chengdu, Sichuan, China

Zhiming M. Wang
University of Electronic Science
and Technology of China
Chengdu, Sichuan, China

ISSN 2195-2159 ISSN 2195-2167 (electronic)
Lecture Notes in Nanoscale Science and Technology
ISBN 978-3-030-35812-9 ISBN 978-3-030-35813-6 (eBook)
https://doi.org/10.1007/978-3-030-35813-6

This Springer imprint is published by the registered company Springer Nature Switzerland AG
The registered company address is: Gewerbestrasse 11, 6330 Cham, Switzerland

Preface

Nanoscale semiconductor devices have been extensively studied as next-generation alternatives to conventional devices owing to the achievable high integration and functionality. Semiconductor quantum dot (SQD) is a quasi-zero-dimensional structure composed of a small number of atoms. The carriers are three-dimensionally confined and the exciton Bohr radius is highly crowded, allowing quantum dots to exhibit unique and unusual physical, optical, and electronic properties that are absent in larger samples of a semiconductor material. Due to the quantum confinement effect, the energy level of quantum dots is similar to that of atoms with discontinuous energy level structure. The quantum size effect, quantum tunneling effect, Coulomb blocking effect, quantum interference effect, multi-body correlation, nonlinear optical effect, *etc.*, demonstrated by SQDs are of immense significance to the study of basic optical and electrical properties of quantum-confined structures and novel optoelectronic devices [1, 2].

SQD-based novel optoelectronic devices hold promise for current and future information technology applications. Research on quantum dot optoelectronics has extensively evolved over the past three decades alongside the ever-growing technological demands, with huge advancements in solar cell energy conversion [3], laser [4], single photon source [5], photodetector [6], and LED [7]. The goal of this book is to present a comprehensive overview of the current state of the art in quantum dot-based optoelectronics covering a broad range of applications and related technologies.

The main body of the book comprises contributions that focus on the quantum dot and quantum dot-based optoelectronics, including quantum dot-based solar cells, single photon emitters, LEDs, photodetectors, nanolasers, memristor, quantum bit applications, quantum communication, and nonlinear optical applications. Specifically, Chap. 1 provides a comprehensive perspective on quantum dot-based thin film III–V solar cells. Unlike solar cells using III–V and Si materials, solution-processed colloidal quantum dots can overcome the Shockley–Queisser limit through multiple exciton generation. Chapter 2 offers a detailed introduction to chemically synthetized colloidal quantum dots for highly efficient solar cells. Quantum dots are widely considered as the best candidate for quantum emitters due to their advantages on the controllability, purity, brightness, indistinguishability, and coherence [5]. Chapter 3 presents a general review of the development of electrically and optically driven quantum emitters based on SQDs. While the quantum dots, as mentioned above, are obtained by epitaxy and chemical methods, Chap. 4 introduces gate-defined quantum dots obtained by etching and their applications on quantum computing via electrical pumping. In Chap. 5, a progress report on quantum communication with quantum dot-based devices is discussed. This chapter is particularly important for optoelectronic integration. Light-matter interaction is critical for optoelectronics based on quantum dots. In Chap. 6, a comprehensive introduction to cesium lead halide perovskite quantum dots is provided. Chapter 7 deals with recent developments of InAs/GaAs quantum dots and their applications to high-speed and ultrafast lasers. Chapter 8 introduces studies of bioresource derived graphene quantum dots and their applications on optoelectronic and biological applications. Chapter 9 presents influences of quantum confinement effect on the optical and electronic properties of quantum dots and their application for efficient memristor design and performance.

The editors thank all the contributors of this book for their remarkable chapters. We owe many thanks to Dr. David Packer, executive editor at Springer, for supporting this book and Mr. Nirmal Selvaraj, production editor at Springer, for ably managing the production process. Last but not least, we appreciate Mr. Hezhuang Liu, who provided indispensable editorial assistance and support. The editors acknowledge the financial support of UESTC Shared Research Facilities of Electromagnetic Wave and Matter Interaction.

Chengdu, People's Republic of China Peng Yu
Zhiming M. Wang

References

1. Coe-Sullivan, S. (2009). Quantum dot developments. *Nature Photonics, 3*, 315–316.
2. Wang, Z. M. (2007). *Self-assembled quantum dots* (Vol. 1). New York: Springer Science & Business Media.
3. Wu, J., & Wang, Z. M. (2014) *Quantum dot solar cells*. Berlin: Springer.
4. Ellis, B., et al. (2011). Ultralow-threshold electrically pumped quantum-dot photonic-crystal nanocavity laser. *Nature Photonics, 5*, 297.
5. Michler, P., et al. (2000). A quantum dot single-photon turnstile device. *Science, 290*, 2282–2285.
6. Konstantatos, G., et al. (2006). Ultrasensitive solution-cast quantum dot photodetectors. *Nature, 442*, 180.
7. Jang, E., et al. (2010). White-light-emitting diodes with quantum dot color converters for display backlights. *Advanced Materials, 22*, 3076–3080.

Contents

Chapter 1
Quantum Dot-Based Thin-Film III–V Solar Cells

F. Cappelluti, A. Tukiainen, T. Aho, F. Elsehrawy, N. Gruginskie, M. van Eerden, G. Bissels, A. Tibaldi, G. J. Bauhuis, P. Mulder, A. Khalili, E. Vlieg, J. J. Schermer, and M. Guina

Abstract In this work, we report our recent results in the development of thin-film III–V solar cells fabricated by epitaxial lift-off (ELO) combining quantum dots (QD) and light management structures. Possible paths to overcome two of the most relevant issues posed by quantum dot solar cells (QDSC), namely, the degradation of open circuit voltage and the weak photon harvesting by QDs, are evaluated both theoretically and experimentally. High open circuit voltage QDSCs grown by molecular beam epitaxy are demonstrated, both in wafer-based and ELO thin-film configuration. This paves the way to the implementation in the genuine thin-film structure of advanced photon management approaches to enhance the QD photocurrent and to further optimize the photovoltage. We show that the use of light trapping is essential to attain high-efficiency QDSCs. Based on transport and rigorous electromagnetic simulations, we derive design guidelines towards light-trapping enhanced thin-film QDSCs with efficiency higher than 28% under unconcentrated light, ambient temperature. If photon recycling can be fully exploited, 30% efficiency is deemed to be feasible. Towards this goal, results on the development and integration of optimized planar and micro-patterned mirrors, diffractive gratings and broadband antireflection coatings are presented.

Keywords Solar cell · III–V semiconductors · Thin-film · Epitaxial lift-off · Quantum dot · Light trapping

---

F. Cappelluti (✉) · F. Elsehrawy · A. Tibaldi · A. Khalili
Department of Electronics and Telecommunications, Politecnico di Torino, Torino, Italy
e-mail: federica.cappelluti@polito.it

A. Tukiainen · T. Aho · M. Guina
Optoelectronics Research Centre, Physics Unit, Faculty of Engineering and Natural Sciences, Tampere University, Tampere, Finland

N. Gruginskie · M. van Eerden · G. J. Bauhuis · P. Mulder · E. Vlieg · J. J. Schermer
Institute for Molecules and Materials, Radboud University, Nijmegen, The Netherlands

G. Bissels
tf2 devices B.V., Nijmegen, The Netherlands

P. Yu, Z. M. Wang (eds.), *Quantum Dot Optoelectronic Devices*,
Lecture Notes in Nanoscale Science and Technology 27,
https://doi.org/10.1007/978-3-030-35813-6_1

1.1 Introduction

III–V semiconductors enable the highest efficiency solar cells demonstrated to date owing to their excellent optical and transport properties and a very mature technology. Single-junction GaAs solar cells hit a record efficiency of 29.1% [1], very close to the Shockley-Queisser (SQ) limit [2] of 33.5% under the terrestrial AM1.5G sun spectrum. The SQ limit sets the upper efficiency attainable by a single-junction solar cell only based on the energy bandgap of the semiconductor material and the incident spectrum. The bandgap limits in fact at the same time the spectral sensitivity of the cell—i.e. the minimum energy of photons that can be converted into free charge carriers— and the maximum energy with which those carriers can be extracted from the cell, the excess energy being lost through carriers' thermalization processes. Figure 1.1 shows the calculated SQ limit of single-gap solar cells for representative air-mass spectra in space (AM0), terrestrial (AM1.5G), and concentration photovoltaics (AM1.5D). One concept—the most successful to date—to improve the efficiency beyond the SQ limit relies on stacking different semiconductor junctions which selectively absorb a fraction of the incoming sunlight, extending the harvested spectrum range and keeping the thermalization losses low [4]. The resulting multi-junction solar cell has a maximum theoretical efficiency of 67% under the AM1.5G spectrum [5]. The most common implementation relies on a two-terminal monolithic device wherein the subcells are series-connected through tunnel diodes. Thus, the current through the solar cell is limited by the subcell producing the lowest current, and the key requirement is to divide the sun spectrum absorption among semiconductors with the right bandgap—to provide similar photocurrent—and that can be epitaxially grown together with high quality [6, 7]. Due to the richness of III–V materials in terms of bandgaps well matched to the sun

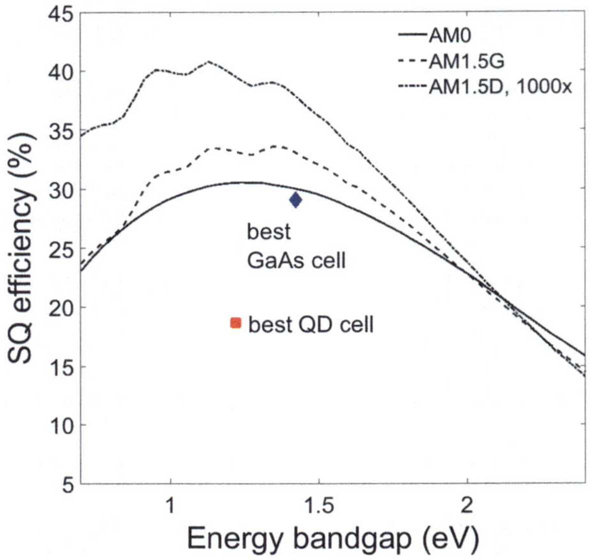

Fig. 1.1 Shockley-Queisser limiting efficiency curves for single-junction solar cells under AM0, AM1.5G and 1000 × AM1.5D spectra. For the sake of reference, also the record GaAs [1] and QD solar cell [3] are shown

spectrum and the relative ease to combine them, the conversion efficiency of multi-junction III–V solar cells has been increasing continuously for decades. Lattice-matched, metamorphic and wafer-bonded approaches have been developed [8]. The present record is over 46% [1, 9], and there are ample opportunities for further improvements.

While the present terrestrial photovoltaic industry is dominated by silicon cells, III–Vs are the predominant technology for space and concentrator cells. For spacecraft applications, III–V solar cells have an unchallenged position owing to their high power-to-weight ratio and resistance against extreme conditions such as electron, proton and harsh UV radiation as well as cyclic temperature changes between plus and minus 100 °C. In spacecraft applications (i.e. under AM0), the industry's standard is provided by triple-junction III–V cells with a typical efficiency of 30%. Since they can handle large thermal loads and feature high open circuit voltages (V_{oc}), III–V solar cells are also highly suited for concentrator photovoltaic (CPV) systems. Such systems use cost-efficient concentrating optics to collect the sunlight incident over a large area and focus this on a solar cell with a much smaller surface area to reduce the cost per unit output power. Typical concentration ratios are 500 to 1000 [10]. Commercially available multiple-junction III–V solar cells dedicated for CPV applications reach efficiencies up to 40% at these concentration ratios. With such high-efficiency cells, CPV modules can have a lower levelized cost of electricity (LCOE) than regular silicon solar modules in areas with high direct normal irradiation (DNI) levels [11].

Worldwide research in III–V solar cells aims today at increasing their power/cost ratio by improving the output power and reducing the production costs. In fact, the high performance of III–Vs comes at the price of a high cost, mainly dominated by the epitaxial growth and the wafer. In particular, multiple-junction cells are the most complex to manufacture and inherently the most expensive, since they involve thick, complex epitaxial layers and, in some approaches, the use of multiple wafers. On the other hand, III–V semiconductors are generally optically thick, meaning that a few micrometres of active material are sufficient to harvest all the incident light. Therefore, a drastic cost reduction is possible if the epitaxial layers can be separated from their wafer, which only serves during the growth process, and the wafer can be reused several times. Thin-film III–V technologies based on epitaxial lift-off (ELO) [12–14] are among the most promising approaches to attain III–V cells with high efficiency/cost ratio. Besides the remarkable reduction of cost, the thin-film architecture enables the integration of mirrors to implement photon trapping and recycling mechanisms that are essential to reach the highest efficiency [15–17]. As a matter of fact, the world record single-junction cell is an ELO thin-film GaAs cell [1, 14]. Besides the development of a useful wafer separation and reuse technology, the reduction of chemical vapour deposition costs—by increasing the cell deposition rate and reducing the cell thickness—and the development of alternative cell deposition technologies are key research areas towards the development of low-cost III–V photovoltaics [18]. In this regard, light trapping plays again an important role because enabling electrically thin layer (with respect to the diffusion length) may relax the requirements in terms of crystal quality. The improvement of power output of multiple-junction cells encompasses (a) optimization of the efficiency of each individual subcell; (b) development of alternative subcells with a better response to the dominant illumination condition; (c) adding more junctions (subcells) to the device. With respect to (b), a point of attention for series-connected multi-junction cells is related to the large spectral sensitivity to changes of the solar spectrum that impacts their annual energy production capacity [7, 19–21]. Therefore, alternative architectures such as multiterminal devices with independently connected subcells are receiving renewed attention [22, 23].

Nanostructured absorbers such as quantum well (QW) and quantum dot (QD) materials offer a complementary path to further developing III–V photovoltaics. With respect to the research directions (a)–(c), bandgap engineering in QWs [24] and QDs [25] is an attractive solution for the spectral tuning and optimization of multi-junction cells [26, 27]. Moreover, single-junction nanostructured solar cells can match the optimum bandgap at high concentration—around 1.2 eV at 1000 sun (see Fig. 1.1)—and achieve 40% efficiency with a simpler and possibly less-expensive technology and more stable performance under spectral fluctuations [27, 28]. In space applications, QDs may bring additional benefits such as enhanced radiation tolerance [26]. Finally, owing to their tunable optoelectronic characteristics, quantum-confined materials are among the most promising material platforms for introducing next-generation photovoltaic concepts. In particular, QD solar cells (QDSCs) have been extensively investigated both theoretically and experimentally for the implementation of the intermediate band solar cell (IBSC) [29, 30].

As well-known, QDs introduce 3D potential wells within the host semiconductor, confining carriers in energy states within the host semiconductor forbidden energy band. Under sun illumination, charge carriers are photogenerated in the extended and bound states through interband and intraband optical absorption. Depending on the effectiveness of thermalization of photogenerated carriers from the extended to the bound energy states, two different situations arise in the photovoltaic process:

- If intraband carrier thermalization is slow with respect to radiative recombination (and conversely, thermal escape is slow with respect to photon-assisted escape), a non-equilibrium condition exists between the continuum and the confined states sustained by a two-step photon absorption process [29]. The result is an enhanced short-circuit current that does more than compensate for the small reduction of V_{oc} due to the reduced bandgap of the nanostructured material. The SQ limit is therefore overcome, breaking the trade-off between spectral harvesting and thermalization loss of conventional single-junction cells.
- If intraband carrier thermalization is instead faster than radiative recombination, the continuum and confined states are in thermal equilibrium, and the solar cell operates somewhat similarly to a single-junction cell of bandgap lower than the host semiconductor. In this case, the limiting efficiency is arguably still provided by the SQ limit. We refer to this second class of devices as thermally limited QDSC.

From the experimental standpoint, even though the ground principles of the IB concept have been proven [30, 31], QDSCs usually fall in the thermally-limited case. A clear enhancement of the infrared spectral response is observed in these devices, but the maximum demonstrated efficiency (18.7% at 1 sun for a InAs/GaAs QDSC) [3] lags well behind that of state-of-art GaAs single-junction cells, not to mention the gap with respect to the efficiency predicted by the IB theory ($\approx$ 36% for the InAs/

GaAs material system at 1 sun). The large discrepancy between demonstrated and theoretical efficiencies is often ascribed to the fact that reported devices work in the thermally limited regime. However, under such conditions, a conservative benchmark value for the attainable efficiency is given by that of a single-junction cell with bandgap comparable to the optical transition energy of the QD ground state. Assuming a bandgap of 1 eV (representative of the InAs/GaAs QD system), the SQ limit efficiency is above 30% under 1 sun, a value remarkably higher than the demonstrated one.

There are two main problems to overcome for achieving high-efficiency QDSCs, regardless if they fall in the IB or thermally limited class. First, the small optical absorption cross-section of QDs and their reduced volume fraction within the absorber material make the extended absorptivity provided by the QDs quite low, whereas the SQ theory assumes complete absorptivity for above-gap photons. The requirements are even more demanding to approach the IB operating regime, since very high—and comparable—interband and intraband photon absorption [31] would be needed. Based on detailed balance calculations, InAs/GaAs IBSCs require a QD density larger than 5×10^{13} cm^{-2} [32], but present QDSCs usually have a few tens of layers and in-plane density about 5×10^{10} cm^{-2}. The second fundamental issue is related to the reduction of V_{oc}. While this is in part an *intrinsic* effect, because the large electrical confinement of electrons and holes leads to enhanced radiative recombination at the QD sites (or equivalently the cell effective bandgap is narrowed by the QDs), reported devices are often plagued by further *extrinsic* loss due to defect-mediated recombination. Such defects are induced by the crystal strain inherent to the QD growth. Strain compensation techniques are therefore demanded during the epitaxial growth and have led to the demonstration of high V_{oc} QDSCs with low extrinsic loss [33].

To enhance the QD photogeneration, significant research efforts are devoted to grow high-quality crystals with increased QD sheet density [34–37] and number of QD layers [38, 39]. However, the intrinsic V_{oc} penalty also increases with the number and density of QD layers. Theoretical calculations show that such efforts will give frustrating results in terms of cell efficiency improvement, unless they are combined with clever ways to further increase the QD absorptivity [40, 41]. This calls for the development of light-trapping approaches that can be implemented within a thin-film solar cell architecture [28, 42, 43]. ELO thin-film InAs/GaAs QD cells with planar rear mirror were first reported in [44, 45]. In [40] we have demonstrated a twofold increase of QD photocurrent in an InAs/GaAs QDSC fabricated by ELO; however, the cells—either QD or bulk ones—suffered a high V_{oc} degradation due to a non-sufficient quality of the epilayers.

In this chapter, we report our most recent results in the development of high-efficiency thin-film III–V QD solar cells, with high V_{oc} and enhanced QD photon harvesting. Section 1.2 reviews the ELO method and discusses the optimum junction configuration for thin-film III–V cells approaching the radiative limit operation. In Sect. 1.3, we study extrinsic and intrinsic V_{oc} loss with the help of device-level physics-based simulations and report our latest achievements in the molecular beam epitaxy growth of high-voltage QDSCs. The implication of the intrinsic V_{oc} penalty and the demand for light-trapping approaches are discussed in Sect. 1.4, where we summarize our work on the development of photon management approaches in thin-film III–V QDSCs. After the demonstration of high V_{oc} ELO QD cells with planar reflectors, Sect. 1.4 discusses the design and fabrication of back surface reflectors and nanostructured antireflection coating for effective light trapping. Finally, an approach for enhanced photon recycling in thin-film cells by a highly reflective patterned rear contact/mirror is presented.

1.2 Epitaxial Lift-Off for Genuine Thin-Film III–V Devices

Most III–V materials are direct bandgap semiconductors with high absorption coefficients. Therefore, the active structure in a III–V solar cell only needs to be several microns thick to absorb all the light that the cell can convert into electricity. Nevertheless, III–V solar cells are referred to as a wafer-based technology because in order to produce high-quality single crystal III–V structures, state-of-the-art low defect single crystal InP, GaAs or Ge wafers are required as a production platform. After deposition of the cell structure, the wafer is of no further use. However, in the present fabrication techniques, the thin active structure is processed together with its passive substrate to become a wafer-based solar cell (see Fig. 1.2 left).

Compared to the manufacturing methods used nowadays, the amount of precious semiconductor material used to generate a certain output power can be reduced by two orders of magnitude by the utilization of a thin-film separation technology that

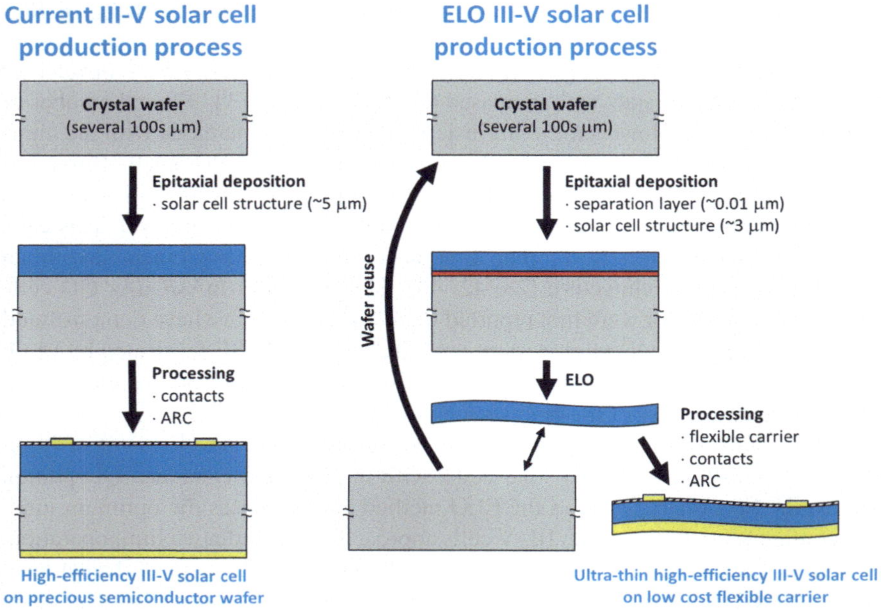

Fig. 1.2 Schematic representation (not to scale) of the current production process for wafer-based III–V cells (left) and thin-film cells via the epitaxial lift-off technology which facilitates reuse of the wafers (right)

allows to reuse the native wafer (which is about 100 times thicker than the active cell structure on top of it) as a production platform. The epitaxial lift-off (ELO) method offers the opportunity to separate thin-film III–V structures from their original wafer substrate [46–49] and transfer them to a low-cost foreign carrier like glass or plastic for stability during further processing and application. In order to apply the ELO method, a typically 10-nm-thick $Al_xGa_{1-x}As$ ($x > 0.6$) release layer has to be deposited before the actual III–V device structure (see Fig. 1.2 right). This sacrificial layer allows for the separation of the device structure from its substrate by selective wet etching using an aqueous hydrofluoric (HF) solution. Owing to the large selectivity ($>10^6$) of the HF solution for etching of AlGaAs over GaAs [50, 51], the original substrate is hardly affected. ELO reaction residues, mainly elemental arsenic, form As_2O_3 crystallites over time [52]. However, with an appropriate surface re-preparation strategy (e.g. a polishing etch or removable protection layer), the substrate can be reused [13, 53–56].

The first attempts to separate III–V devices from their substrates using the extreme selectivity of an aqueous HF solution for AlGaAs over GaAs were described as early as 1978 [46]. A wax layer was applied to support the circa 30-μm-thick fragile films during the process which at that time was referred to as 'peeled film technology'. Almost a decade later, it was noted that if the film structures have a thickness in the range of a few micrometres, the tension induced by the wax support layer causes the III–V films to curl up as they become undercut [47]. This was concluded to be beneficial for removal of the etch products during the process. As a result, the lateral etch rate of the AlGaAs release layer with a typical thickness in the 10–100 nm range increased to about 0.3 mm/h [57]. After etching of the sacrificial AlGaAs layer, the surface morphology of the thin-film device structures typically shows an increased surface roughness and the presence of reaction residues [52, 58]. This has no consequences for the performance of the final devices as the epi-layer structure can be designed in such a way that the outer layer affected by the substrate removal process is noncritical (e.g. a contact layer) or is etched away completely during subsequent device processing.

Using this process, now referred to as 'epitaxial lift-off technique', many III–V devices such as solar cells [59–61], photodiodes [62, 63], LEDs [64, 65], LASERs [66, 67], HEMTs [68], FETs [69] and THz antennas [70] transferred to silicon, silicon oxide, sapphire and glass plates were demonstrated. However, because the etch rate still was fairly low, the demonstrated devices were generally limited from several millimetres up to a square centimetre size.

1.2.1 Rapid Release of Full Wafer-Sized Thin-Film Structures for Photovoltaic Applications

At Radboud University, the parameters influencing the ELO etch process were systematically investigated [48, 58, 71, 72], and a model to describe the etch process was developed [73, 74]. By making some essential modifications like mounting a temporarily flexible carrier/handle on top of the III–V epi-structure and applying a controlled force to bend the carrier with the epi-structure away to allow the etchant to reach the etch front, the lateral etch rate increased more than 100 times to well over 30 mm/h [74–76]. At the same time, the flexible foil and increased etch rate offered the possibility to increase the lateral size of the released III–V structures from typically a square centimetre to full wafers ranging from 2 in. up to 6 in. in diameter [12, 13, 49, 77, 78]. Besides structures from GaAs wafers, ELO has also been successfully demonstrated for Ge and InP wafers, as well as multiple-times reused wafers. The method has successfully been demonstrated for a whole range of solar cell structures, such as single- and multi-junction, inverted and upright, lattice-matched and metamorphic structures as well as structures containing quantum dots. Both III–V epi-structures produced by metal organic vapour phase epitaxy (MOVPE) and by molecular beam epitaxy (MBE) can be processed by ELO. Automated ELO production equipment has also been developed, as shown in Fig. 1.3. Examples of thin films released from 2-in. and 4-in. diameter wafers are shown in Fig. 1.4.

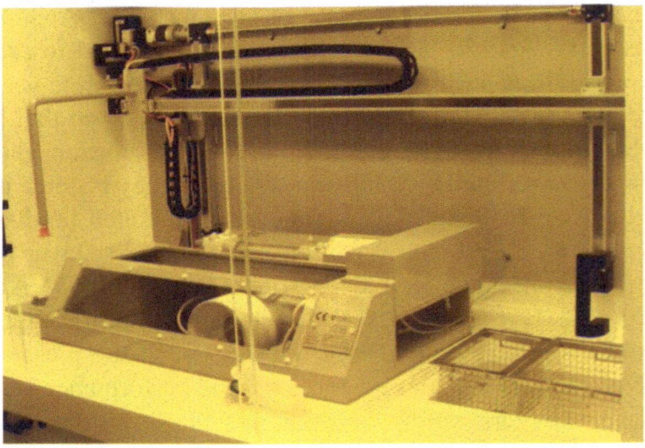

Fig. 1.3 Automated ELO processing equipment developed at tf2 devices

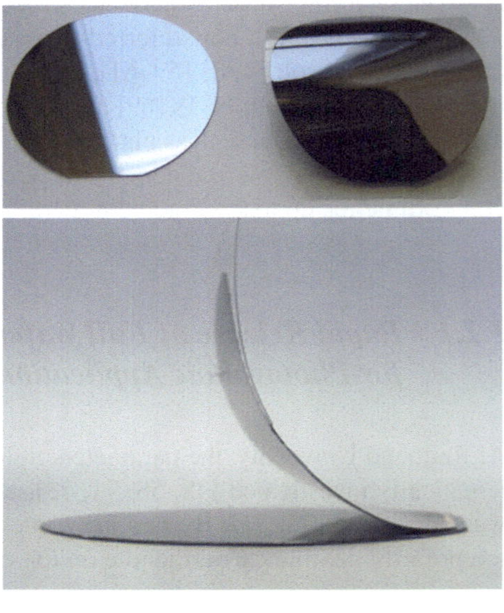

Fig. 1.4 Wafer-sized thin-film structures retrieved from 2-in. (upper) and 4-in. (lower) diameter GaAs substrates by epitaxial lift-off using a flexible carrier/handle

After ELO, the III–V solar cell structure can, by flipping from one temporary carrier to another, be processed on both sides. Together with the final carrier to be applied, this allows for new device structures which can be optimized for different purposes. The carrier can be a lightweight material for space applications [61, 79], a flexible carrier to cover bended surfaces [80, 81], a heat-conducting material for concentrator applications or any type of low bandgap cell to form a mechanically stacked multi-junction solar cell [49, 82, 83]. In case of a transparent carrier, the thin-film cell can be processed with a grid contact pattern on both sides. In this way semitransparent or bifacial III–V solar cells are obtained [84, 85]. Alternatively, an arbitrary carrier can be applied to support a thin-film cell with a full back contact. If applied correctly, this back contact might at the same time act as a mirror which reflects any photons that reach the rear side of the cell back into the cell structure [85, 86]. Figure 1.5 shows some examples of solar cells produced from thin-film III–V structures that were released from 4-in. wafers.

The deposition of III–V epi-structures for ELO thin-film solar cells typically is performed by MOVPE [13, 87], but structures produced by MBE [40, 45] work equally well. It is however essential to utilize epi-structures of sufficient quality. Owing to photon recycling, i.e. the excitation of electrons resulting from reabsorption of photons generated by radiative recombination of electrons and holes [88–91], for state-of-the-art epi-structures, thin-film ELO solar cells typically outperform their substrate-based equivalent cells [92]. This is clearly demonstrated by the present power conversion efficiency world records for single-junction GaAs solar cells being

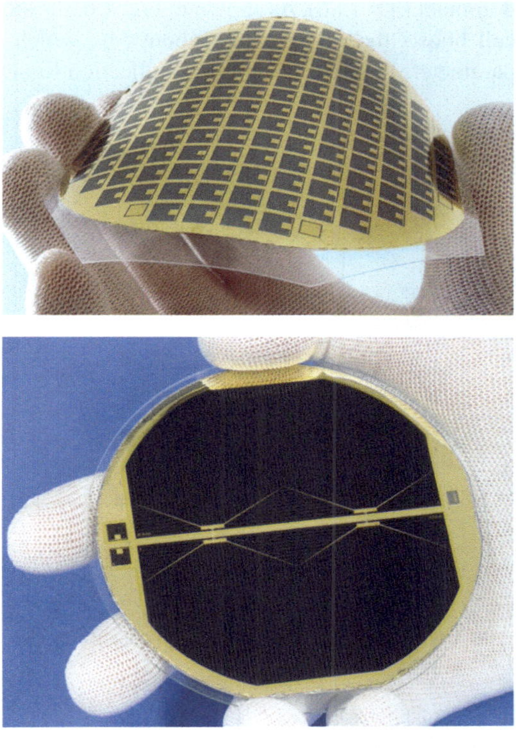

Fig. 1.5 Thin-film III–V solar cells produced by *tf2 devices* using 4-in. diameter GaAs substrates as a production platform followed by ELO and thin-film cell processing. Top: array of 5×5 mm^2 ELO solar cells on a flexible carrier, bottom: two 30 cm^2 ELO solar cells on a rigid carrier

27.8% for a substrate-based and 29.1% for the thin-film ELO cell [1]. On the other hand, epi-structures of lesser quality might suffer from the ELO etch process, resulting in a reduced cell performance of the ELO thin-film cells compared to their substrate-based equivalent. In case of inferior epi-structures, the difference in cell performance will be huge and directly apparent from the electroluminescence (EL) analysis of the devices [40]. For regular state-of-the-art epi-structure quality, the EL signals of thin-film- and substrate-based cells typically show no defects whatsoever, and differences in cell performance can only be judged from the *I–V* and EQE characteristics.

1.2.2 Shallow- and Deep-Junction Configurations

III–V solar cells are typically produced in a n-on-p shallow-junction (SJ) geometry, i.e. with a thin highly doped n-type emitter and a thick p-type basis. The shallow-junction geometry ensures that most of the light is absorbed close to the p-n junction, thus minimizing the constraints on the free carrier lifetimes. However, with the high quality of the epi-layers that are currently obtained by MOVPE and MBE, carrier lifetimes are sufficiently high to allow for n-on-p cells in a deep-junction (DJ) configuration [93–95]. In a comparative study, substrate-based DJ GaAs and InGaP cells demonstrated to perform better than their SJ counterparts [96]. This is because at the maximum power point, the DJ cells operate mainly in the radiative recombination regime (ideality factor $n = 1.22$), while in the SJ cells, non-radiative recombination is dominant ($n = 1.86$). As shown in Fig. 1.6, the steeper slope of the IV curve of the DJ cell boosts the fill factor by about 5%, which is thereby the most improved cell parameter. In order to minimize collection losses in the upper part of the solar cell,

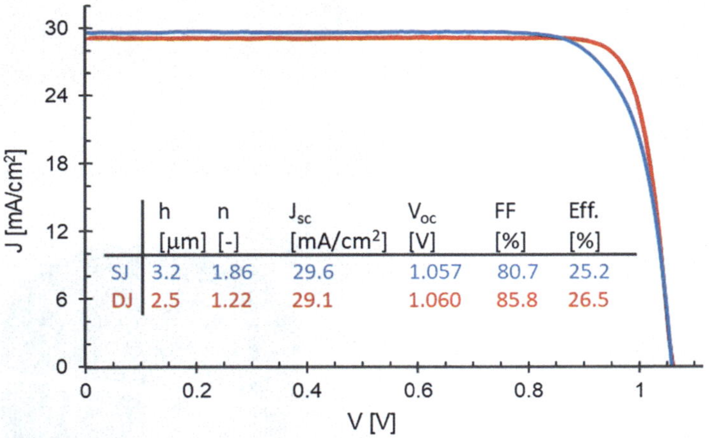

Fig. 1.6 *J–V* characteristics of 1 cm² substrate-based shallow-junction and deep-junction GaAs cells. The cells were processed with optimal antireflective coatings and a front grid that covers only 1.5% of the cells' surface

the optimal thickness (h) of the emitter plus base of the GaAs DJ cell is only 2.5 μm compared to 3.2 μm for the SJ cell. Nevertheless, the associated lower cell current is more than compensated by the higher fill factor and open circuit voltage, and as a result, the efficiency of the cell is raised from 25.2% for the SJ GaAs cell to 26.5% for the DJ cell. In the thinner InGaP cell, the absence of current loss resulted in an even larger efficiency benefit for the DJ geometry (18.2% as compared to 16.6%) [96].

1.3 High V_{oc} InAs/GaAs QD Solar Cells by MBE Growth

InAs QDs embedded in GaAs exhibit a small bandgap and induce a large electrical confinement for electrons and holes. The strong carrier confinement leads to enhanced carrier recombination at the QD sites, which results in an intrinsic reduction of the V_{oc}. Also, every grown InAs QD layer results in excessive compressive strain which eventually leads to formation of lattice defects manifesting as dislocation lines and a further extrinsic reduction of the V_{oc}. Recently, it has been demonstrated that by using metal-organic chemical vapour deposition (MOCVD), it is possible to fabricate InAs/GaAs QD solar cells with V_{oc} close to 1 V and only a marginal V_{oc} reduction compared to GaAs reference solar cells [33, 97]. On the other hand, using molecular beam epitaxy (MBE), another viable and mass-production capable epitaxy method, such as high V_{oc} values for InAs/GaAs QD solar cells, was not published for a long time. The published V_{oc} values for MBE-grown InAs/GaAs QD solar cells have typically been about 0.2 V lower than the V_{oc} values of high-quality GaAs reference solar cells [3, 40, 98, 99]. This has been partially due to larger QDs used in the structures or due to increased recombination at the QD regions. However, when comparing the works done using MOCVD or MBE, it is important to understand that for both techniques, the V_{oc} values degrade for larger QDs in very similar manner [3]. Recently, the authors have demonstrated that InAs/GaAs QD solar cells with high V_{oc}, approaching 0.95 V, can be fabricated also by MBE [100].

In this section, we present an assessment of the typical intrinsic and extrinsic V_{oc} reduction mechanisms in QDSCs based on the study of experimental devices by quantum-corrected transport simulations. Then, we review the fabrication of MBE-grown high-V_{oc} InAs QD solar cells and discuss the characteristics of deep-junction and shallow-junction structures [100].

1.3.1 Causes for V_{oc} Reduction in QDSCs

1.3.1.1 QD-Corrected Transport Model

Models of QDSCs have been traditionally developed based on the hypotheses of IB operation, with detailed balance [101–103] and in some cases with drift-diffusion [104–106] approaches. Recently, a detailed balance model of thermally limited

QDSCs has been reported in [107], suggesting a substantial equivalence between these devices and bulk cells. When it comes to understanding the detailed physics of real devices and providing useful guidelines to their development, transport models become essential tools. However, the accurate device-level modelling of QDSCs is a difficult task since it requires managing details both at the nano-/mesoscopic and at the microscopic scales. Novel quantum kinetic approaches [108] are very well suited for nanostructure modelling, but the available computational resources prevent to date their application to device-level simulations. In [109] an effective medium approach was developed to study the performance of QD multi-junction cells, where the QD recombination and related V_{oc} loss are somehow modelled by describing the QDs as equivalent trap states. Recently, the Polytechnic University of Turin has developed a quantum-corrected drift-diffusion simulator that, including the peculiar physics of the QD material, leads to an accurate description of the interband and intersubband charge transfer processes involved in QDSC operation. The initial works [110, 111] demonstrated the impact of such processes on intrinsic and extrinsic V_{oc} reduction, and a number of follow-up papers have applied the model to the analysis of real solar cells and to the design of high-efficiency QDSCs [40, 43, 112, 113].

To focus the ideas, one can consider the solar cell structure and the associated band diagram in Fig. 1.7b. A DJ n-p solar cell embeds a stack of 20 QD layers with undoped interdot layers. The QD stack is placed next to the junction to ensure optimal carrier collection at short circuit. As sketched in Fig. 1.7c, the three-dimensional confinement of carriers in the QDs introduces sub-bandgap energy states modelled by three discrete energy levels in the conduction and valence band: ground state (GS), excited state (ES) and wetting layer (WL). At each QD layer, possible interband and intraband charge transfer mechanisms include intersubband relaxation and excitation, photogeneration and radiative recombination. Within the present work, we assume that charge transfer between subband energy states occurs only through thermal cascaded capture and relaxation, because at room temperature, for the InAs/GaAs material system, optical intersubband processes are negligible with respect to

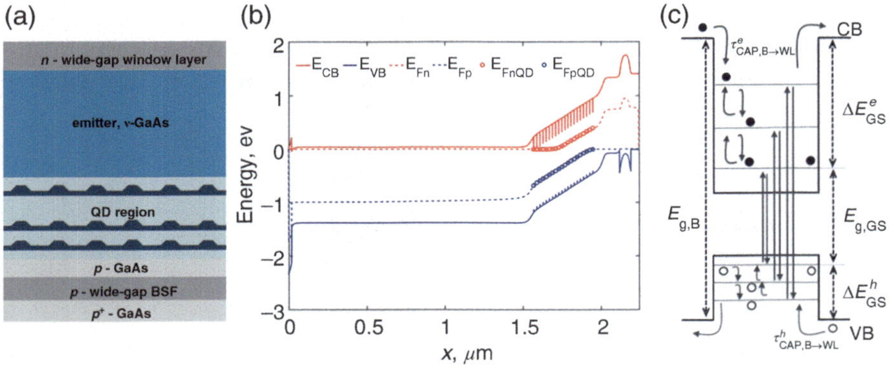

Fig. 1.7 Sketch of the studied GaAs n-p solar cell with embedded QD layers and the corresponding energy band diagram at short-circuit condition under AM1.5G illumination. On the right, QD energy states and interband and intersubband transitions are considered in the model

the thermally activated ones [114]. On the other hand, the model can be extended to deal with intersubband optical transitions in material systems with relatively higher energy band separations (or for lower operating temperature) where thermal transitions are reduced [115, 116].

The QDSC operation can be described as follows: electron and holes are photogenerated in the barrier and QD states by above and below energy gap photons, respectively. Bulk carriers move by drift-diffusion, and when they fall within the QD interaction range (a few nm), they can either be emitted through the QD layer or captured in QD states. Capture and escape mechanisms connecting the bulk and the bound states are mediated by the WL states. To account for this, 3D carriers continuity equations and Poisson equation for the electrostatic potential are corrected with terms accounting for capture/escape processes and QD occupation:

$$\frac{\partial n}{\partial t} = \frac{1}{q}\frac{\partial J_n}{\partial x} - U_n^{\mathrm{B}} + G_{\mathrm{ph}}^{\mathrm{B}} - \sum_i \left(R_{n,\mathrm{cap}}^{\mathrm{B}\to WL,i} - R_{n,esc}^{WL,i\to \mathrm{B}} \right) \delta\left(x - x_i\right) \tag{1.1}$$

$$\frac{\partial^2 \phi}{\partial x^2} = -\frac{q}{\epsilon}\left(p - n + N + \sum_{i,k} \left(p_{k,i} - n_{k,i} \right) \delta\left(x - x_i\right) \right) \tag{1.2}$$

where i identifies the QD layer, n the electron density, J_n the electron current density, $G^{\mathrm{B}}{}_{\mathrm{ph}}$ the photogeneration rate and U_n^{B} the net recombination rate due to radiative and non-radiative recombination, modelled according to Shockley-Read-Hall (SRH) theory. An equation analogous to (1.1) holds for holes.

The charge transfer between bulk and the QDs is treated locally near the i-th QD layer by carrier capture ($R_{n,\mathrm{cap}}^{\mathrm{B}\to WL,i}$) and escape ($R_{n,esc}^{WL,\mathrm{i}\to \mathrm{B}}$) rates. The modified Poisson equation accounts for the local charge contribution of the QD states ($p_{k,i}$ and $n_{k,i}$ with k identifying the GS, ES and WL state).

Intersubband charge transfer is governed by rate equations for each QD state describing the subband population in the i-th QD layer as follows:

$$\frac{\partial n_{WL,i}}{\partial t} = R_{n,\mathrm{cap}}^{\mathrm{B}\to WL,i} - R_{n,esc}^{WL,i\to \mathrm{B}} - R_{n,\mathrm{cap}}^{WL,i\to ES,i} + R_{n,esc}^{ES,i\to WL,i} - U^{WL,i} + G$$

$$\frac{\partial n_{ES,i}}{\partial t} = R_{n,\mathrm{cap}}^{WL,i\to ES,i} - R_{n,esc}^{ES,i\to WL,i} - R_{n,\mathrm{cap}}^{ES,i\to GS,i} + R_{n,esc}^{GS,i\to ES,i} - U^{ES,i} + G$$

$$\frac{\partial n_{GS,i}}{\partial t} = R_{n,\mathrm{cap}}^{ES,i\to GS,i} - R_{n,esc}^{GS,i\to ES,i} - U^{GS,i} + G$$

where $n_{\mathrm{GS},i}$, $n_{\mathrm{ES},i}$ and $n_{\mathrm{WL},i}$ are the electron sheet densities in the GS, ES and WL. $U^{k,i}$ and $G^{k,i}{}_{\mathrm{ph}}$ are net band-to-band recombination and photogeneration rates in the k-th state. Capture and escape processes depend on the occupation probability in each state and characteristic scattering times. In particular, the detailed balance at thermal equilibrium implies that escape and capture times are related by

$$\tau_{n,esc}^{k-1,i} = \frac{N_{k-1}}{N_k} \tau_{n,\mathrm{cap}}^{k-1,i} \exp\left(\frac{E_k - E_{k-1}}{k_{\mathrm{B}} T} \right)$$

N_k being the effective density of states of state E_k.

QD electronic structure, optical properties and scattering rates for the charge transfer mechanisms can be derived from quantum mechanical models or from experimental data. In the following examples, carrier capture and relaxation times between the QD states range on a scale of 100 fs for holes and 1 ps for electrons, while 1 ns radiative lifetime is used for all the QD levels [110].

Finally, the electrical model is coupled to a suitable optical model for the calculation of the spatially resolved photogeneration rates [40, 112].

1.3.1.2 Intrinsic and Extrinsic V_{oc} Loss

To clarify the issue of V_{oc} degradation in QDSCs, it is helpful to analyse a real-life example with the help of the quantum-corrected model described in Sect. 1.3.1.1. Figure 1.8 reports the EQE and *J–V* characteristics of two InAs/GaAs QDSCs with 20 and 50 QD layers, respectively, and a reference GaAs cell [40]. The cells have an n-p+ structure, with the junction located close to the rear-side contact and emitter thickness of 2 μm. In the QDSC samples, a portion of the emitter is replaced by an intrinsic stack of QD layers with interdot spacing of 20 nm and areal density of 8×10^{10} cm^{-2}. QDs were grown by a Sb-mediated technique to achieve high density without using any strain balancing compensation [36].

From the analysis of the photoluminescence (PL) spectra of precursor samples with one QD layer and of the sub-bandgap EQE, the GS, ES and WL optical bandgaps result as 1.17 eV, 1.25 eV and 1.37 eV, 80% of which is attributed to the electron confinement energy in the conduction band [117]. In the QD layers, the optical absorption associated with each interband transition is modelled by a Gaussian function with absorption peaks on the order of 10^3 cm^{-1} for GS and ES states and 10^4 cm^{-1} for WL state. The full list of QD and bulk material parameters used in this study is summarized in [40].

The experimental results in Fig. 1.8 point out a clear reduction of the collection efficiency of the 50 × QD cell with respect to the 20 × QD one and the reference cell, well visible from the EQE spectra, and a large and almost identical V_{oc} penalty in the QDSCs with respect to the reference cell. From the simulation study, it turns out that the main factors affecting the cell operation are the minority carrier lifetime in the emitter and the non-radiative recombination in the intrinsic barrier layers of the QD stack. In particular, SRH recombination in the emitter significantly impairs the short-circuit current (which in such rear-side junction cell sustains most of the short-circuit current), while it has a lower impact on the V_{oc}. Vice versa, a decrease of the SRH lifetime in the intrinsic interdot layers of the QD stack has a limited impact on carrier collection but affects strongly the recombination current and thus is responsible for the observed low V_{oc} in the QDSCs. Figure 1.9 shows the predicted behaviour of the QDSCs V_{oc} as a function of the SRH carrier lifetime in the GaAs barrier layers. A range of SRH lifetimes in the low-doped portion of the emitter is considered, which provides a reasonable fit to the EQE spectra of the 50 × QDSC (about 3 ns) and of the 20 × QDSC (15–200 ns depending on the specific sample). It is worth noticing that these cells were grown without any strain management.

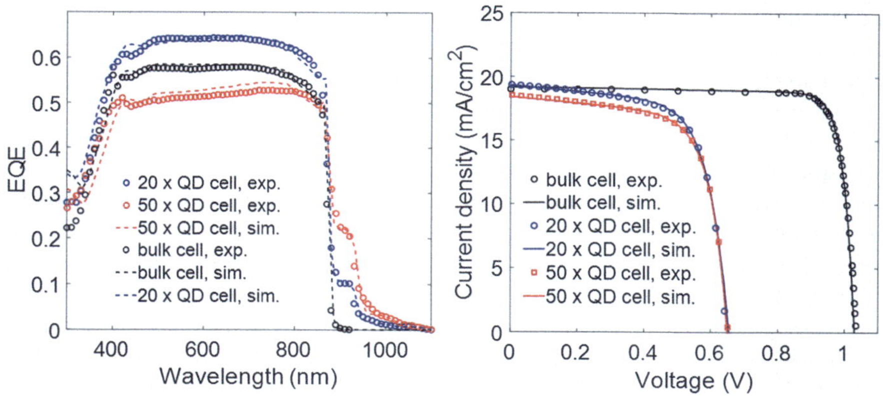

Fig. 1.8 Measured and simulated EQE spectra (left) and *J*–V characteristics (right) of a GaAs regular cell, a 20 × QD cell and a 50 × QD cell. (Reprinted from Sol. Energy Mater. Sol. Cells, 181, F. Cappelluti et al., Light-trapping enhanced thin-film III-V quantum dot solar cells fabricated by epitaxial lift-off, pp. 83–92, Copyright (2018), with permission from Elsevier)

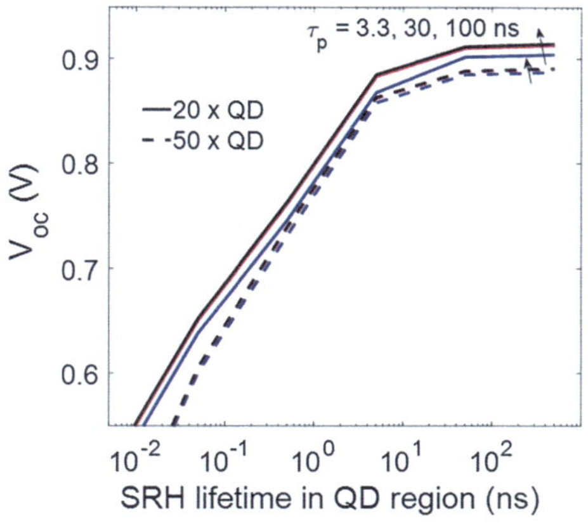

Fig. 1.9 Behaviour of V_{oc} for the 20 × (solid lines) and 50 × (dashed lines) QD cells as a function of the SRH lifetime in the interdot layers of the QD stack and for different values of SRH hole lifetime in the doped portion of the emitter (τ_p). (Reprinted from Sol. Energy Mater. Sol. Cells, 181, F. Cappelluti et al., Light-trapping enhanced thin-film III-V quantum dot solar cells fabricated by epitaxial lift-off, pp. 83–92, Copyright (2018), with permission from Elsevier)

The reduction of the SRH lifetime in the emitter of the 50 × cell with respect to the 20 × one suggests a higher defect density correlated to the larger number of QD layers and the higher strain accumulation during the epitaxial growth. As long as the SRH lifetime in the QD stack is significantly larger than the QD radiative lifetime (1 ns), the V_{oc} remains high with calculated limiting values (for negligible SRH recombination) on the order of 900 mV—denoting an intrinsic penalty of about 150 mV with respect to the reference cell due to the radiative recombination through the QD states. Therefore, in the samples under study, the observed large reduction of V_{oc} can be attributed to crystal quality degradation, with an estimated SRH lifetime in the QD stack on the order of 100 ps.

Concerning the intrinsic V_{oc} loss, Fig. 1.10 shows some calculated trend in terms of V_{oc} scaling with the number of QD layers (N_S), QD areal density (N_{QD}) and GS emission energy under the hypothesis of negligible non-radiative recombination. As shown in Fig. 1.10 (left), the intrinsic V_{oc} scales linearly with the natural logarithm of the number of layers, with a slope of 26 mV. At open circuit, $V_{oc} = V_T \ln(J_{sc}/J_0)$, where J_0 is the reverse saturation current, J_{sc} the short-circuit current density and $V_T = 26$ mV. Since J_{sc} is weakly affected by the number of QD layers, the linear scaling of V_{oc} with $\ln(N_S)$ denotes an almost linear increase of the radiative recombination with N_S. Similar considerations can be drawn for the V_{oc} scaling with the areal density N_{QD}. In Fig. 1.10 (right), the intrinsic V_{oc} is analysed for different GS emission energies. This demonstrates that the linear scaling with $-V_T \ln(N_S)$ holds regardless of the confinement energy of carriers, since V_{oc} increases almost linearly with the GS bandgap. The reason is that QDs act as trap centres whose activation energy increases as the confinement energy decreases, i.e. as thermal escape becomes more effective. From the experimental standpoint, a linear trend of V_{oc} with the E_{GS} was verified in [3] and more recently in [100] for high-quality QDSCs, with a promising GS bandgap-voltage offset ($W_{oc} = E_{GS}/q - V_{oc}$) as low as 0.3 V, comparable to the highest efficiency thin-film GaAs cells. For reference, the theoretical offset is 0.36 V for radiative-limited GaAs cells [118], while 0.4 V is the value of state-of-art wafer-based GaAs cells. However, the results in Fig. 1.10 show that the bandgap offset in nanostructured absorbers significantly depends on the QD volume density and changes from about 0.15 V for the case with lowest volume fraction (low density and low number of layers) to about 0.35 V for the case with highest volume fraction.

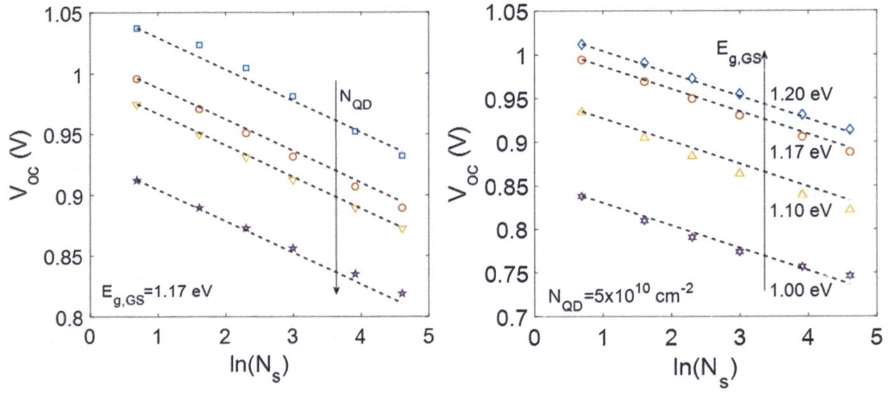

Fig. 1.10 Left: Scaling of V_{oc} with the number of QD layers (N_S) and QD areal density ($N_{QD} = 1 \times 10^{10}, 5 \times 10^{10}, 1 \times 10^{11}, 1 \times 10^{12}$ cm^{-2}) for GS bandgap of 1.17 eV. Right: Scaling of V_{oc} with the number of QD layers and the GS bandgap for fixed areal density $N_{QD} = 5 \times 10^{10}$ cm^{-2}. Symbols are results of simulations, and dashed lines are linear fit according to $V_{oc} = V_T \ln(J_{sc}/J_0) \propto - V_T \ln(N_S)$, with $V_T = 26$ mV

In fact, as the QD volume occupation increases, the onset of significant QD absorption becomes closer to the GS bandgap, and W_{oc} approaches the typical value found for III–V bulk solar cells with radiative recombination only. On the other hand, in samples with low volume occupation, the offset is significantly lower. In this case, as noted in [119], the W_{oc} is no more a useful figure of merit to predict the efficiency potential of the cell, because the hypothesis of almost full absorptivity above the bandgap energy does not hold anymore. Nevertheless, as done in Sect. 1.3.2.5, the calculated intrinsic W_{oc} still provides a useful benchmark to quantify in real devices the extrinsic V_{oc} loss due to non-radiative recombination.

1.3.2 *Molecular Beam Epitaxy-Grown InAs/GaAs Quantum Dot Solar Cells with High Open Circuit Voltage*

1.3.2.1 Experimental Details

The samples were grown on p-GaAs(100) substrates using either a V90 or a V80H MBE system equipped with effusion cells for group III elements and dopants and valved cracker sources for group V elements. Silicon and beryllium were used as n-type and p-type dopants, respectively. The nominal layer structures for DJ and SJ architectures are shown in Fig. 1.11. Growth temperature for the p-GaAs, n-GaAs and p-AlGaAs BSF layers was kept at about 580 °C. The lattice-matched n-AlInP window layer was grown at 490 °C, and the InAs QDs were grown at 465–475 °C. The nominal amount of deposited InAs, determined by simple ion gauge flux measurements, was varied between 1.6 monoatomic layer (ML) and 2.2 ML to adjust the QD size. The QD density was targeted to be over 5×10^{10} cm^{-2}. A set of samples was grown without any strain management, and in some samples thin, 2-monolayer-thick,

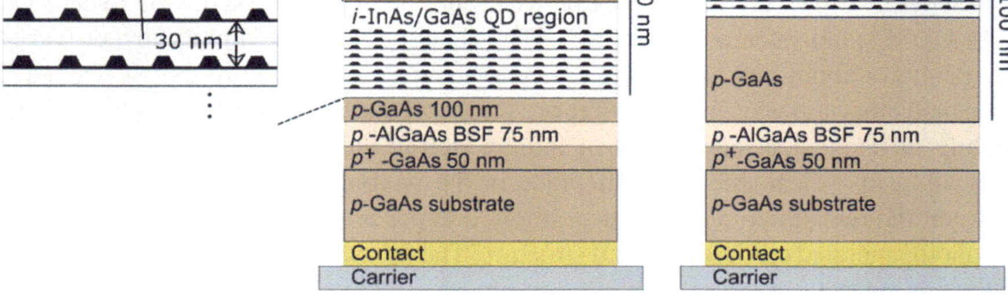

Fig. 1.11 The schematic layer structures for the deep (left)- and shallow (right)-junction InAs QD solar cells with the nominal layer thicknesses

GaAsP layers were inserted within the undoped GaAs barrier layers between the QD sheets for managing the accumulating compressive strain [33]. The number of QD sheets was varied between 0 and 20 to study the effects of increasing the active InAs QD volume on the photogeneration and structural behaviour of the solar cells.

A set of samples was also grown in inverted manner to allow thin-film processing and structures providing light management and photon recycling on the back and front surfaces.

The grown samples were processed to wafer-based solar cells with size of $6 \times 6\ mm^2$ by dicing and using a simple shadow mask process for forming front contact fingers. The backsides of the substrates were metallized as p-contacts. Most of the cells were left without antireflection coating. The front contact finger pattern was formed using Ni/Au, and Ti/Au metallization was used for the bottom contacts. The contact metals were deposited by electron beam evaporation.

Optical emission properties were characterized using an automated Accent RPM2000 photoluminescence (PL) mapping tool with ~25 mW diode laser excitation at 785 nm. The PL emission was recorded either using a CCD detector array or an InGaAs photodiode array, depending on the emission wavelength. The symmetric GaAs (004) x-ray diffraction (XRD) Ω-2θ scans, to compare the crystal properties of DJ and SJ QD structures, were recorded using an X'Pert Pro MRD high-resolution XRD system. Surface quality and QD density properties were investigated using light microscopy (Nikon L200) and atomic force microscopy (AFM) (Veeco DI3100 AFM system), respectively. The external quantum efficiencies (EQE) were measured in the wavelength range from 380 nm to 1100 nm using a self-made measurement system based on a 250 W QTH lamp (Oriel), a DK240 monochromator (Digikröm), appropriate long-pass and short-pass optical filters and NIST-calibrated Si and Ge reference photodiodes. The electrical parameters of the solar cells were deduced from light-biased current-voltage (*I-V*) measurements using an OAI TriSol solar simulator adjusted to AM1.5D conditions (the spectrum was normalized to 1000 W/m^2).

1.3.2.2 Photoluminescence Characteristics

The room temperature PL of two sets of InAs QD cells are shown in Fig. 1.12. Figure 1.12a shows the differences between luminescence properties of SJ and DJ InAs QDSCs and reference GaAs SC. The DJ cell emitting at slightly longer wavelength shows about one half of the emission intensity compared to the SJ structure. The small emission wavelength difference for the SJ and DJ structures originates from slightly different deposition rates of indium during the growths on different days. There is also a clear intensity difference in the wetting layer emission at 920 nm between the two structures. This may imply a larger non-radiative recombination of the photogenerated carriers in the DJ structure. The full-width at half-maximum values of the ground state emission peaks are ~64 nm and ~67 nm for SJ and DJ, respectively. The rather wide emission peak width indicates that either excited states take part to the emission or the broadness arises at least partially from QDs with different sizes.

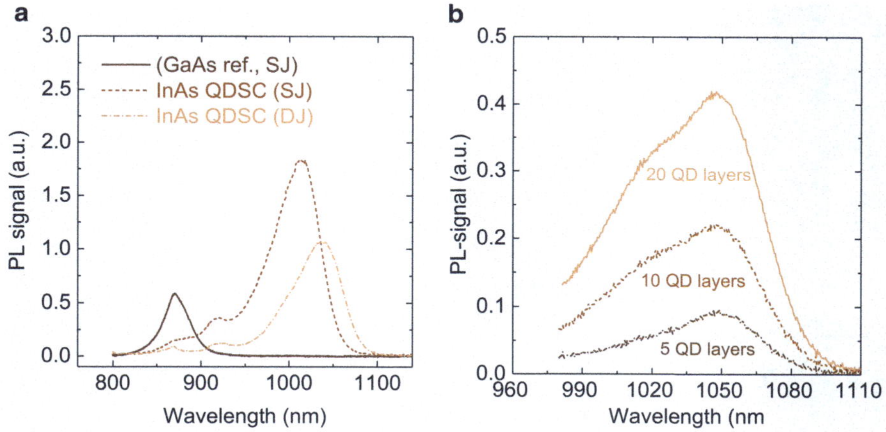

Fig. 1.12 (**a**) PL of GaAs, DJ and SJ InAs QDSCs. (**b**) PL from QDSCs with 5, 10 and 20 QD layers

In Fig. 1.12b, the PL spectra from InAs QDSCs with different number of QD layers are shown. We varied the number of InAs QD layers in a DJ structure from 5 to 20 to study the effects of the number of QD layers on the optical emission. The PL peak intensity increases roughly linearly with the number of QD layers indicating that at least from the optical emission point of view, the QDs maintain their activity even with 20 QD layers. On the high energy side of the ground state PL peak, one also observes a PL component from excited states or from a population of smaller-sized QDs.

1.3.2.3 Surface Characteristics

The InAs QD density was measured using AFM on samples that were separately grown on GaAs(100). Typically, the measured QD density in our samples was higher than 5×10^{10} cm^{-2}. For the growth conditions used in [100], the QD density was measured to be ~7×10^{10} cm^{-2}. AFM images also indicated that the size distribution of the QDs at these growth conditions is at least bimodal, suggesting that there are different sizes of QD contributing to the photogeneration.

The accumulation of strain in the InAs QD solar cells along the number of QD layers is also another concern. The consequence of not applying strain management is demonstrated in microscopy images shown in Fig. 1.13. The samples with 10 QD layers show only a few surface defects at magnifications of 10× and 50×, but otherwise the surface seems to be free of dislocation lines. For samples with 20 QD layers, a number of dislocation lines have appeared already at 10× magnification, which indicates that for 20 InAs QD layers, the accumulated compressive strain exceeds the threshold for dislocation formation.

Additionally, we find that the inter-QD-layer distance drastically affects the crystalline quality of the InAs QD solar cells—especially when no strain management is

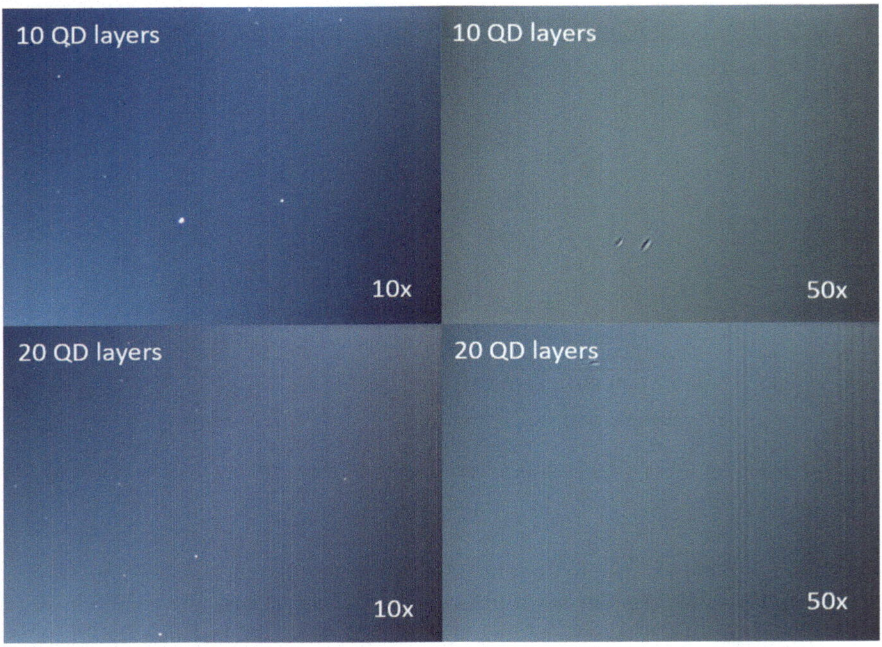

Fig. 1.13 Microscopy images for InAs QD solar cell structures with 10 and 20 QD layers

used. Figure 1.14 demonstrates the effect of the inter-QD-layer distance on surface morphology of the structures with 20 QD layers. For inter-QD-layer distances of 50 nm, no dislocation lines are visible at 10× magnification. However, when the inter-QD-layer distance becomes 30 nm or smaller, dislocation lines appear in microscopy images. For 30 nm inter-QD-layer distance, only a few lines are visible, but for 26 nm, a dense dislocation line pattern is observed. Such observations are in line with earlier observations on multi-QD-layer structures in which the accumulating strain in one QD layer affects the growth of the next QD layer when the inter-QD-layer distance becomes too thin [120]. Therefore, to maintain pseudomorphic growth and keep the misfit dislocation line density at minimum, it is essential to manage the accumulating strain within the structures using tensile-strained layers added to the structures.

1.3.2.4 X-Ray Diffraction Characteristics

We have adopted the strain management approach published by Hubbard's group on MOCVD-grown InAs/GaAs QD solar cells [33] in which a thin layer of GaAsP is grown inside the GaAs inter-QD-layer between the QD layers. To study the effects of insertion of GaAsP layers, XRD measurements are essential to determine the grown structure. The fitting to the measured XRD patterns was done using a commercial dynamical diffraction theory-based software RADS Mercury [121].

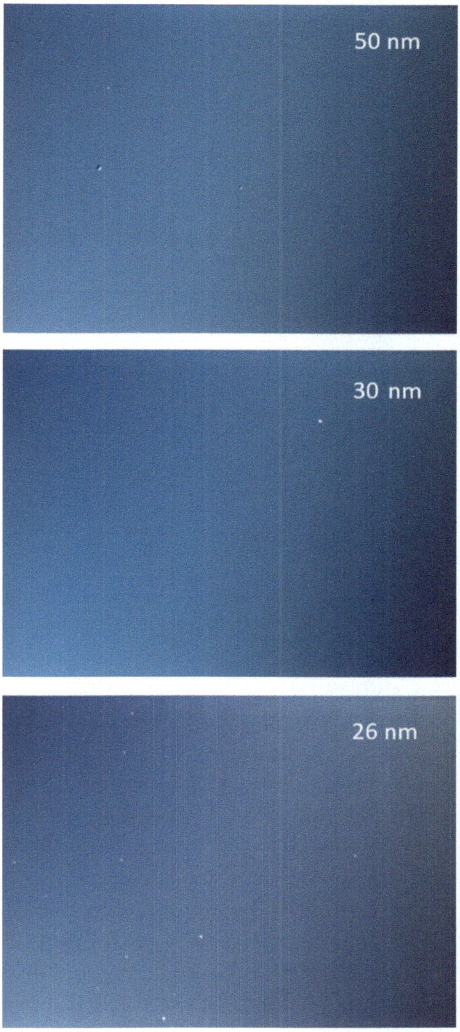

Fig. 1.14 Microscopy images (10× magnification) taken from DJ InAs QD solar cell structures with different inter-QD-layer distances varying from 50 nm down to 26 nm

Below, we shortly discuss the differences observed in XRD measurements between InAs QD solar cells with and without strain management. Figure 1.15 compares typical symmetric Ω-2θ XRD scans taken over the GaAs (004) reflection on InAs QD solar cells with and without strain management. In the upper pane of Fig. 1.15, XRD of a solar cell structure without strain management is shown, while in the lower pane, the solar cell has thin GaAsP strain management layers inserted between the QD layers. Both structures show the main peak corresponding to GaAs (004) reflection and periodic satellite peaks on both sides of the main peak. The satellite peaks for the sample without strain management are slightly shifted towards compressive side of the main peak compared to the sample with strain management.

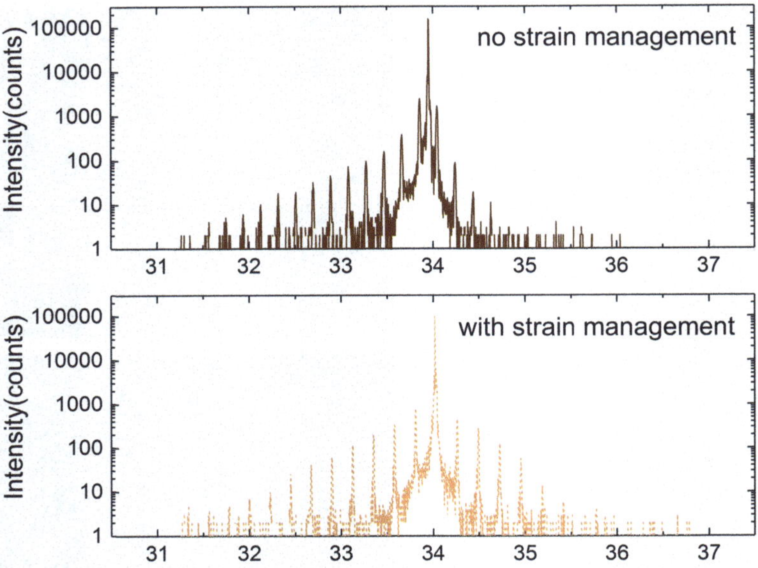

Fig. 1.15 Omega-2Theta XRD scans over the GaAs (004) reflection from InAs QD solar cells with (bottom) and without (top) GaAsP strain-management layers

This shift is related to the strain in the superlattice structure and can be used for further analysis of the effect of GaAsP layers used for strain management. In fact, one of the satellite peaks is masked by the GaAs (004) peak in the sample with strain management. Another difference between the samples can be seen on the right-hand side of the GaAs (004) diffraction peak. For the strain-managed structure, the periodic satellite peaks are higher and more of them are detected, especially on the tensile strain side of the GaAs (004) diffraction peak suggesting higher crystal quality. The main difference between the samples can be seen on the right-hand side of the GaAs (004) diffraction peak. For the strain-managed structure, the periodic satellite peaks are higher and more of them are detected, especially on the tensile strain side of the GaAs (004) diffraction peak. The inter-QD-layer distance can also be deduced from the distance of the superlattice (SL) satellite peaks in the XRD measurement graph using the relation $1/\Lambda = (2\sin(\theta_n) - 2\sin(\theta_{n-1}))/\lambda$, where Λ is the period of the SL, θ_n and θ_{n-1} are the adjacent SL peak angles and λ is the x-ray wavelength (1.54056 Å for Cu $K_{\alpha 1}$) [122].

1.3.2.5 InAs/GaAs QD Solar Cell I–V Characteristics

The room temperature *I–V* measurement results for best performing SJ and DJ solar cells are shown in Fig. 1.16. The electrical parameters of the best DJ and SJ InAs QD solar cells without antireflection coating are reported in Table 1.1. For the DJ structures, there is an inverse linear relation between the V_{oc} and the number of QD layers. For zero layers, i.e. GaAs reference cell, the V_{oc} is 0.917 V, whereas for a cell with 20 QD layers, the V_{oc} is reduced to 0.852 V. The decrease of 2.5 mV per QD

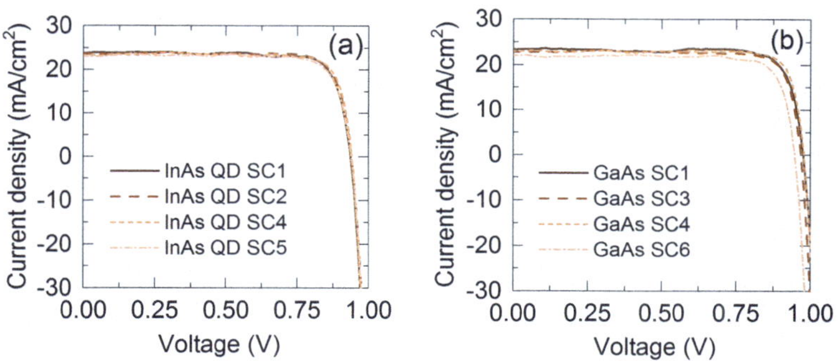

Fig. 1.16 (**a**) *I–V* for the four best wafer-based SJ InAs QD SCs. (**b**) *I–V* for the four best wafer-based SJ GaAs SCs

Table 1.1 Electrical parameters of the best DJ and SJ solar cells [100]

Structural elements				Electrical parameters			
Material		QD layers	Junction type	V_{oc}(V)	J_{sc}(mA/cm^2)	FF (%)	Efficiency (%) (active area)
R2262	GaAs	0	DJ	0.917	19.5	80.2	14.3
R2272	InAsQD	10	DJ	0.877	22.6	81.7	16.2
R2270	InAsQD	20	DJ	0.852	21.9	79.2	14.8
R2275	GaAs	0	SJ	0.979	21.7	83.8	17.8
R2274	InAsQD	10	SJ	0.940	22.1	83.9	17.4

layer between 10 and 20 QD layers is close to 26 mV × ln(20/10), i.e. 1.8 mV per QD layer predicted in [40]. Our interpretation is that the origin of the reduction in the V_{oc} is most likely related to increased recombination in the QD region [100]. The SJ structures seem to work better in our case: for a wafer-based GaAs reference cell, V_{oc} of 0.979 V was obtained, while an InAs QD cell with 10 QD layers exhibited V_{oc} of 0.94 V, which is only 0.39 mV less than for the GaAs reference cell. The small V_{oc} penalty upon insertion of the QD layers indicates that the quality of the InAs QD layers is very good. Also, the QD cells show slightly increased short-circuit current densities for both DJ and SJ structures, which indicates that indeed there is an increased photocurrent from the QD layers.

Most of the increased photocurrent of the wafer-based cells originated from the thin InAs wetting layer. This can be seen in the EQEs of Fig. 1.17a, which shows that at the wavelengths corresponding to the wetting layer photogeneration, the EQE for 10 QD layers is about 7%, whereas the contribution of the electronic states of the QD at longer wavelengths is marginal. For 20 QD layers, the contribution of the wetting layer increases to 10%, and those of the QD ground and excited states to the photogeneration become more visible.

Also, it was found, similarly to MOCVD-grown InAs QD solar cells [3], that there is an inverse relationship between the V_{oc} values and QD PL emission wavelength of the cells. In Fig. 1.17b, the simulated V_{oc} under radiative limit for a solar cell with 10 QD layers is also drawn for comparison. The best obtained experimental V_{oc} values are ~ 40–60 mV below the calculated radiative limit. The experimental

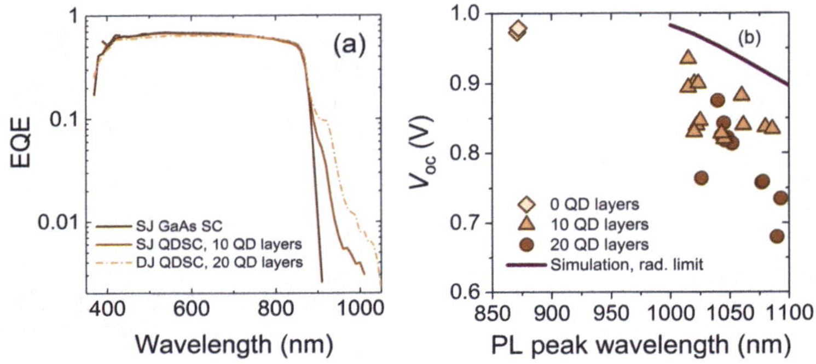

Fig. 1.17 (**a**) Examples of EQEs of the SJ GaAs and InAs QDSCs with 10 and 20 InAs QD layers. (**b**) V_{oc} values versus PL peak emission wavelength for wafer-based GaAs and InAs QDSCs with either 10 or 20 QD layers. The simulated radiative limit for the V_{oc} is also shown as continuous line

and simulated results indicate that, the larger an InAs QD gets, the deeper the associated potential well becomes, leading to higher recombination probability at the QD sites. A high V_{oc} can be obtained by reducing the QD size. The downside of this is that the decreased QD size also leads to decreased added photocurrent. To increase photogeneration in the QD layers, one would thus need to use higher number of QD layers, which again eventually would lead to reduced V_{oc}. To circumvent the problems arising from increasing the number of QD layers, one shall use smart photon management within the structures to allow for higher absorption at the QD layers and higher photogeneration. The next section will concentrate on the photon management aspects of the thin-film InAs QDSCs.

1.4 Photon Management Approaches for High-Efficiency Thin-Film III–V Solar Cells

Overcoming the weak optical absorption of QDs to compensate for the intrinsic V_{oc} loss is a mandatory path to demonstrate high-efficiency QDSCs. The development of light-trapping structures in QD solar cells involves two components: the fabrication of the back surface reflector and the structuring of the antireflection coating. With planar back surface reflectors, the optical length in absorptive layers can be effectively doubled. Furthermore, the optical length of the photons can be enhanced even more with diffractive gratings with reflector in the back surface of the solar cell. Diffractive gratings together with structured antireflection coatings (ARCs) create an optical system inducing photons to experience multiple round trips across the cell, increasing the absorption of light into the active layers.

In this section, we discuss the potential of light-trapping enhanced QD cells based on the realistic device model introduced in Sect. 1.2.1 and the associated requirements in terms of QD current gain for the light-trapping structures. ELO thin-film cells with almost doubled photocurrent and preserved V_{oc} are demonstrated. Different planar reflectors and their suitability to be used in III–V solar cells are discussed. More specifically, we address the design and development of diffractive gratings and the development of structured antireflection coatings. Finally, the role of a micro-patterned rear contact design to optimize the photon-recycling effect is discussed.

1.4.1 Light-Trapping Enhanced QD Solar Cells

The genuine thin-film structure obtained by ELO is an excellent platform for the implementation of effective photon management strategies—as described in Sect. 1.5—to maximize photon trapping and minimize reflection loss. Figure 1.18 analyses the potential in terms of achievable short-circuit current and efficiency of QDSCs with (a) a wafer-based configuration; (b) a thin-film structure with planar rear mirror—double pass configuration; and (c) a thin-film structure with textured rear surface—light-trapping configuration.

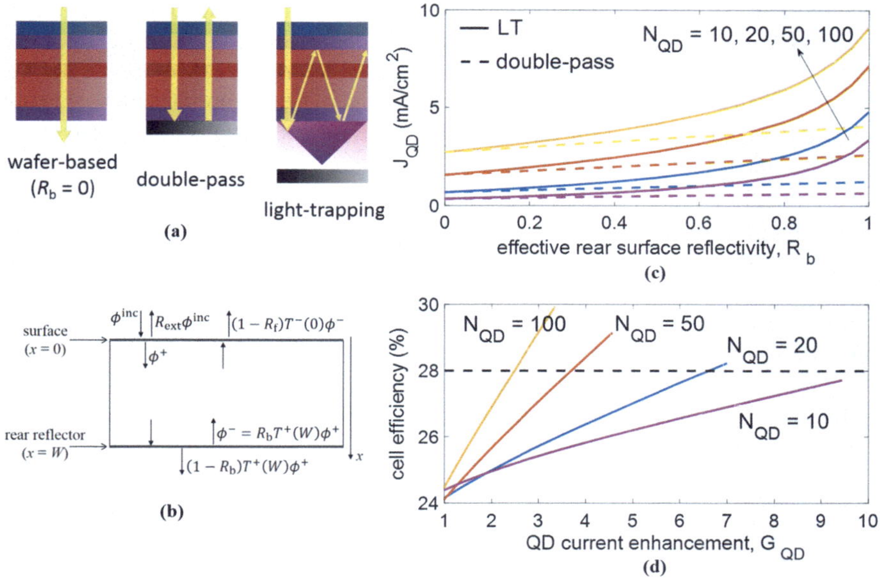

Fig. 1.18 (**a**) Basic InAs/GaAs QD solar cell structure implementing a single-pass configuration, double-pass configuration using a planar reflector and full light trapping using a periodic diffraction grating and a reflector. (**b**) Conceptual scheme of light trapping. (**c**) QD current density as a function of the effective rear surface reflectivity for thin-film cell with planar mirror—double-pass configuration—and thin-film cell with textured mirror, light-trapping configuration, using different numbers of QD layers. (**d**) Predicted cell efficiency as a function of the QD current enhancement with respect to the wafer-based configuration. The $R_b = 0$ point identifies the wafer-based configuration. (Reprinted from Sol. Energy Mater. Sol. Cells, 181, F. Cappelluti et al., Light-trapping enhanced thin-film III-V quantum dot solar cells fabricated by epitaxial lift-off, pp.83–92, Copyright (2018), with permission from Elsevier)

The optical flux in the cell is described by an incoherent multiple reflection model [123, 124], as depicted in Fig. 1.18b, with

- $R_f = 0$ and $R_b = 0$ for the wafer-based configuration.
- $R_f = 0$ and arbitrary $0 < R_b \leq 1$ for the double-pass configuration.
- $R_f = 1 - (1 - R_{ext})/n^2$, n being the semiconductor refractive index and R_{ext} the reflectance at the illuminated surface; $0 < R_b \leq 1$ and angular independent for the Lambertian-like light-trapping configuration. Here R_b is intended as an equivalent parameter modelling the reflectance of the specific textured structure implemented at the cell rear side.

We take as reference the epilayer structure exploited for the 50 × QD solar cell analysed in Sect. 1.3.1.2, with 10, 20, 50 or 100 QD layers uniformly distributed in the ≈ 1-μm-thick undoped part of the emitter. The analysis is carried out by assuming radiative recombination only and neglecting photon recycling. Moreover, it is assumed that the power loss due to the cumulative effect of reflection and shadowing is limited to 5%.

Figure 1.18c reports the QD short-circuit current contribution as a function of the rear surface reflectivity. J_{QD} is estimated by integrating the EQE in the QD wavelength range ($\lambda > 880$ nm) over the AM1.5G solar spectrum. The light-trapping (LT) scheme provides a remarkable increase of the QD current contribution compared to the wafer-based configuration ($R_b = 0$) and to the planar (double-pass) one. Defining the QD current gain, G_{QD}, as the ratio between J_{QD} at a certain value of R_b and the J_{QD} provided by the wafer-based cell ($R_b = 0$), we find that the maximum achievable gain is about 9.3, 7, 4.5 and 3.4 for the 10×, 20×, 50×, and 100× QD cell, respectively.

The predicted power conversion efficiency as a function of G_{QD} is shown in Fig. 1.18d. It is worth noticing that when $G_{QD} = 1$—i.e. in the wafer-based cell—there is no practical advantage to increase the number of QD layers, because the J_{sc} increase produced by the QDs is barely sufficient, in the best case, to compensate for the intrinsic V_{oc} loss. Therefore, improvements in the QD growth—either in terms of layers or sheet density—must be combined with light trapping to achieve high efficiency. Similar considerations were drawn for IB-QDSCs [41] on the basis of detailed-balance calculations. A tenfold or sevenfold increase of the QD current is demanded for thin-film 10 × and 20 × QD solar cells, as those investigated in this chapter, to approach 28% efficiency.

1.4.2 *ELO QD Cells with Planar Rear Mirror*

Figure 1.19 shows a 3-in. diameter thin-film QD GaAs solar cell structure that was processed into an array of individual 5 × 5 mm² solar cells. The structure containing 10 InAs QD layers was produced by MBE at Tampere University, ELO, and thin-film cell processing was performed by Radboud University and tf2 devices. As a reference, an equivalent structure was produced and processed while it remained on its native wafer (substrate-based cell).

Figure 1.20 indicates that both cells suffer from a relatively small V_{oc} penalty in the range of 150–170 mV (as shown in Fig. 1.6 where GaAs cells without QDs have

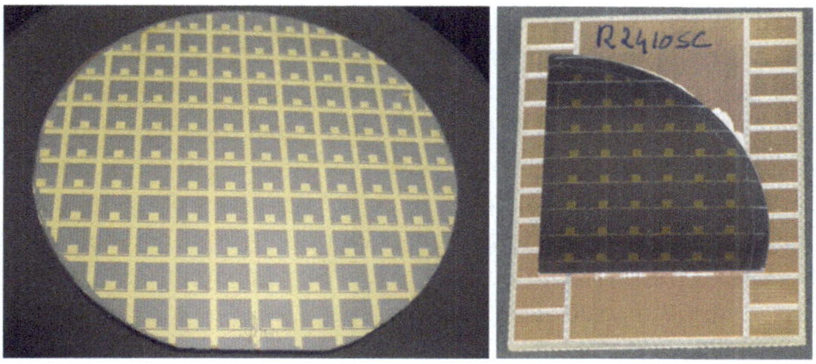

Fig. 1.19 Array of 5 × 5 mm² thin-film quantum dot (TFQD) solar cells, released from a 3-in. diameter GaAs wafer (left). Array of 5 × 5 mm² substrate-based quantum dot reference cells, on a section of a 3 in. wafer (right)

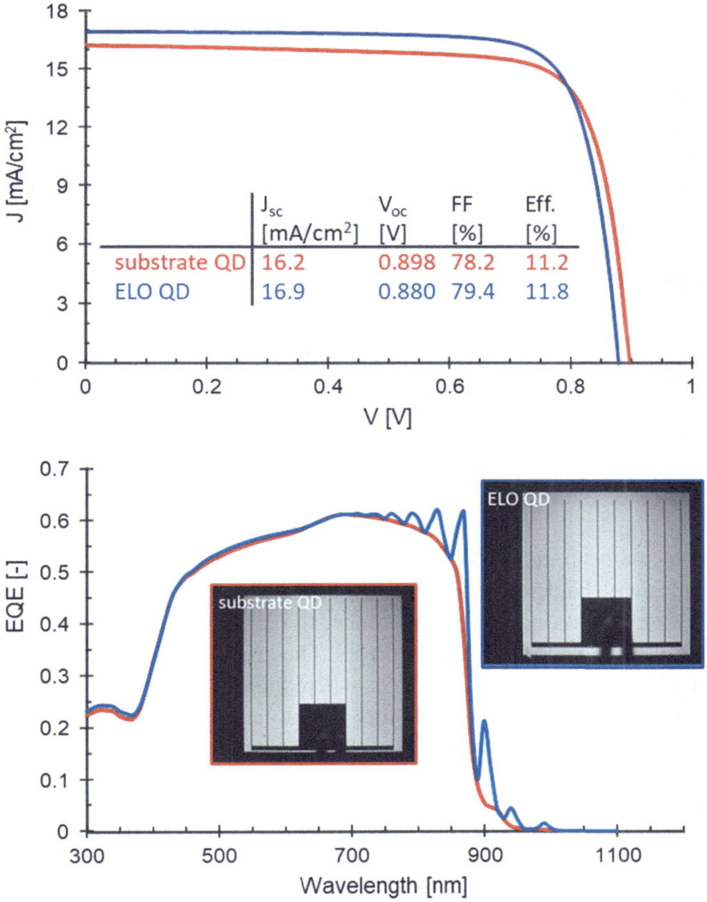

Fig. 1.20 *J–V* (upper figure), EQE (lower figure) and EL (insets) characteristics of 5 × 5 mm² ELO QD and substrate-based QD solar cells. Note that neither the ELO nor the substrate-based cells were equipped with antireflection coatings

a V_{oc} of about 1.05 V). The V_{oc} penalty in the ELO cell is slightly larger (18 mV), but this difference is more than compensated for by the larger J_{sc} and fill factor (FF) of the ELO cell as compared to the substrate-based cell, resulting in an 0.4% higher cell efficiency. The EQE shown in Fig 1.20 shows that the increased J_{sc} of 0.7 mA/cm^2 results from an enhanced absorption in the wavelength range above the GaAs cut-off wavelength of 875 nm owing to the QD layers as well as in the wavelength range just below 875 nm owing to an enhanced photon confinement. The EL signal of the cells (insets in Fig 1.20) does not reveal any apparent defects neither in the substrate-based nor in the thin-film cells.

1.4.3 Structures for Photon Management

1.4.3.1 Comparison of Different Metals for Planar Reflectors

Planar back surface reflectors such as Ag, Au and Cu provide almost ideal reflectivity at wavelengths longer than 800 nm [125]. Au is known to be a rather expensive material, and therefore the use of other highly conductive metals, such as Ag and Cu, would reduce the solar cell fabrication costs. Ag back reflectors have been demonstrated in III–V solar cells [126, 127], and Cu has also been used as a part of the back contact. However, Cu is known to diffuse into III–V semiconductors and thus to reduce the V_{oc} of the solar cell [128–130]. This could be alleviated by applying a double-layer Ag/Cu reflector. The purpose of Ag is to act as a diffusion barrier for Cu and to provide high reflectance in combination with Cu layer. The Cu layer acts as a current spreader and conductor. In addition, Ni is required between the Au and Cu layers as an adhesion layer, and Au acts as a protective layer and bonding surface.

The performance of planar back surface reflectors incorporating different metal layers were investigated with single-junction GaInNAs solar cells that were grown by MBE [8]. The metal reflectors were deposited using an electron beam evaporator, and the samples' details are given in Table 1.1.

EQE results of the solar cells with different metal layers as back reflectors are presented in Fig. 1.21a. The optical properties of the reflectors were characterized by reflectance measurements. To this end, double-side polished semi-insulating (SI) GaAs samples with back reflectors were used, and the results are shown in Fig. 1.21b. Compared to conventional Ti/Au reference, the EQE was improved for the solar cells with Au, Ag/Cu and Cu back reflectors. The greatest improvement in EQE was observed near the band edge of GaInNAs, where the absorption coefficient decreases and transmission increases – we note that this situation resembles the behaviour of QD solar cells, both in terms of wavelength range and low value of absorption. The highest EQE was obtained for the samples with Ag/Cu reflector, which could be the result of higher reflectance of the back reflector. This means that the photons that are not absorbed during the first pass are reflected back. The results between the 10 nm, 30 nm and 50 nm Ag/Cu reflectors are identical, meaning that a 10 nm layer of Ag is already enough to preserve high reflectivity. Furthermore, thermal annealing had no impact on the EQE of the 10 nm Ag/Cu reflector, permitting the processing conditions at 200 °C [131].

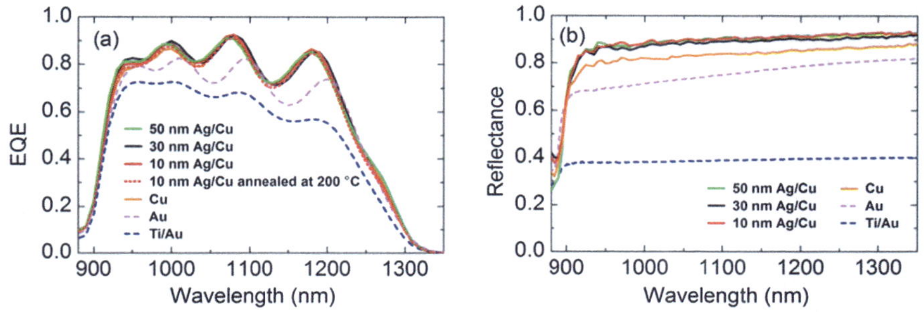

Fig. 1.21 (**a**) The EQE of the III–V solar cells with different back reflectors and (**b**) the reflectance of double-side polished SI-GaAs samples with back surface reflectors [131]. (Reproduced from T. Aho et al., Enhancement of photocurrent in GaInNAs solar cells using Ag/Cu double-layer back reflector, Appl. Phys. Lett. 109, 251,104 (2016), with the permission of AIP)

Table 1.2 Description of the fabricated samples

Back surface reflector	Metal layers	Thickness (nm)
Ti/Au	Ti/Au	50/100
Au	Au	100
Cu	Cu	100
10 nm Ag/Cu	Ag/Cu/Ni/Au	10/100/10/50
30 nm Ag/Cu	Ag/Cu/Ni/Au	30/100/10/50
50 nm Ag/Cu	Ag/Cu/Ni/Au	50/100/10/50

The J_{sc} values at AM1.5D (1000 W/m^2) spectral conditions [132] were calculated from the EQE results by integrating over the given spectrum and assuming a thick GaAs filter on top of GaInNAs cell. The results are collected in Table 1.2. The solar cells with Ag/Cu reflectors exhibited J_{sc} values of ~14 mA/cm^2, which is more than 25% higher compared to the reference solar cell with Ti/Au reflector.

Back surface reflectors employed as ohmic contacts require a low contact resistivity. Conventional ohmic contacts are made of alloyed metals wherein annealing is shown to lower reflectivity due to the reduction of interface sharpness when compared to non-alloyed contacts [92]. On the other hand, non-alloyed contacts usually suffer from higher contact resistivity [133], yet high doping levels can be employed in the contact GaAs layer to reach low contact resistivity values [128]. Typically, contact resistivity values for p-GaAs are of the order of 10^{-5} Ω·cm^2 and 10^{-7} Ω·cm^2 for non-alloyed and for alloyed metal contacts, respectively [133, 134]. The contact resistivity obtained for the different non-alloyed layers used in our study deposited on a p-type GaAs layer with a doping level of ~10^{20} cm^{-3} is presented in Table 1.2; the measurements were made using the transmission line method (TLM). As presented in Table 1.3, all the reflectors showed low contact resistivity of the order of 10^{-6} Ω·cm^2, which is lower than the reported contact resistivity of non-alloyed contacts [134]. These results show that an ohmic contact is formed even without annealing due to high doping in the GaAs layer, and high reflectivity of the metal reflector is retained. In addition, the Ag/Cu reflector had good adhesion to III–V semiconductor, and a thin Ag layer acts as a diffusion barrier, preventing Cu diffusion into semiconductor structure even when the cell is subjected to thermal annealing [131].

Table 1.3 Measured contact resistivity results and J_{sc} values at AM 1.5D [131]

Back surface reflector	Contact resistivity (Ω cm^2)	J_{sc} (mA/cm^2)
Ti/Au	3×10^{-6}	10.7
Au	1×10^{-6}	12.4
Cu	4×10^{-6}	13.4
10 nm Au/Cu	4×10^{-6}	13.6
10 nm Ag/Cu[a]	–	13.3
30 nm Ag/Cu	4×10^{-6}	13.7
50 nm Ag/Cu	5×10^{-6}	13.7

[a]Annealed at 200 °C for 90 s

1.4.3.2 Gratings for Light Diffraction in InAs/GaAs QDSC

Periodically textured reflectors patterned on the cell rear side can be exploited to enhance the QD absorptivity. ELO thin-film QD cells with textured back surface reflectors were reported in [42], demonstrating a 30% increase of QD current contribution compared to a cell with planar reflector. Effective light trapping by backside periodic grooves has also been demonstrated in multiple quantum well solar cells in [135], attaining a fivefold increase of the sub-bandgap optical path length.

Here, we report the design and experimental development of diffractive rear mirrors for thin-film QD cells. These structures are designed with a period larger than the incident wavelength, so that the reflector excites high-order diffractive modes that propagate outside the escape cone and get reflected multiple times through the absorber material.

Possible designs with uni-periodic gratings with triangular cross-section and bi-periodic pyramidal gratings directly imprinted on the widegap (AlInP) back surface field layer are shown in Fig. 1.22. A planarizing polymer layer (SU-8) is used to alleviate the parasitic losses due to unwanted surface plasmon effect that could arise at the large interface between the microstructured semiconductor and the metallic (Ag) mirror [137, 138]. The planar ARC can be composed of several materials, including a single layer of Si_3N_4 or a double layer of SiO_2 and TiO_2. Alternatively, a nanostructured pyramidal ARC could be patterned directly on the AlInP window layer [139].

The structures were simulated with the rigorous coupled-wave analysis (RCWA) method [140] in order to identify the grating configuration that maximizes the QD current contribution, $J_{sc,QD}$. This is estimated integrating over the QD wavelength range the absorbed photon density in the active region. As seen in Fig. 1.22b, it turns out that the most relevant design parameter is the height/period aspect ratio of the grating, with optimum values between 0.32 and 0.38. The same optimum was also found for gratings realized in low index materials [141].

Figure 1.23 shows examples of calculated absorbance spectra for cells with a pyramidal grating with height of 900 nm and period of 2400 nm, exploiting a planar ARC and a nanostructured AlInP ARC, respectively. For reference, also the absorbance of an unpatterned cell (equivalent to a wafer-based configuration) and of a cell with planar reflector and ARC is shown. The enhancement due to light trapping is evident close to the GaAs bandgap and in the QD absorption region (λ = 890 –1200 nm).

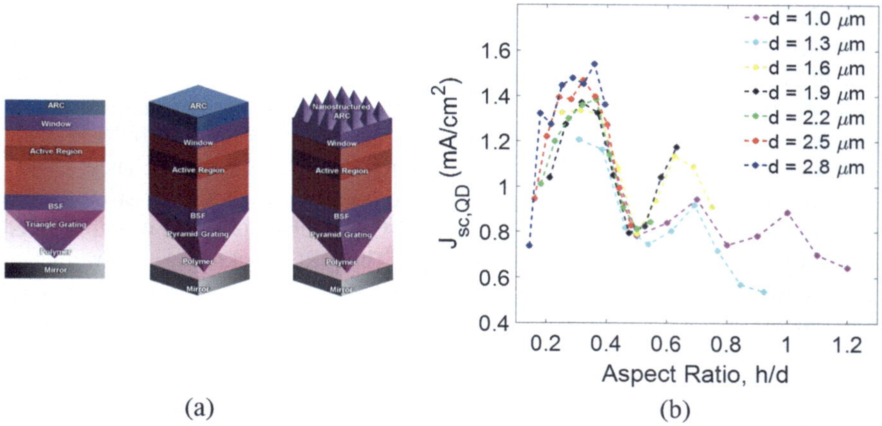

Fig. 1.22 (**a**) Sketch of thin-film solar cells integrating: planar ARC and triangular grating, planar ARC and pyramidal grating, nanostructured ARC and pyramidal grating. The active region is composed of GaAs emitter, a stack of 10 InAs/GaAs QD layers and GaAs base. (**b**) QD short-circuit current contribution as a function of the aspect ratio for a uni-periodic triangular grating without metallic reflector. (Adapted with permission from [136] © The Optical Society)

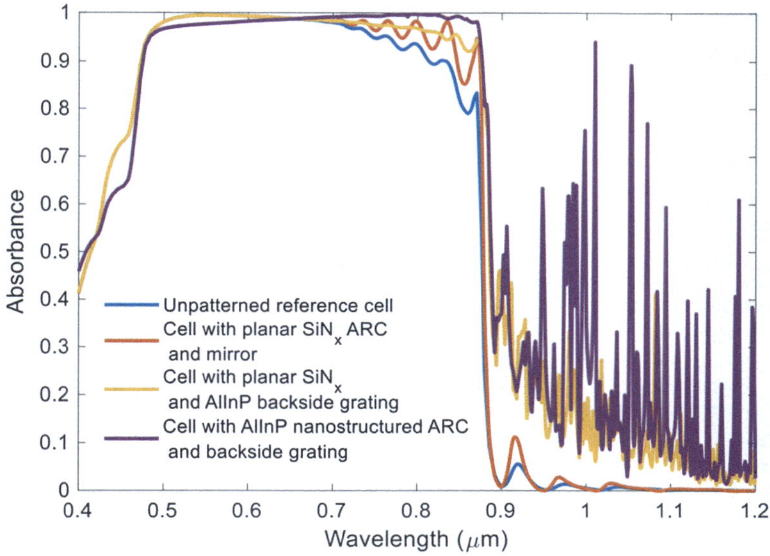

Fig. 1.23 Absorbance comparison between the planar reference cell, cell with planar ARC with mirror, cell with planar ARC and backside grating and cell with nano-ARC and backside grating. (Adapted with permission from [142] © The Optical Society)

From these spectra, we estimate that the cell with planar ARC and pyramidal backside grating provides a photocurrent increase of 16 times with respect to the unpatterned cell. In the double-side patterned cell, the combined action of the front ARC and backside diffraction grating provides more effective light trapping with a current increase of about 27. Such values are very promising when compared to the needs in terms of QD current gain set in Fig. 1.18 to surpass the 28% target efficiency.

Diffraction gratings test structures with the optimum aspect ratio of 0.32–0.38 were fabricated using a double-side polished SI-GaAs substrate which was embedded with single-layer SiN_x antireflection coatings. The ARC thicknesses and refractive indexes are presented in Fig. 1.24, which introduces the three investigated geometries, namely, half-sphere, blazed and pyramid gratings, and the planar reference.

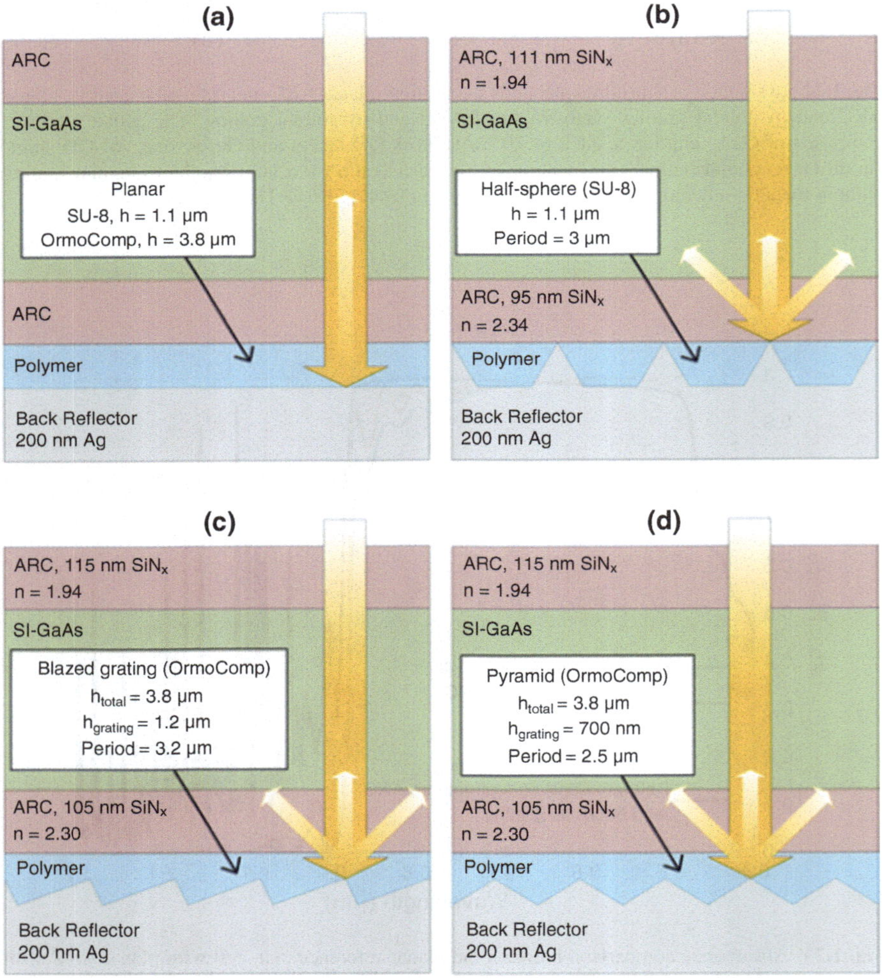

Fig. 1.24 The schematic drawing of the diffraction structures. (**a**) Planar, (**b**) half-sphere grating, (**c**) blazed grating and (**d**) pyramidal grating. (Reprinted with permission from [143] © The Optical Society)

The half-sphere gratings were fabricated by photolithography using a commercial epoxy-based negative photoresist, SU-8 (MicroChem Corp.), which is commonly used in the electronic industry. To obtain the desired half-sphere structure, photolithography process parameters such as spin coating speed, exposure time and baking temperatures were optimized. For blazed grating and pyramidal grating, a commercial OrmoComp nanoimprint lithography (NIL) photoresist (micro resist technology GmbH) was used in the NIL processes. The gratings were transferred from NIL master into the NIL photoresist. Figure 1.25 shows the detailed images of the three experimental structures. The top of the half-sphere appeared slightly flattened, as shown in the optical profilometer measurement in Fig. 1.25a. The scanning electron microscopy (SEM) images of the pyramid grating, presented in Fig. 1.25c, d, are taken from the back surface of the sample where 200 nm Ag is evaporated, whereas the blazed grating image in Fig. 1.25b is the cross-section of the polymer grating without Ag.

The specular reflectance at 8° incident angle and the total reflectance were measured with spectrophotometer. All the measurements were carried out through the SI-GaAs wafer, and the performance of the back reflectors was evaluated at the wavelengths for which SI-GaAs is transparent. The reflectance results of the half-sphere grating are presented in Fig. 1.26 (left), where the difference between the total and the specular reflectance represents the diffuse reflectance. A local maximum in the specular reflectance of the half-sphere is observed, decreasing the diffuse reflectance at the wavelength range of 950–1150 nm. The reflectance results of the blazed grating and the pyramid grating are compared in Fig. 1.26 (right). According to the results, the pyramid grating shows the highest diffuse reflectance when compared to the blazed grating or the half-sphere structure. In addition, the total reflectance of the blazed grating was lower compared to the pyramid grating, revealing some extra losses in the structure, probably originating from the surface plasmon resonance (SPR) in the metal reflector.

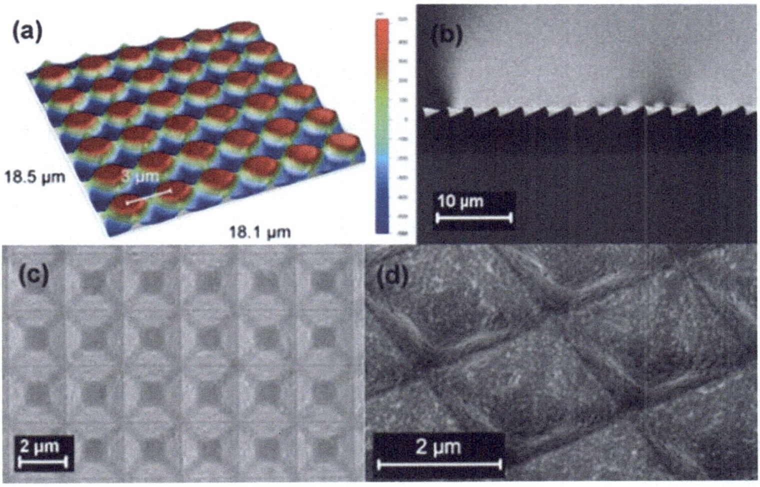

Fig. 1.25 (**a**) Optical profilometer image of the half-sphere structure, (**b**) SEM image of the blazed grating, (**c**) and (**d**) SEM image of the pyramidal grating. (Reprinted with permission from [143] © The Optical Society)

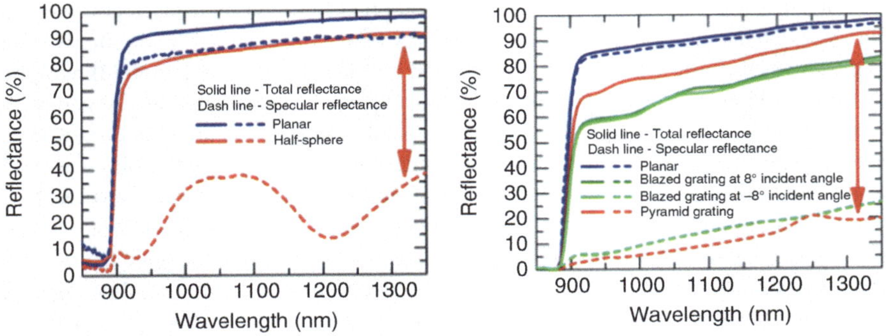

Fig. 1.26 Left: Total and specular reflectance of the half-sphere grating and the planar reference. Right: the blazed grating, pyramidal grating and planar reference. (Reprinted with permission from [143] © The Optical Society)

In order to understand the suitability of polymer grating architectures for space applications, an electron irradiation (1 MeV) study was performed to the samples with the OrmoComp polymer gratings. The reflectance spectra were measured after the electron irradiation. The reflectance results of the irradiated samples were identical up to doses of 1×10^{15} e^-/cm^2 when compared to the sample without irradiation in Fig. 1.26 (right), indicating that the polymer could be suitable for diffractive gratings in space solar cells.

1.4.3.3 Antireflection Coatings Using Nano-Gratings

To fully exploit light trapping in QD solar cells, nanostructured ARCs need to be considered. Tommila et al. have demonstrated NIL process to fabricate nanocones into AlInP window layer with the period of 300 nm [139]. The reflectance results of the nanocone ARC are presented in Fig. 1.27, where the red line is 300 nm period cone and the black line is the reference sample without any structuring. The 300 nm cone structure resulted in less than 5% average reflectance in the wide wavelength range of 400–1300 nm. As observed in the SEM image of the cone structure, presented in Fig. 1.27, the grating showed high uniformity.

To produce nanocone structures with smaller period (< 300 nm), a process simpler than the 300 nm cones process was devised and utilized. Based on simulations (not reported here), gratings with a period of 180 nm can be designed to achieve comparable reflectance lowering the height of the cones. This simplifies the processing of the nanocones. The SEM images of AlInP cones with the period of 180 nm are presented in Fig. 1.28. In addition, 40 nm SiN_x layer with refractive index of 2.0 was evaporated on top of the cones to further lower the reflectance.

The reflectance results to the samples before and after the SiN_x evaporation are presented in Fig. 1.29. The results show that the reflectance remains less than 5% from wavelength of 320 nm to ~950 nm for the sample without SiN_x layer and to

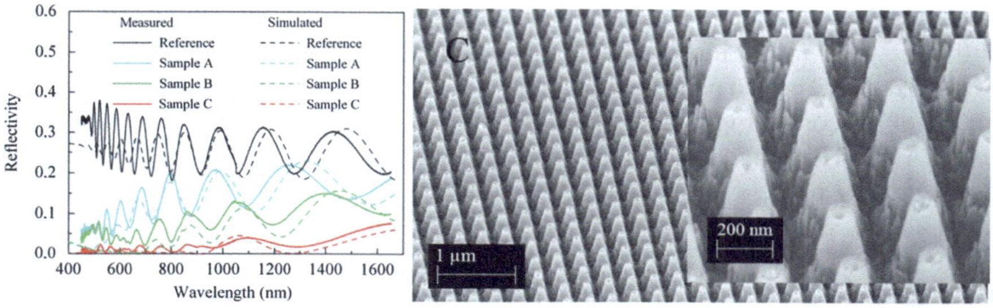

Fig. 1.27 Left: Simulated and measured reflectivity results of 300 nm period cones in AlInP (red line). Right: SEM images of 300 nm period cones [139]. (Reprinted from Solar Energy Mater. Solar Cells, 94, J. Tommila et al., Nanostructured broadband antireflection coatings on AlInP fabricated by nanoimprint lithography, pp. 1845–1848, Copyright (2010), with permission from Elsevier)

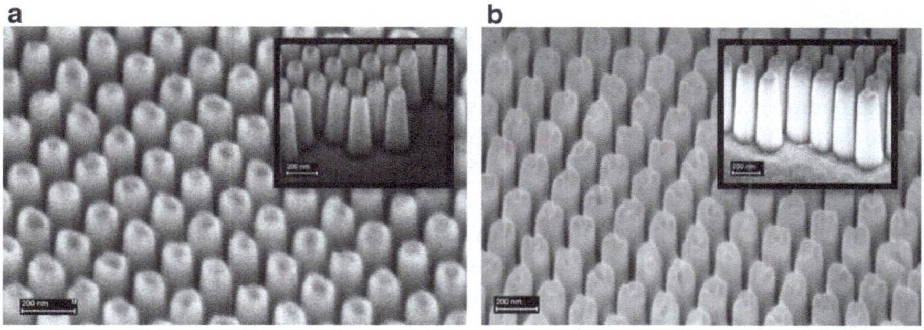

Fig. 1.28 (**a**) SEM images of 180 nm period cone structure in AlInP, (**b**) 40 nm SiN_x layer of top of the AlInP cones

Fig. 1.29 Reflectance results of 180 nm AlInP cone structure. The red line represents the 5% reflectance

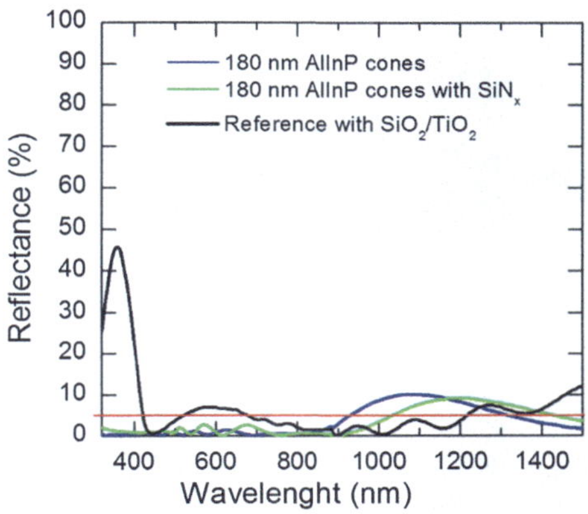

~1000 nm for the sample with SiN_x layer. The SiN_x layer on top of the AlInP cones shifts the peak to the longer wavelengths. As a reference, a GaAs solar cell with SiO_2/TiO_2 ARC is also presented in Fig. 1.29. The 180 nm cones have rough surface on top of the cones (Fig. 1.28), which might have an effect on the optical properties of the ARC. The rough top surface could possibly be avoided with the optimization of the etching process of the cones.

1.4.3.4 Patterned Rear Mirror/Contact for Enhanced Photon Confinement and Recycling

For solar cells where the non-radiative recombination is sufficiently reduced, radiative recombination becomes the dominant loss mechanism. The performance of cells operating in this regime can be strongly enhanced by utilizing the mechanism of photon recycling. Whereas in wafer-based cells a large part of the regenerated photons might escape to the substrate, the back contact geometry of a thin-film cell can be designed to optimize its reflectivity and maximally benefit from photon recycling [14, 16, 86, 94, 95, 144]. Figure 1.30 shows the results of an approach in which the initially fully metalized back contact of a single-junction GaAs cell is reduced to a discrete number of contact points. In between the highly doped GaAs contact layer of the cell is etched away and replaced by a dielectric (ZnS) layer. In combination with the Au-coated copper carrier, the ZnS layer yields a strongly increased rear-side reflectivity (R) from 63.5% to 96.5% in the relevant wavelength range (840–870 nm) [145]. Therefore, by reducing the size of the rear contact points, the fraction of the rear contact surface coverage (Cr) also decreases, resulting in an increased reflectivity from $R = 63.5\%$ for $\mathrm{Cr} = 100\%$ up to $R = 93.2\%$ for $\mathrm{Cr} = 10\%$. Figure 1.31a shows a 2-in. diameter thin-film single-junction GaAs cell structure that was processed into multiple individual $5 \times 5\ \mathrm{mm}^2$ solar cells with a different rear contact coverage in each quadrant. The EL signal of the cells (Fig. 1.31) does not reveal any apparent defect in the thin-film cells but clearly shows the decreasing size of the rear

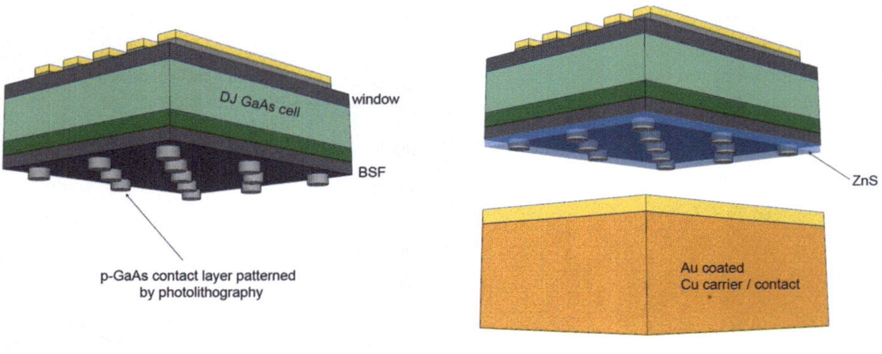

Fig. 1.30 Schematic representation (not to scale) of a partial contact approach to optimize the rear-side reflectivity of a III–V thin-film cell

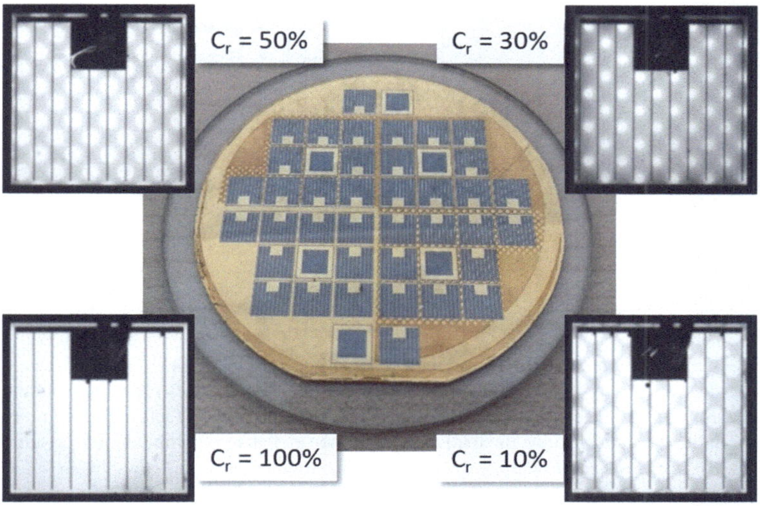

Fig. 1.31 A 2-in. diameter thin-film GaAs cell structure containing multiple 5 × 5mm² solar cells with a different rear contact coverage (Cr) in each quadrant with the electroluminescence image of a cell from each section

contact points. For the ease of contacting, the cells have a large front coverage of 16.6% which is about ten times larger than that applied in our record cells [92]. Nevertheless, *I-V* analysis shows that the cells have efficiencies around 22% which corresponds to well over 25% in case a low coverage grid would be applied.

Related to the increasing distance between the rear contact points, the series resistance of the cells increases with decreasing Cr. As a result, the fill factor of the cells initially decreases gradually with decreasing Cr and later quite abruptly for Cr = 10%. On the other hand, decreasing Cr results in an increased rear-side reflectivity which according to a theoretical study by Miller et al. [16] should result in a linear increase in short-circuit current density and a more than linear increase in open circuit voltage. Figure 1.32 shows that the experimentally determined trend in J_{sc} and V_{oc} confirms the theoretical predictions. The increased J_{sc} with R is simply the result of the increased reflection and subsequent absorption of photons in the active layers of the cell. On the other hand, the more than linear increase in V_{oc} with R is indicative for effective photon recycling in the solar cell. In follow-up experiments, the rear-side reflectivity of the cell was increased from 93.2% to 96% by further decreasing the diameter of the rear contact points to 40 μm, while at the same time, their mutual distance was decreased to minimize resistive losses and maintain a high fill factor. The V_{oc} of these cells was found to be 1.071 V providing further evidence for the rapid increase in cell voltage as the reflectivity approaches 100%. The present world record thin-film GaAs solar cell was reported to have a V_{oc} of 1.1272 V and a fill factor of 0.867 [1]. Although the exact structure of this cell is not published, the high V_{oc} and fill factor values suggest a DJ geometry with a near-perfect back reflector.

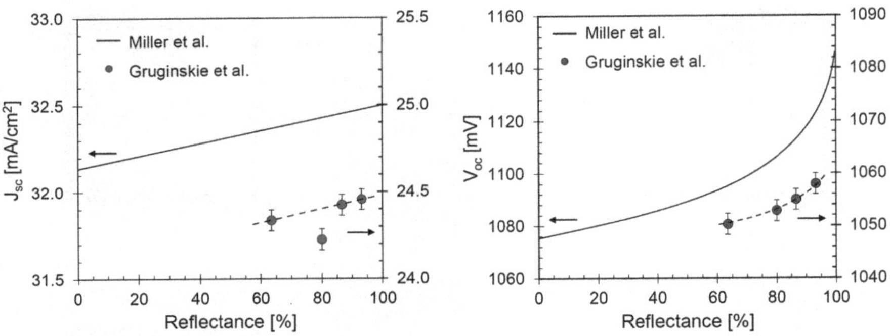

Fig. 1.32 Theoretically predicted [16] and experimentally determined [145] J_{sc} and V_{oc} as a function of the rear-side reflectivity in thin-film GaAs cells. Note that the J_{sc} values resulting from the theory are significantly higher since the reflection of light at the ARC and grid contacts of the cell is neglected. Moreover, in the theoretical study, the total absence of non-radiative recombination is assumed

In a more recent study [146], a highly doped AlGaAs back contact layer was applied which between the localized contact points was not fully removed but partially etched using a NaOH/H_2O_2 solution. The induced surface roughness facilitated the application of a diffusive back reflector which allowed for a strong reduction of the solar cell thickness. Due to the increased photon confinement, ultra-thin (300 nm) GaAs cells with such textured mirrors were demonstrated to have a highly increased current density yielding efficiency up to 21.4%, whereas the efficiency in ultra-thin cells with planar mirrors was limited to 18.5%.

1.5 Summary and Outlook

Thin-film III–V cells combined with light management structures are a very promising concept as alternative to the costly multi-junction cell route, both for space and terrestrial uses. QDs may provide additional advantages such as higher radiation hardness in space and better matching to the sun spectrum in CPV, provided that the issues of large V_{oc} reduction and small QD photogeneration are solved.

In this contribution, we have analysed the V_{oc} reduction due to intrinsic and extrinsic mechanisms. Intrinsic loss causes a linear dependence of the V_{oc} with the QD emission wavelength as verified experimentally by a large set of fabricated samples and theoretically by quantum-corrected drift-diffusion simulations. MBE-grown 10 × InAs/GaAs QD solar cells with high in-plane density QDs demonstrate V_{oc} higher than 0.95 V, a record value for MBE-grown QDSCs, in line with the record value for MOCVD-grown QDSCs.

Numerical simulations demonstrate that light management is essential to improve the QD absorption and the cell efficiency. Incidentally, our study evidences that increasing the number of QD layers or the QD areal density is useful to achieve high-efficiency QDSCs only if combined with light trapping. Towards this goal, successful ELO of MBE-grown layer structures from 3 in. wafers is reported. ELO QDSCs show a twofold increase of the QD photocurrent and a high V_{oc} of 0.88 V, the highest ever reported for thin-film III–V QD cells. Several photon management structures have been discussed. A planar double-layer Ag/Cu back contact reflector—withstanding annealing temperatures of at least 200 °C—was demonstrated, resulting in 28% higher calculated J_{sc} than that of the reference Ti/Au back contact and low contact resistivity. Back reflector structures enabling light diffraction were designed and developed by employing different geometries. Efficient diffraction of light is observed for all the structures, indicating potential for improving the absorption and hence the current generation in the thin-film solar cells. Pyramid gratings showed the highest diffuse reflectance in the wavelengths corresponding to the photogeneration in InAs/GaAs QDSCs. The nanostructured ARCs fabricated into AlInP window layer showed less than 5% reflectance in the wavelength range of 400–1300 nm. These ARC nanostructures together with diffraction back surface reflectors will enable light trapping in QDSCs. The fabrication techniques of the light-trapping structures are applicable also to other types of III–V solar cells by adjusting the dimensions of the gratings to be optimal for the specific wavelength range.

As the quality of the QD solar cell structures approaches that of regular solar cells, ELO QD cells can achieve higher V_{oc} than their substrate-based counterparts by utilizing the photon confinement and recycling mechanisms, as already demonstrated in conventional single-junction cells. A novel rear contact/mirror design aimed at optimizing photon recycling has been reported. A selective patterning technique of the bottom contact layer implemented in thin-film GaAs cells yields a clear non-linear increase of V_{oc} well correlated to the increase of reflectance as the contact layer coverage is reduced from 100% to 10%. In addition, application of a diffusive rear mirror was demonstrated to significantly increase the current density and efficiency of ultra-thin GaAs cells with an active layer thickness of only 300 nm.

The promising experimental results and the guidelines provided by the numerical simulations presented in this work demonstrate that high-efficiency QD-based single- and multi-junction cells are feasible in the next future. In the longer term, the achieved high material quality and V_{oc} and the technology of photonic structures for light trapping and photon recycling will be instrumental to overcome the challenges posed by the implementation of intermediate band QDSCs.

Based on its efficiency improvements, III–V-based photovoltaic technologies have good prospects for a stepwise expansion from the current niche application areas. With the present wafer-based production technologies, this might gradually proceed via CPV applications in areas with lower DNI. However, once the III–V photovoltaics industry succeeds in maturing the cost-effective ELO production technology for genuine thin-film devices from reusable wafers, the cost of III–V space solar cells can drop significantly, while their power-to-weight ratio can increase substantially. Both factors will ease the realization of satellite constellations. More importantly, it would enable CPV to become competitive in areas with much lower DNI, thereby substantially expanding the CPV and III–V photovoltaics markets. The utilization of III–V cells might then be further accelerated by the access of new market segments as well. Such markets might include unmanned aerial vehicles [147, 148], motor vehicles [149] and portable electronics [147]. For ultimate large-scale application in non-concentrating III–V solar modules in the build environment, besides ELO, a low-cost deposition technology is required for the production of the cell structures. Despite the huge challenges, developments are already being made in this direction [150].

Acknowledgements The work was partly funded by the European Union Horizon 2020 projects TFQD (Grant Agreement No. 687253) and ERC AdG project AMETIST (Grant Agreement No. ERC-2015-AdG 695116). The authors would also like to thank Dr. Arto Aho for valuable discussions, Dr. Jari Lyytikäinen and Mr. Eero Halonen for MBE-related actions and Ms. Marianna Raappana and Mrs. Ninja Kajas for sample processing support. The authors would also like to thank Prof. Huiyun Liu for his help with starting the QD growths.

References

1. Green, M. A., Emery, K., Hishikawa, Y., Warta, W., & Dunlop, E. D. (2015). Solar cell efficiency tables (Version 45). *Progress in Photovoltaics: Research and Applications, 23*, 1–9.
2. Shockley, W., & Queisser, H. J. (1961). Detailed balance limit of efficiency of p-n junction solar cells. *Journal of Applied Physics, 32*, 510–519.
3. Tanabe, K., Guimard, D., Bordel, D., & Arakawa, Y. (2012). High-efficiency InAs/GaAs quantum dot solar cells by metalorganic chemical vapor deposition. *Applied Physics Letters, 100*, 193905.
4. Jackson, E. D. (1955). *Transactions of the conference on the use of solar energy* (pp. 122–126). Tucson, AZ:University of Arizona Press.
5. Ekins-Daukes, N. J. (2014) III-V solar cells. In G. Conibeer, & A. Willoughby (Eds.), *Solar cell materials: Developing technologies* (pp. 113–143). Wiley.
6. Marti, A., & Araújo, G. L. (1996). Limiting efficiencies for photovoltaic energy conversion in multigap systems. *Solar Energy Materials & Solar Cells, 43*, 203–222.
7. Brown, A. S., & Green, M. A. (2002). Limiting efficiency for current-constrained two-terminal tandem cell stacks. *Progress in Photovoltaics: Research and Applications, 10*, 299–307.
8. Dimroth, F., & Kurtz, S. (2007). High-efficiency multijunction solar cells. *MRS Bulletin, 32*, 230–235.
9. Dimroth, F., Tibbits, T. N., Niemeyer, M., Predan, F., Beutel, P., Karcher, C., Oliva, E., Siefer, G., Lackner, D., Fuß-Kailuweit, P., & Bett, A. W. (2015). Four-junction wafer-bonded concentrator solar cells. *IEEE Journal of Photovoltaics, 6*, 343–349.
10. Philipps, S. P., Bett, A. W., Horowitz, K., & Kurtz, S. (2015). Current status of concentrator photovoltaic (CPV) technology (No. NREL/TP-5J00-65130). National Renewable Energy Lab (NREL), Golden, CO (United States).
11. Branker, K., Pathak, M. J. M., & Pearce, J. M. (2011). A review of solar photovoltaic levelized cost of electricity. *Renewable and Sustainable Energy Reviews, 15*, 4470–4482.
12. Tatavarti, R., Hillier, G., Dzankovic, A., Martin, G., Tuminello, F., Navaratnarajah, R., Du, G., Vu, D. P., & Pan, N. (2008). Lightweight, low cost GaAs solar cells on 4″ epitaxial liftoff (ELO) wafers. In 2008 33rd IEEE Photovoltaic Specialists Conference, 1–4.
13. Bauhuis, G. J., Mulder, P., Haverkamp, E. J., Schermer, J. J., Bongers, E., Oomen, G., Köstler, W., & Strobl, G. (2010). Wafer reuse for repeated growth of III–V solar cells. *Progress in Photovoltaics: Research and Applications, 18*, 155–159.
14. Kayes, B. M., Nie, H., Twist, R., Spruytte, S. G., Reinhardt, F., Kizilyalli, I. C., & Higashi, G. S. (2011). 27.6% conversion efficiency, a new record for single-junction solar cells under 1 sun illumination. *37th IEEE Photovoltaic Specialists Conference* (pp. 4–8).
15. Lush, G., & Lundstrom, M. (1991). Thin film approaches for high-efficiency III–V cells. *Solar cells, 30*, 337–344.

16. Miller, O. D., Yablonovitch, E., & Kurtz, S. R. (2012). Strong internal and external luminescence as solar cells approach the Shockley–Queisser limit. *IEEE Journal of Photovoltaics, 2*, 303–311.
17. Wang, X., Khan, M. R., Gray, J. L., Alam, M. A., & Lundstrom, M. S. (2013). Design of GaAs solar cells operating close to the Shockley–Queisser limit. *IEEE Journal of Photovoltaics, 3*, 737–744.
18. National Renewable Energy Laboratory (NREL). (n.d.). Low-cost III-V solar cells. Retrieved from https://www.nrel.gov/pv/low-cost-iii-v-solar-cells.html.
19. Kurtz, S. R., Olson, J. M., & Faine, P. (1991). The difference between standard and average efficiencies of multijunction compared with single-junction concentrator cells. *Solar Cells, 30*, 501–513.
20. Warmann, E. C., & Atwater, H. A. (2016). Energy production advantage of independent subcell connection for multijunction photovoltaics. *Energy Science & Engineering, 4*, 235–244.
21. Villa, J., & Martí, A. (2017). Impact of the spectrum in the annual energy production of multijunction solar cells. *IEEE Journal of Photovoltaics, 7*, 1479–1484.
22. Warren, E. L., Deceglie, M. G., Rienäcker, M., Peibst, R., Tamboli, A. C., & Stradins, P. (2018). Maximizing tandem solar cell power extraction using a three-terminal design. *Sustainable Energy & Fuels, 2*, 1141–1147.
23. Linares, P. G., Antolín, E., & Martí, A. (2019). Novel heterojunction bipolar transistor architectures for the practical implementation of high-efficiency three-terminal solar cells. *Solar Energy Materials & Solar Cells, 194*, 54–61.
24. Barnham, K. W. J., Ballard, I., Connolly, J. P., Ekins-Daukes, N. J., Kluftinger, B. G., Nelson, J., & Rohr, C. (2002). Quantum well solar cells. *Physica E: Low-dimensional Systems and Nanostructures, 14*, 27–36.
25. Aroutiounian, V., Petrosyan, S., Khachatryan, A., & Touryan, K. (2001). Quantum dot solar cells. *Journal of Applied Physics, 89*, 2268–2271.
26. Hubbard, S., Bailey, C., Polly, S., Cress, C. D., Andersen, J., Forbes, D. V., & Raffaelle, R. P. (2009). Nanostructured photovoltaics for space power. *Journal of Nanophotonics, 3*, 031880.
27. Lumb, M. P., Dobbin, A. L., Bushnell, D. B., Lee, K. H., & Tibbits, T. N. D. (2010). Comparing the energy yield of (III–V) multi-junction cells with different numbers of subcells. *AIP Conference Proceedings 1277*, 299–302.
28. Welser, R. E., Sood, A. K., Laghumavarapu, R. B., Huffaker, D. L., Wilt, D. M., Dhar, N. K., & Sablon, K. A. (2015). The physics of high-efficiency thin-film III-V solar cells. In *Solar cells-new approaches and reviews*. IntechOpen.
29. Luque, A., & Martí, A. (1997). Increasing the efficiency of ideal solar cells by photon induced transitions at intermediate levels. *Physical Review Letters, 78*, 5014.
30. Luque, A., & Martí, A. (2010). The intermediate band solar cell: Progress toward the realization of an attractive concept. *Advanced Materials, 22*, 160–174.
31. Okada, Y., Ekins-Daukes, N. J., Kita, T., Tamaki, R., Yoshida, M., Pusch, A., Hess, O., Phillips, C. C., Farrell, D. J., Yoshida, K., & Ahsan, N. (2015). Intermediate band solar cells: Recent progress and future directions. *Applied Physics Reviews, 2*, 021302.
32. Sakamoto, K., Kondo, Y., Uchida, K., & Yamaguchi, K. (2012). Quantum-dot density dependence of power conversion efficiency of intermediate-band solar cells. *Journal of Applied Physics, 112*, 124515.
33. Hubbard, S. M., Cress, C. D., Bailey, C. G., Raffaelle, R. P., Bailey, S. G., & Wilt, D. M. (2008). Effect of strain compensation on quantum dot enhanced GaAs solar cells. *Applied Physics Letters, 92*, 123512.
34. Zhou, D., Sharma, G., Thomassen, S., Reenaas, T., & Fimland, B. (2010). Optimization towards high density quantum dots for intermediate band solar cells grown by molecular beam epitaxy. *Applied Physics Letters, 96*, 061913.
35. Fujita, H., Yamamoto, K., Ohta, J., Eguchi, Y., & Yamaguchi, K. (2011). In-plane quantum-dot superlattices of inas on gaassb/gaas (001) for intermediate band solar-cells. *Proceedings of the 37th IEEE Photovoltaic Specialists Conference (PVSC)* (pp. 2612–2614).

36. Tutu, F. K., Wu, J., Lam, P., Tang, M., Miyashita, N., Okada, Y., Wilson, J., Allison, R., & Liu, H. (2013). Antimony mediated growth of high-density InAs quantum dots for photovoltaic cells. *Applied Physics Letters, 103*, 043901.
37. Sameshima, K., Sano, T., & Yamaguchi, K. (2016). Self-formation of ultrahigh-density (1012 cm-2) inas quantum dots on inassb/gaas (001) and their photoluminescence properties. *Applied Physics Express, 9*, 075501.
38. Akahane, K., Yamamoto, N., & Kawanishi, T. (2011). Fabrication of ultra-high-density inas quantum dots using the strain-compensation technique. *Physica Status Solidi (a), 208*, 425–428.
39. Sugaya, T., Numakami, O., Oshima, R., Furue, S., Komaki, H., Amano, T., Matsubara, K., Okano, Y., & Niki, S. (2012). Ultra-high stacks of InGaAs/GaAs quantum dots for high efficiency solar cells. *Energy & Environmental Science, 5*, 6233–6237.
40. Cappelluti, F., Kim, D., van Eerden, M., Cédola, A. P., Aho, T., Bissels, G., Elsehrawy, F., Wu, J., Liu, H., Mulder, P., Bauhuis, G., Schermer J. J., Niemi T., & Guina M. (2018). Light-trapping enhanced thin-film III-V quantum dot solar cells fabricated by epitaxial lift-off. *Solar Energy Materials & Solar Cells, 181*, 83–92.
41. Mellor, A., Luque, A., Tobías, I., & Martí, A. (2014). The feasibility of high-efficiency InAs/GaAs quantum dot intermediate band solar cells. *Solar Energy Materials & Solar Cells, 130*, 225–233.
42. Smith, B. L., Slocum, M. A., Bittner, Z. S., Dai, Y., Nelson, G. T., Hellstroem, S. D., Tatavarti, R., Hubbard, S. M. (2016). Inverted growth evaluation for epitaxial lift off (ELO) quantum dot solar cell and enhanced absorption by back surface texturing. *IEEE 43rd Photovoltaic Specialists Conference (PVSC)* (pp. 1276–1281).
43. Cappelluti, F., Gioannini, M., Ghione, G., Khalili, A. (2016). Numerical study of thin-film quantum-dot solar cells combining selective doping and light-trapping approaches. *IEEE 43rd Photovoltaic Specialists Conference (PVSC)* (pp. 1282–1286).
44. Bennett, M. F., Bittner, Z. S., Forbes, D. V., Tatavarti, S. R., Phillip Ahrenkiel, S., Wibowo, A., Pan, N., Chern, K., & Hubbard, S. M. (2013). Epitaxial lift-off of quantum dot enhanced gaas single junction solar cells. *Applied Physics Letters, 103*, 213902.
45. Sogabe, T., Shoji, Y., Mulder, P., Schermer, J., Tamayo, E., & Okada, Y. (2014). Enhancement of current collection in epitaxial lift-off InAs/GaAs quantum dot thin film solar cell and concentrated photovoltaic study. *Applied Physics Letters, 105*, 113904.
46. Konagai, M., Sugimoto, M., & Takahashi, K. (1978). High efficiency GaAs thin film solar cells by peeled film technology. *Journal of Crystal Growth, 45*, 277–280.
47. Yablonovitch, E., Gmitter, T., Harbison, J. P., & Bhat, R. (1987). Extreme selectivity in the lift-off of epitaxial GaAs films. *Applied Physics Letters, 51*, 2222–2224.
48. Schermer, J. J., Bauhuis, G. J., Mulder, P., Meulemeesters, W. J., Haverkamp, E., Voncken, M. M. A. J., & Larsen, P. K. (2000). High rate epitaxial lift-off of InGaP films from GaAs substrates. *Applied Physics Letters, 76*, 2131–2133.
49. Schermer, J. J., Mulder, P., Bauhuis, G. J., Voncken, M. M. A. J., Van Deelen, J., Haverkamp, E., & Larsen, P. K. (2005). Epitaxial lift-off for large area thin film III/V devices. *Physica Status Solidi (a), 202*, 501–508.
50. Wu, X. S., Coldren, L. A., & Merz, J. L. (1985). Selective etching characteristics of HF for AlxGa1-xAs/GaAs. *Electronics Letters, 21*, 558–559.
51. Voncken, M. M. A. J., Schermer, J. J., Bauhuis, G. J., Mulder, P., & Larsen, P. K. (2004). Multiple release layer study of the intrinsic lateral etch rate of the epitaxial lift-off process. *Applied Physics A: Materials Science & Processing, 79*, 1801–1807.
52. Smeenk, N. J., Engel, J., Mulder, P., Bauhuis, G. J., Bissels, G. M. M. W., Schermer, J. J., Vlieg, E., & Kelly, J. J. (2013). Arsenic formation on GaAs during etching in HF solutions: Relevance for the epitaxial lift-off process. *ECS Journal of Solid State Science and Technology, 2*, P58–P65.
53. Lee, K., Zimmerman, J. D., Xiao, X., Sun, K., & Forrest, S. R. (2012). Reuse of GaAs substrates for epitaxial lift-off by employing protection layers. *Journal of Applied Physics, 111*, 033527.

54. Horng, R. H., Tseng, M. C., Wu, F. L., Li, C. H., Wu, C. H., & Yang, M. D. (2012). Thin film solar cells fabricated using cross-shaped pattern epilayer lift-off technology for substrate recycling applications. *IEEE Transactions on Electron Devices, 59*, 666–672.
55. Adams, J., Elarde, V., Hains, A., Stender, C., Tuminello, F., Youtsey, C., Wibowo, A., & Osowski, M. (2012). Demonstration of multiple substrate reuses for inverted metamorphic solar cells. *IEEE 38th Photovoltaic Specialists Conference (PVSC)* (pp. 1–6).
56. Lee, K., Zimmerman, J. D., Hughes, T. W., & Forrest, S. R. (2014). Non-destructive wafer recycling for low-cost thin-film flexible optoelectronics. *Advanced Functional Materials, 24*, 4284–4291.
57. Yablonovitch, E., Hwang, D. M., Gmitter, T. J., Florez, L. T., & Harbison, J. P. (1990). Van der Waals bonding of GaAs epitaxial liftoff films onto arbitrary substrates. *Applied Physics Letters, 56*, 2419–2421.
58. Van Niftrik, A. T., Schermer, J. J., Bauhuis, G. J., van Deelen, J., Mulder, P., & Larsen, P. K. (2007). The influence of in x Ga1–x as and GaAs1–y P y layers surrounding the AlAs release layer in the epitaxial lift-off process. *Crystal Growth and Design, 7*, 2472–2480.
59. Omnes, F., Guillaume, J. C., Nataf, G., Jager-Waldau, D., Vennegues, P., & Gibart, P. (1996). Substrate free GaAs photovoltaic cells on Pd-coated silicon with a 20% AM1.5 efficiency. *IEEE Transactions on Electron Devices, 43*, 1806–1811.
60. Lee, X. Y., Goertemiller, M., Boroditsky, M., Ragan, R., & Yablonovitch, E. (1997). Thin film GaAs solar cells on glass substrates by epitaxial liftoff. *AIP Conference Proceedings 394*, 719–727.
61. Schermer, J. J., Mulder, P., Bauhuis, G. J., Larsen, P. K., Oomen, G., & Bongers, E. (2005). Thin-film GaAs epitaxial lift-off solar cells for space applications. *Progress in Photovoltaics: Research and Applications, 13*, 587–596.
62. Schumacher, H., Gmitter, T. J., LeBlanc, H. P., Bhat, R., Yablonovitch, E., & Koza, M. A. (1989). High-speed InP/GaInAs photodiode on sapphire substrate. *Electronics Letters, 25*, 1653–1654.
63. Kobayashi, F., & Sekiguchi, Y. (1992). GaAs Schottky photodiode fabricated on glass substrate using epitaxial lift-off technique. *Japanese Journal of Applied Physics, 31*, L850.
64. Schnitzer, I., Yablonovitch, E., Caneau, C., Gmitter, T. J., & Scherer, A. (1993). 30% external quantum efficiency from surface textured, thin-film light-emitting diodes. *Applied Physics Letters, 63*, 2174–2176.
65. Sasaki, Y., Katayama, T., Koishi, T., Shibahara, K., Yokoyama, S., Miyazaki, S., & Hirose, M. (1999). High-speed GaAs epitaxial lift-off and bonding with high alignment accuracy using a sapphire plate. *Journal of the Electrochemical Society, 146*, 710–712.
66. Yablonovitch, E., Kapon, E., Gmitter, T. J., Yun, C. P., & Bhat, R. (1989). Double heterostructure GaAs/AlGaAs thin film diode lasers on glass substrates. *IEEE Photonics Technology Letters, 1*, 41–42.
67. Pollentier, I., Buydens, L., Van Daele, P., & Demeester, P. (1991). Fabrication of a GaAs-AlGaAs GRIN-SCH SQW laser diode on silicon by epitaxial lift-off. *IEEE Photonics Technology Letters, 3*, 115–117.
68. Shah, D. M., Chan, W. K., Caneau, C., Gmitter, T. J., Song, J. I., Hong, B. P., Micelli, P. F., & De Rosa, F. (1995). Epitaxial lift-off GaAs HEMT's. *IEEE Transactions on Electron Devices, 42*, 1877–1881.
69. Morf, T., Biber, C., & Bachtold, W. (1998). Effects of epitaxial lift-off on the DC, RF, and thermal properties of MESFET's on various host materials. *IEEE Transactions on Electron Devices, 45*, 1407–1413.
70. Georgiou, G., Tyagi, H. K., Mulder, P., Bauhuis, G. J., Schermer, J. J., & Rivas, J. G. (2014). Photo-generated THz antennas. *Scientific Reports, 4*, 3584.
71. Voncken, M. M. A. J., Schermer, J. J., Bauhuis, G. J., Van Niftrik, A. T. J., & Larsen, P. K. (2004). Strain-accelerated HF etching of AlAs for epitaxial lift-off. *Journal of Physics: Condensed Matter, 16*, 3585.
72. Bauhuis, G. J., Mulder, P., & Schermer, J. J. (2014). Thin-film iii–v solar cells using epitaxial lift-off. In X. Wang & Z. Wang (Eds.), *High-efficiency solar cells* (pp. 623–643). Cham: Springer.

73. Voncken, M. M. A. J., Schermer, J. J., Van Niftrik, A. T. J., Bauhuis, G. J., Mulder, P., Larsen, P. K., Peters, T. P. J., De Bruin, B., Klaassen, A., & Kelly, J. J. (2004). Etching AlAs with HF for epitaxial lift-off applications. *Journal of the Electrochemical Society, 151*, G347–G352.
74. Van Niftrik, A. T. J., Schermer, J. J., Bauhuis, G. J., Mulder, P., Larsen, P. K., & Kelly, J. J. (2007). A diffusion and reaction related model of the epitaxial lift-off process. *Journal of the Electrochemical Society, 154*, D629–D635.
75. Voncken, M. M. A. J., Schermer, J. J., Maduro, G., Bauhuis, G. J., Mulder, P., & Larsen, P. K. (2002). Influence of radius of curvature on the lateral etch rate of the weight induced epitaxial lift-off process. *Materials Science and Engineering: B, 95*, 242–248.
76. Van Niftrik, A. T. J., Schermer, J. J., Bauhuis, G. J., Mulder, P., Larsen, P. K., Van Setten, M. J., Attema, J. J., Tan, N. C. G., & Kelly, J. J. (2008). HF species and dissolved oxygen on the epitaxial lift-off process of GaAs using AlAsP release layers. *Journal of the Electrochemical Society, 155*, D35–D39.
77. Tatavarti, R., Hillier, G., Martin, G., Wibowo, A., Navaratnarajah, R., Tuminello, F., Hertkorn, D., Disabb, M., Youtsey, C., McCallum, D., & Pan, N. (2009). Lightweight, low cost InGaP/GaAs dual-junction solar cells on 100 mm epitaxial liftoff (ELO) wafers. *34th IEEE Photovoltaic Specialists Conference (PVSC)* (pp. 2065–2067).
78. Youtsey, C., Adams, J., Chan, R., Elarde, V., Hillier, G., Osowski, M., McCallum, D., Miyamoto, H., Pan, N., Stender, C., & Tatavarti, R. (2012). Epitaxial lift-off of large-area GaAs thin-film multi-junction solar cells. *Proceedings of the CS MANTECH Conference*.
79. Van Leest, R. H., Mulder, P., Gruginskie, N., van Laar, S. C., Bauhuis, G. J., Cheun, H., Lee, H., Yoon, W., van der Heijden, R., Bongers, E., & Vlieg, E. (2017). Temperature-induced degradation of thin-film III–V solar cells for space applications. *IEEE Journal of Photovoltaics, 7*, 702–708.
80. Farah, J. (2012). Dry-epitaxial lift-off, integration, interconnect and encapsulation of foldable/rollable high efficiency solar cell modules. *38th IEEE Photovoltaic Specialists Conference* (pp. 2868–2873).
81. Trautz, K. M., Jenkins, P. P., Walters, R. J., Scheiman, D., Hoheisel, R., Tatavarti, R., Chan, R., Miyamoto, H., Adams, J. G., Elarde, V. C., & Grimsley, J. (2012). Mobile solar power. *IEEE Journal of Photovoltaics, 3*, 535–541.
82. Essig, S., Steiner, M. A., Allebé, C., Geisz, J. F., Paviet-Salomon, B., Ward, S., Descoeudres, A., LaSalvia, V., Barraud, L., Badel, N., & Faes, A. (2016). Realization of GaInP/Si dual-junction solar cells with 29.8% 1-sun efficiency. *IEEE Journal of Photovoltaics, 6*, 1012–1019.
83. Hannappel, T., Sagol, B. E., Seidel, U., Szabo, N., Schwarzburg, K., Bauhuis, G. J., & Mulder, P. (2008). Measurement of an InGaAsP/InGaAs tandem solar cell under GaAs. *33rd IEEE Photovoltaic Specialists Conference* (pp. 1–3).
84. Essig, S., Allebé, C., Remo, T., Geisz, J. F., Steiner, M. A., Horowitz, K., Barraud, L., Ward, J. S., Schnabel, M., Descoeudres, A., & Young, D. L. (2017). Raising the one-sun conversion efficiency of III–V/Si solar cells to 32.8% for two junctions and 35.9% for three junctions. *Nature Energy, 2*, 17144.
85. Schermer, J. J., Bauhuis, G. J., Mulder, P., Haverkamp, E. J., Van Deelen, J., Van Niftrik, A. T. J., & Larsen, P. K. (2006). Photon confinement in high-efficiency, thin-film III–V solar cells obtained by epitaxial lift-off. *Thin Solid Films, 511*, 645–653.
86. Bauhuis, G. J., Schermer, J. J., Mulder, P., Voncken, M. M. A. J., & Larsen, P. K. (2004). Thin film GaAs solar cells with increased quantum efficiency due to light reflection. *Solar Energy Materials & Solar Cells, 83*, 81–90.
87. Tatavarti, R., Wibowo, A., Elarde, V., Tuminello, F., Pastor, R., Giannopoulos, T., Osowski, M., Chan, R., Youtsey, C., Hillier, G., & Pan, N. (2011). Large-area, epitaxial lift-off, inverted metamorphic solar cells. In *37th IEEE Photovoltaic Specialists Conference* (pp. 1941–1944).
88. Asbeck, P. (1977). Self-absorption effects on the radiative lifetime in GaAs-GaAlAs double heterostructures. *Journal of Applied Physics, 48*, 820–822.
89. Ahrenkiel, R. K., Dunlavy, D. J., Keyes, B., Vernon, S. M., Dixon, T. M., Tobin, S. P., Miller, K. L., & Hayes, R. E. (1989). Ultralong minority-carrier lifetime epitaxial GaAs by photon recycling. *Applied Physics Letters, 55*, 1088–1090.

90. Parrott, J. E. (1993). Radiative recombination and photon recycling in photovoltaic solar cells. *Solar Energy Materials & Solar Cells, 30*, 221–231.
91. Lush, G. B. (2009). B-coefficient in n-type GaAs. *Solar Energy Materials & Solar Cells, 93*, 1225–1229.
92. Bauhuis, G. J., Mulder, P., Haverkamp, E. J., Huijben, J. C. C. M., & Schermer, J. J. (2009). 26.1% thin-film GaAs solar cell using epitaxial lift-off. *Solar Energy Materials & Solar Cells, 93*, 1488–1491.
93. Geisz, J. F., Steiner, M. A., Garcia, I., Kurtz, S. R., & Friedman, D. J. (2013). Enhanced external radiative efficiency for 20.8% efficient single-junction GaInP solar cells. *Applied Physics Letters, 103*, 041118.
94. Steiner, M. A., Geisz, J. F., Garcia, I., Friedman, D. J., Duda, A., & Kurtz, S. R. (2013). Optical enhancement of the open-circuit voltage in high quality GaAs solar cells. *Journal of Applied Physics, 113*, 123109.
95. Lumb, M. P., Steiner, M. A., Geisz, J. F., & Walters, R. J. (2014). Incorporating photon recycling into the analytical drift-diffusion model of high efficiency solar cells. *Journal of Applied Physics, 116*, 194504.
96. Bauhuis, G., Mulder, P., Hu, Y. Y., & Schermer, J. (2016). Deep junction III–V solar cells with enhanced performance. *Physica Status Solidi (a), 213*, 2216–2222.
97. Bailey, C. G., Forbes, D. V., Raffaelle, R. P., & Hubbard, S. M. (2011). Near 1 V open circuit voltage InAs/GaAs quantum dot solar cells. *Applied Physics Letters, 98*, 163105.
98. Sablon, K. A., Little, J. W., Olver, K. A., Wang, Z. M., Dorogan, V. G., Mazur, Y. I., Salamo, G. J., & Towner, F. J. (2010). Effects of AlGaAs energy barriers on InAs/GaAs quantum dot solar cells. *Journal of Applied Physics, 108*(74305).
99. Lam, P., Wu, J., Tang, M., Jiang, Q., Hatch, S., Beanland, R., Wilson, J., Allison, R., & Liu, H. (2014). Submonolayer InGaAs/GaAs quantum dot solar cells. *Solar Energy Materials & Solar Cells, 126*, 83–87.
100. Tukiainen, A., Lyytikäinen, J., Aho, T., Halonen, E., Raappana, M., Cappelluti, F., & Guina, M. (2018). Comparison of 'shallow' and 'deep' junction architectures for MBE-grown InAs/GaAs quantum dot solar cells. *IEEE 7th World Conference on Photovoltaic Energy Conversion (WCPEC)(A Joint Conference of 45th IEEE PVSC, 28th PVSEC & 34th EU PVSEC)* (pp. 2950–2952).
101. Levy, M. Y., & Honsberg, C. (2008). Solar cell with an intermediate band of finite width. *Physical Review B, 78*, 165122.
102. Cuadra, L., Martí, A., & Luque, A. (2004). Influence of the overlap between the absorption coefficients on the efficiency of the intermediate band solar cell. *IEEE Transactions on Electron Devices, 51*, 1002–1007.
103. Strandberg, R., & Reenaas, T. (2011). Optimal filling of the intermediate band in idealized intermediate-band solar cells. *IEEE Transactions on Electron Devices, 58*, 2559–2565.
104. Yoshida, K., Okada, Y., & Sano, N. (2010). Self-consistent simulation of intermediate band solar cells: Effect of occupation rates on device characteristics. *Applied Physics Letters, 97*, 133503.
105. Lin, A. S., & Phillips, J. D. (2009). Drift-diffusion modeling for impurity photovoltaic devices. *IEEE Transactions on Electron Devices, 56*, 3168–3174.
106. Strandberg, R., & Reenaas, T. W. (2011). Drift-diffusion model for intermediate band solar cells including photofilling effects. *Progress in Photovoltaics: Research and Applications, 19*, 21–32.
107. Zhu, L., Akiyama, H., & Kanemitsu, Y. (2018). Intrinsic and extrinsic drops in open-circuit voltage and conversion efficiency in solar cells with quantum dots embedded in host materials. *Scientific Reports, 8*, 11704.
108. Aeberhard, U. (2013). Simulation of nanostructure-based high-efficiency solar cells: Challenges, existing approaches, and future directions. *IEEE Journal of Selected Topics in Quantum Electronics, 19*, 1–11.
109. Walker, A. W., Thériault, O., & Hinzer, K. (2014). The dependence of multijunction solar cell performance on the number of quantum dot layers. *IEEE Journal of Quantum Electronics, 50*, 198–203.

110. Gioannini, M., Cedola, A. P., Di Santo, N., Bertazzi, F., & Cappelluti, F. (2013). Simulation of quantum dot solar cells including carrier intersubband dynamics and transport. *IEEE Journal of Photovoltaics, 3*, 1271–1278.
111. Cappelluti, F., Gioannini, M., & Khalili, A. (2016). Impact of doping on InAs/GaAs quantum-dot solar cells: A numerical study on photovoltaic and photoluminescence behavior. *Solar Energy Materials & Solar Cells, 157*, 209–220.
112. Cédola, A. P., Kim, D., Tibaldi, A., Tang, M., Khalili, A., Wu, J., Liu, H., & Cappelluti, F. (2018). Physics-based modeling and experimental study of si-doped InAs/GaAs quantum dot solar cells. *International Journal of Photoenergy, 2018*, 7215843.
113. Khalili, A., Tibaldi, A., Elsehrawy, F., Cappelluti, F. (2019). Multiscale device simulation of quantum dot solar cells. In *Physics, simulation, and photonic engineering of photovoltaic devices VIII* (Vol. 10913, p. 109131N). International Society for Optics and Photonics.
114. Jolley, G., Fu, L., Lu, H. F., Tan, H. H., & Jagadish, C. (2013). The role of intersubband optical transitions on the electrical properties of ingaas/gaas quantum dot solar cells. *Progress in Photovoltaics: Research and Applications, 21*, 736–746.
115. Khalili, A., & Cappelluti, F.(2018). Modeling of type-ii quantum dot intermediate band solar cells accounting for thermal and optical intersubband transitions. *International Conference on Numerical Simulation of Optoelectronic Devices (NUSOD)* (pp. 139–140).
116. Hwang, J., Martin, A. J., Millunchick, J. M., & Phillips, J. D. (2012). Thermal emission in type-ii gasb/gaas quantum dots and prospects for intermediate band solar energy conversion. *Journal of Applied Physics, 111*, 074514.
117. Williamson, A. J., Wang, L. W., & Zunger, A. (2000). Theoretical interpretation of the experimental electronic structure of lens-shaped self-assembled InAs/GaAs quantum dots. *Physical Review B, 62*, 12963.
118. King, R. R., Bhusari, D., Boca, A., Larrabee, D., Liu, X. Q., Hong, W., Fetzer, C. M., Law, D. C., & Karam, N. H. (2011). Band gap-voltage offset and energy production in next-generation multijunction solar cells. *Progress in Photovoltaics: Research and Applications, 19*, 797–812.
119. Ekins-Daukes, N., Pusch, A. (2018). The use and abuse of W_{oc} as a figure of merit. *Proceedings of PVSEC 2018.*
120. Nuntawong, N., Tatebayashi, J., Wong, P. S., & Huffaker, D. L. (2007). Localized strain reduction in strain-compensated In As/Ga As stacked quantum dot structures. *Applied Physics Letters, 90*, 163121.
121. Bowen, D. K., & Tanner, B. K. (2005). *High resolution X-ray diffractometry and topography.* Boca Raton, FL: CRC press.
122. Fewster, P. F. (1986). X-ray diffraction from multiple quantum well structures. *Philips Journal of Research, 41*, 268–289.
123. Gee, J. M. (1988). The effect of parasitic absorption losses on light trapping in thin silicon solar cells. *Proceedings of the 2016 IEEE Conference Record of the Twentieth IEEE Photovoltaic Specialists Conference 1*, 549–554.
124. Green, M. A. (2002). Lambertian light trapping in textured solar cells and light-emitting diodes: Analytical solutions. *Progress in Photovoltaics: Research and Applications, 10*, 235–241.
125. Weber, M. J. (2002). *Handbook of optical materials*. Boca Raton, FL: CRC press.
126. Tsai, C., Liu, G., Fan, G., & Lee, Y. (2010). Substrate-free large gap InGaN solar cells with bottom reflector. *Solid-State Electronics, 54*, 541–544.
127. Vandamme, N., Hung-Ling, C., Gaucher, A., Behaghel, B., Lemaitre, A., Cattoni, A., Dupuis, C., Bardou, N., Guillemoles, J., & Collin, S. (2015). Ultrathin GaAs solar cells with a silver back mirror. *IEEE Journal of Photovoltaics, 5*, 565–570.
128. Baca, A. G., & Ashby, C. I. (2005). *Fabrication of GaAs devices*. London: IET.
129. Van Leest, R. H., Bauhuis, G. J., Mulder, P., van der Heijden, R., Bongers, E., Vlieg, E., & Schermer, J. J. (2015). Effects of copper diffusion in gallium arsenide solar cells for space applications. *Solar Energy Materials & Solar Cells, 140*, 45–53.

130. Van Leest, R. H., de Kleijne, K., Bauhuis, G. J., Mulder, P., Cheun, H., Lee, H., Yoon, W., van der Heijden, R., Bongers, E., Vlieg, E., & Schermer, J. J. (2016). Degradation mechanism (s) of GaAs solar cells with Cu contacts. *Physical Chemistry Chemical Physics, 18*, 10232–10240.
131. Aho, T., Aho, A., Tukiainen, A., Polojärvi, V., Salminen, T., Raappana, M., & Guina, M. (2016). Enhancement of photocurrent in GaInNAs solar cells using Ag/Cu double-layer back reflector. *Applied Physics Letters, 109*, 251104.
132. ASTM Standard (2008). *Standard tables for reference solar spectral irradiances: Direct normal and hemispherical on 37° tilted surface* (ASTM International, Technical Report 12).
133. Stareev, G., Künzel, H., & Dortmann, G. (1993). A controllable mechanism of forming extremely low-resistance nonalloyed ohmic contacts to group III-V compound semiconductors. *Journal of Applied Physics, 74*, 7344–7356.
134. Akdogan, I. G., & Parker, M. A. (2005). Au-Cu ohmic contacts for p+ GaAs. *Electrochemical and Solid-State Letters, 8*, G106–G108.
135. Inoue, T., Watanabe, K., Toprasertpong, K., Fujii, H., Sugiyama, M., & Nakano, Y. (2015). Enhanced light trapping in multiple quantum wells by thin-film structure and backside grooves with dielectric interface. *IEEE Journal of Photovoltaics, 5*, 697–703.
136. Musu, A., Cappelluti, F., Aho, T., Polojärvi, V., Niemi, T., & Guina, M. (2016). Nanostructures for light management in thin-film GaAs quantum dot solar cells. *OSA Light, Energy and the Environment Congress* (pp. JW4A-45).
137. Mokkapati, S., & Catchpole, K. R. (2012). Nanophotonic light trapping in solar cells. *Journal of Applied Physics, 112*, 101101.
138. Palanchoke, U., Jovanov, V., Kurz, H., Obermeyer, P., Stiebig, H., & Knipp, D. (2012). Plasmonic effects in amorphous silicon thin film solar cells with metal back contacts. *Optics Express, 20*, 6340–6347.
139. Tommila, J., Polojärvi, V., Aho, A., Tukiainen, A., Viheriälä, J., Salmi, J., Schramm, A., Kontio, J., Turtiainen, A., & Niemi, T. (2010). Nanostructured broadband antireflection coatings on AlInP fabricated by nanoimprint lithography. *Solar Energy Materials and Solar Cells, 94*, 1845–1848.
140. RSoft Design Group. (n.d.). DiffractMOD. Retrieved from https://www.synopsys.com/optical-solutions/rsoft/passive-device-diffractMOD.html.
141. Elsehrawy, F., Cappelluti, F., Aho, T., Niemi, T., Polojärvi, V., Guina, M. (2017). Back grating optimization for light trapping in thin-film quantum dot solar cells. *19th Italian National Conference on Photonic Technologies 34*.
142. Elsehrawy, F., Aho, T., Niemi, T., Guina, M., & Cappelluti, F. (2018, November). Improved light trapping in quantum dot solar cells using double-sided nanostructuring. In *Optics and photonics for energy and the environment* (pp. JM4A-5). Optical Society of America.
143. Aho, T., Guina, M., Elsehrawy, F., Cappelluti, F., Raappana, M., Tukiainen, A., Alam, A. K., Vartiainen, I., Kuittinen, M., & Niemi, T. (2018). Comparison of metal/polymer back reflectors with half-sphere, blazed, and pyramid gratings for light trapping in III-V solar cells. *Optics Express, 26*, A331–A340.
144. Ganapati, V., Steiner, M. A., & Yablonovitch, E. (2016). The voltage boost enabled by luminescence extraction in solar cells. *IEEE Journal of Photovoltaics, 6*, 801–809.
145. Gruginskie, N., van Laar, S. C. W., Bauhuis, G., Mulder, P., van Eerden, M., Vlieg, E., & Schermer, J. J. (2018). Increased performance of thin-film GaAs solar cells by rear contact/mirror patterning. *Thin Solid Films, 660*, 10–18.
146. van Eerden, M., Bauhuis, G., Mulder, P., Gruginskie, N., Passoni, M., Andreani, L. C., Vlieg, E., & Schermer, J. J. (2019). A facile light trapping approach for ultra-thin GaAs solar cells using wet chemical etching. Progress in Photovoltaics. Submitted for publication.
147. Stender, C. L., Adams, J., Elarde, V., Major, T., Miyamoto, H., Osowski, M., Pan, N., Tatavarti, R., Tuminello, F., Wibowo, A., & Youtsey, C. (2015). Flexible and lightweight epitaxial lift-off GaAs multi-junction solar cells for portable power and UAV applications. *IEEE 42nd Photovoltaic Specialist Conference (PVSC)* (pp. 1–4).

148. Scheiman, D., Hoheisel, R., Edwards, D. J., Paulsen, A., Lorentzen, J., Jenkins, P., Caruthers, S., Carter, S., & Walters, R. (2016). A path toward enhanced endurance of a UAV using IMM solar cells. *IEEE 43rd Photovoltaic Specialists Conference (PVSC)* (pp. 1095–1100).
149. PV Magazine. (n.d.). Audi, Hanergy unit to jointly develop PV for vehicles. Retrieved from http://pv-magazine.com/2017/08/23/audi-hanergy-unit-to-jointly-develop-pv-for-vehicles.
150. US Department of Energy, Office of Energy Efficiency and Renewable Energy. (n.d.). High-efficiency, low-cost, one-sun, III-V photovoltaics. Retrieved from http://energy.gov/eere/solar/project-profile-high-efficiency-low-cost-one-sun-iii-v-photovoltaics.

Chapter 2
Colloidal Quantum Dots for Highly Efficient Photovoltaics

Jiantuo Gan and Liang Qiao

Abstract Owing to strong quantum confinement, solution-processed colloidal quantum dots (CQDs) provide a unique route for fabrication of highly efficient photovoltaics to overcome the Shockley-Queisser limit through multiple exciton generation (MEG). Also, the CQDs PVs are cost-effective due to fabrication at low temperatures. Despite the high quantum limit of 42–44%, the highest experimental power conversion efficiency reported so far in PbS CQD PVs still remains at ~11.2%. Recent studies have shown that the performance of CQD solar cells is mainly limited by defects through different recombination channels. In this work, we identify the reasons for a large open circuit voltage (V_{OC}) deficit, associated with short diffusion length and lifetime of minority carriers by different types of defect recombination pathways in the devices. We also summarize recent progress in improvement of device efficiency through treatment of defect. Surface modification is primarily intended to passivate surface defects in CQDs. Whereas, interface defects can be treated by engineering of transport layers or device architecture through, e.g., ionic doping or additional layers, entailing fast dissociation and smooth transport of charge carriers at junctions.

Keywords Colloidal quantum dots · Multiple exciton generation · Photovoltaics · Defect recombinations · Passivation · Electron transport layer

J. Gan · L. Qiao (✉)
School of Physics, University of Electronic Science and Technology of China, Chengdu, P. R. China
e-mail: liang.qiao@uestc.edu.cn

P. Yu, Z. M. Wang (eds.), *Quantum Dot Optoelectronic Devices*,
Lecture Notes in Nanoscale Science and Technology 27,
https://doi.org/10.1007/978-3-030-35813-6_2

2.1 Introduction

Today, renewable technologies are becoming increasingly important for the world's energy supply, and are accounting for around 4–6% (mainly including contribution of power capacities of wind 539 gigawatts, or GWs, solar photovoltaics (PVs) 422 GWs and biopower 122 GWs) in today's energy consumption [1, 2]. Among these, solar PVs are anticipated to be the most prominent means for generating future clean energy [3]. In 1994–2016, they also have shown the fastest capacity growth (calculated by GWs), in comparison with other renewables. According to a recent release, the Photovoltaic Report by Fraunhofer Institute for Solar Energy Systems (ISE) [4], the power generation by solar PV has reached to 443 TWhrs in 2017, which contributes to 1.7% of the world's total energy. In the next period (2017–2023), it will further increase by 170%, reaching 92.3 Million Tonnes of Oil Equivalent (Mtoe). Whereas for wind it only shows 69% of increase [3]. Moreover, there are also not carbon emission during sunlight conversion by solar PVs. Up to now, wafer (c-Si, ~92% in market share) [4] and thin-film (CdTe, $Cu(In,Ga)Se_2$, a-Si:H with 2.3%, 1.3%, 0.9%) [4]-based PV modules are dominant in the market because of their technological maturity; yet for the future, cost-effective and highly efficient solar cells are anticipated to be more compatible [5, 6]. In particular, colloidal quantum dot (CQD)-based solar cells have generated huge interest in the development of future-generation PVs, due to their high potential over conventional solar cells in terms of breaking the Shockley-Queisser limit (S-Q limit) [7]. More importantly, the CQD PVs are featured with its low carbon footprint in production due to the low temperatures (<200 °C and below) required [8–13]. The CQD thin films can also be fabricated using simple techniques, such as layer-by-layer spin coating [6]. In addition, CQD engineering, i.e. via oxygen doping, has enhanced the device power conversion efficiency (PCE) to above 10% in a highly humid environment [14].

Besides, CQDs are attractive to compete with c-Si, owing to their low costs and cross-industry applicability [15–19]. During synthesis, size tuning of CQDs can enable a wide range of sunlight absorption covering almost the entire spectrum of ultra-violet (UV), visible (VIS) and near-infrared (NIR) ranges [20]. Due to the unique quantum-size effect, they also have been widely used as building blocks in the fabrication of high-performance light emitting diodes (LEDs) [21, 22] and photodetectors [23–25]. Nevertheless, CQD solar cells still remain in the research/prototype stage (with laboratory-scale efficiencies of 11.2–16.6%) [26–28], which must be increased further together with their ambient stability, in order to attract the interest of manufacturers and ultimately to make them commercially available.

Since 2010, CQDs have attracted significant research interest for the development of the so-called third-generation solar cells. Instead of the upper limit of 35% calculated by Shockley and Queisser for a single-junction solar cell, higher efficiency, i.e. a quantum limit of 43%, is predicted for solar cells with CQD-based light absorbers under the certain multiple exciton generation (MEG) conditions: the threshold energy (E_{TH}) of $2E_g$ and the slope of $100\%/E_g$ in internal quantum efficiency (IQE) [29, 30]. Here, the MEG effect can help to harvest multiple electron-hole pairs (EHPs) through excitation of a single photon with energy higher than the E_{TH} in quantum dots. In comparison with bulk, their energy bands are discrete, which provide opportunities for formation of a second EHP. The generation of the second EHP is also viewed as the reverse of Auger recombination; that is, due to the photo-excitation by energy released from non-radiative recombination [31, 32]. Hence, it is anticipated to obtain an increased photocurrent (J_{ph}), which compensates efficiency loss through thermalization. It can be seen that the waste heat is dramatically reduced in quantum dots. On the other hand, voltage loss results from non-radiative recombination in the CQD film and at the device interface [19, 33–35] can be avoided by passivating the recombination centers [36]. Enormous work achieved for uncovering the fundamental mechanism of the MEG in reduced dimension nanocrystals are greatly acknowledged and the role of MEG in CQD solar cells are further discussed [18, 20, 37–44].

2.2 Architecture and Fabrication of CQD PVs

2.2.1 Architecture Design

As shown in Fig. 2.1, CQD PVs have three main structural designs [37, 46]. So far, depleted heterojunctions (in Fig. 2.1a), also identified as type-I structure, is widely accepted as the architecture for CQD solar cells [9, 10, 26, 27, 37, 45, 47–55]. In this structure, *n*-type wide-bandgap semiconductors, such as ZnO [26, 27], SnO_2 [55, 56], and TiO_2 [57], are frequently used as the front electrode, owing to their high optical transmittance in the NIR and visible spectral range, at the same time they have high electron mobility, and suitable band alignment with the photoactive CQD layer. The prepared ZnO nanoparticle (NP) solution are often spin-coated to form the

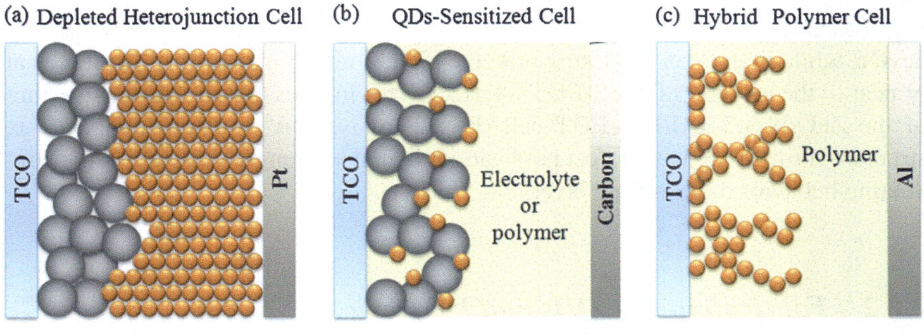

Fig. 2.1 Schematics of three generics of photovoltaic solar cells based on CQDs. (**a**) Depleted-heterojunction cell (type-I); (**b**) QDs sensitized solar cell (type-II) and (**c**) hybrid polymer solar cells (type-III). (Reprinted with permission from Ref. 45)

front electrode layers. Subsequently, the *p*-type CQD absorber layer is also spin-coated, resulting in a ~300 nm compact layer followed by ligand exchange. In this layer, the CQDs are sized uniformly and well aligned to each other, so that a mini-band structure, analogous to a 1D superlattice [46], was obtained for efficient transport of photo-excited carriers. In addition, short dot-to-dot distance (a few nm) enables strong electronic coupling, thus permitting long-range electron transport [58].

In spite of strong electronic coupling, several major issues are frequently investigated, including: the nature of electronic states as a function of inter-dot distance [58, 59], array order versus disorder [37], orientation and shape [60–62], surface states [10] and surface structure/passivation [27, 63]. In addition, band engineering through additional electron transport layers (ETLs) and hole transport layers (HTLs) are proved beneficial in improving carrier transport [9, 64]. Hence, type-II and type-III generic structures are indicated for enhancement of charge transport. In the type-II structure, QDs are chemically adsorbed onto the highly porous TiO_2 nano-crystalline film (in Fig. 2.1b) [37], so that long-wavelength (NIR region) excited electrons can be transferred rapidly to the conduction band of the wide-bandgap semiconductor; whereas, in the type-III structure (in Fig. 2.1c), a conducting polymer blend is used to replace TiO_2, entailing reduced transport length for charge collection. However, the problem remaining with a type-III cell is recombination of EHPs at the enlarged hetero-interface [37]. Hence, in this work we focus mainly on engineering of transport layers and defect investigation, in pursuit of a high performing type-I cell, except when the extraordinary performance is concerned in the last two types.

On the other hand, both regular and inverted structures have been designed to study the mechanisms of charge transport and recombination for the depleted-heterojunction (in Fig. 2.2a, b) [27, 65]. In comparison with ZnO (3.2–3.4 eV), a wider-bandgap *p*-type NiO (~3.5 eV) was used to permit the energetic photons to pass through in the inverted structure. As a result, the device showed a higher J_{SC} of 27.68 mA/cm^2 with PCE of 9.7% comparable to that in a regular device (11.2%) [26, 27, 65]. In Fig. 2.2c, d, the ligand-exchanged photoactive layers (PbS-TBAI, PbS-EDT, or PbS-I^- for CQD surface treatment) are stacked for band-structure engineering in order to transport charge carriers smoothly. Furthermore, from the electron beam-induced current (EBIC) map (see Fig. 2.2e, f), the current is concentrated mainly at the PbS-EDT/Au interface (the red zone) in the PbS-TBAI-only device, while in the PbS-EDT structure, the highest electron current is distributed at or near to the *n*-type ZnO/PbS-EDT interface. This implies that the planar structure of the ZnO/PbS-TBAI/PbS-EDT cell (Fig. 2.2a) is an n^+-n-p, rather than n-i-p, structure [66]. Along with defect passivation, ligand exchanging also alters the conducting behavior of absorbing layer.

2.2.2 *Fabrication of CQDs and PV Devices*

In order to prepare CQDs with high mono-dispersion and narrow size distribution, the underlying mechanisms of the nanocrystals should be elaborated through processing. In particular, processing parameters such as reaction tem-

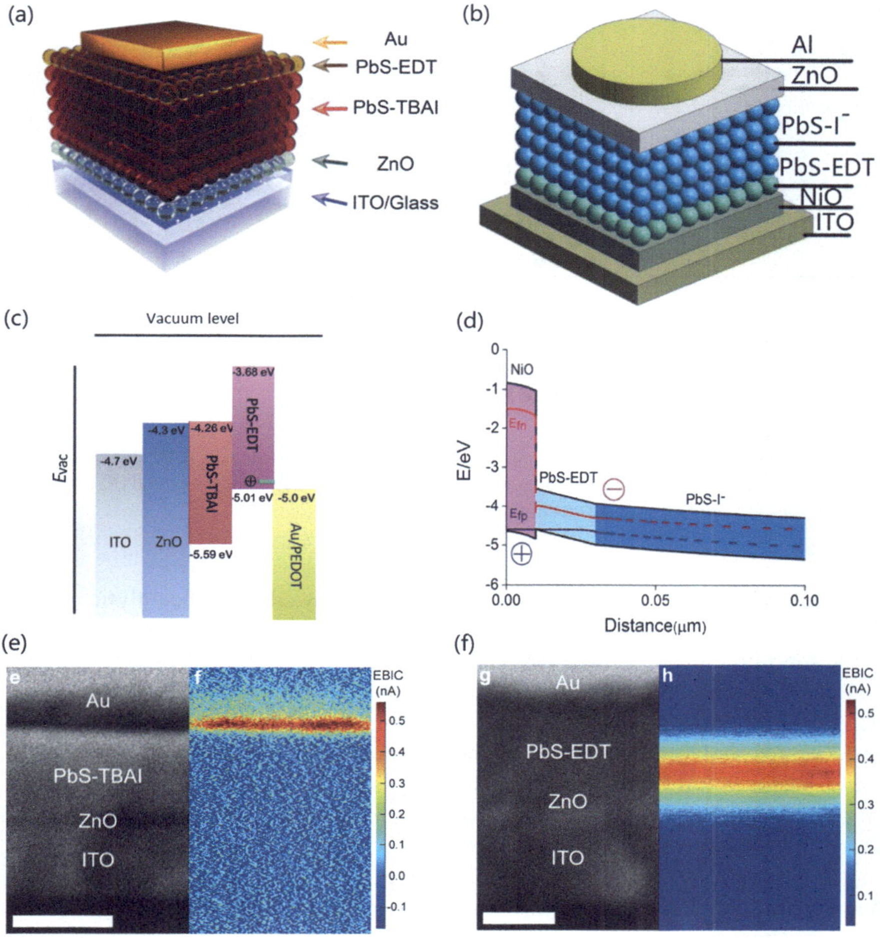

Fig. 2.2 Schematics of type-I CQD solar cell with layer structure of (**a**) regular and (**b**) inverted structure, respectively. (**c**) Band structures and (**d**) band diagram of a type-I CQD solar cell for regular and inverted structures, respectively. (**e**) Cross-sectional SEM image and corresponding EBIC map for planar PbS-TBAI-only structure and (**f**) SEM/EBIC images for PbS-EDT-only cell. (Reprinted with permission from Refs. 27, 65, 66)

perature, degree of super-saturation, reaction time and surface energy etc. [6, 18]. In particular, the surface energy can be adjusted by the choice of precursors [67]. As shown in Fig. 2.3a, in the hot-injection method [69] and/or in the LaMer model [70, 71], the concentration of precursors, i.e., the sulfur precursor for PbS, is increased upon injection, and nucleation begins before it reduces to a low concentration level. After purification, these nanocrystals are dispersed in solvent, forming a CQD solution. On the other hand, Finke-Watzky's theory reveals an

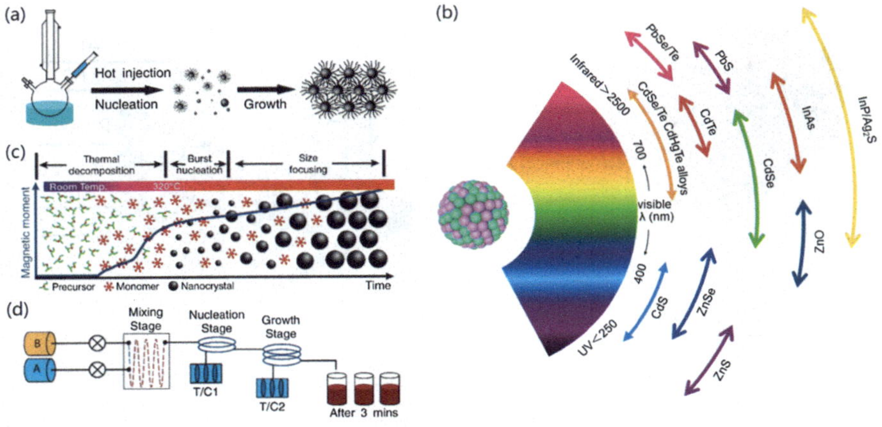

Fig. 2.3 Schematics of CQD synthesis methods: (**a**) hot-injection synthesis, (**b**) CQDs scaled as a function of their emission wavelength in the spectrum, (**c**) non-injection heat-up, and (**d**) flow reactor synthesis. (Reprinted with permission from Refs. 16, 68)

approach where the constant slow nucleation is followed by autocatalytic growth [72]. It means that the nucleation of CQDs and its growth will slow down with reduction in the driving force (the concentration of precursors). In order to control dot size, Ostwald ripening [73] and Lifshitz-Slyozov-Wagner (LSW) theories are used to described CQD size evolution, where the high solubility, large surface area and surface energy of the quantum-size particles allow growth of larger particles [74, 75]. Moreover, the CQD size is also affected by the flux of precursors injected into the particles as descried by Reiss et al. [76] In terms of shape, they vary from isotopic dots to spindle-shaped nanocrystals, or even rods and other elongated structures, by increasing concentration of precursors [77]. As a result, emission range of wavelength can attain to cover the entire spectrum in these CQDs, which indicates their bandgaps can also be tuned for different wavelength sunlight absorption (see Fig. 2.3b) for CQDs based on different types of absorbers. As illustrated in Fig. 2.3c, d, injection heat-up and flow reaction synthesis, are also frequently used for CQD synthesis. In a non-injection heat-up synthesis method, the mixture of all precursors will not react until the temperature reaches a certain point; whereas, in flow reactor synthesis, synthesis phases are separated with a multistage, low-volume, continuous flow synthesis reactor [16, 68, 78, 79]. Among these methods, the hot-injection method is feasible and effective in the preparation of nanocrystal QDs.

For device fabrication, the absorber of the CQD array is fabricated by layer-by-layer spin coating in order to reach the required thickness (Fig. 2.4a). During the process, spinning speed and time are crucial parameters influencing quality of the film. In addition, the volume of CQD solution for each cycle can be adjusted to obtain a continuous and smooth film, followed by ligand-exchange and rinsing. Afterwards, the metal back-contact is deposited by thermal evaporation (the last

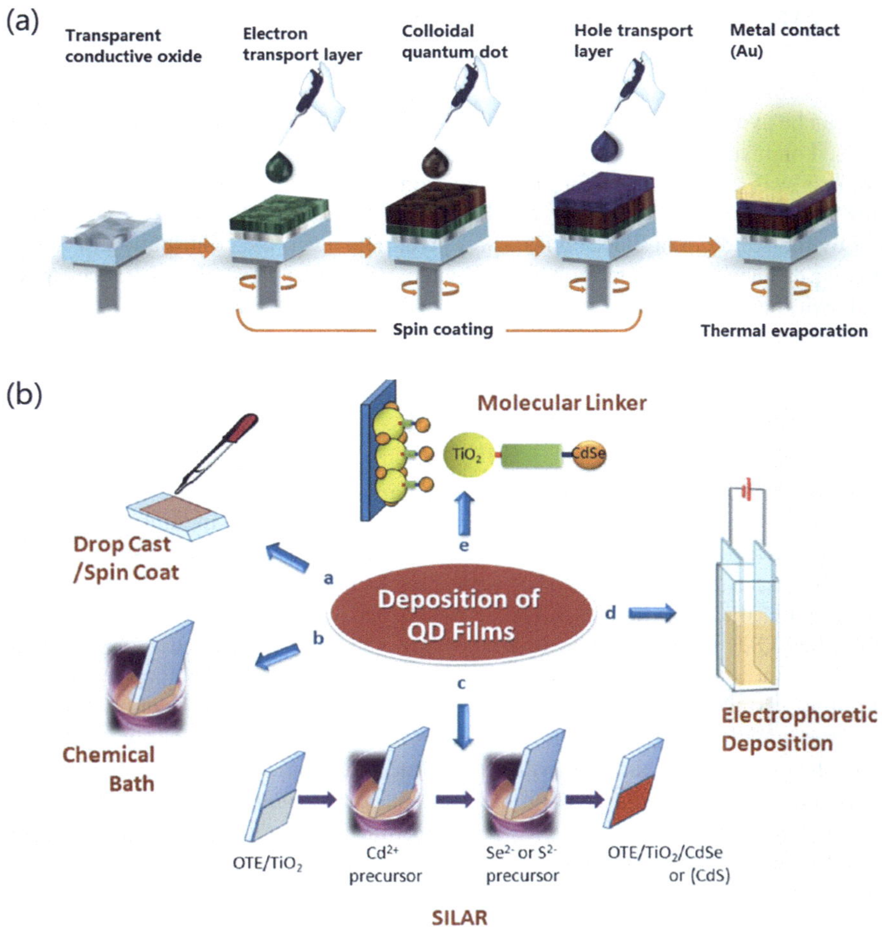

Fig. 2.4 (**a**) Fabrication of quantum dot solar cells through spin coating and thermal evaporation. (**b**) Schematics of various CQD film deposition techniques. (Reprinted with permission from Refs. 80, 81)

procedure in Fig. 2.4a). In addition, there is a large variety of techniques available for deposition of films from solution, as shown in Fig. 2.4b [82]. Typical techniques include drop cast/spin coating, molecular linker [83, 84], electrophoretic deposition [85, 86], successive ionic layer-by-layer adsorption and reaction (SILAR) [87], and chemical bath deposition (CBD) [88, 89], etc. When compared with others, spin-coating/drop-casting is a simpler technique for thin-layer deposition with good quality. Also, it enables large areas of uniform thickness. Furthermore, the desired thickness can also be achieved simply through the number of cycles. Thus, it is widely accepted in fabrication of CQD solar cells.

2.3 MEG and Its Effect on PCE in CQD Solar Cells

Unlike conventional solar cells, energy losses related to charge carrier thermalization can be avoided with the assistance of MEG effect [90]. So far, studies have been focused on investigating quantum yield (QY) in CQDs where the importance of QY progress was demonstrated through reduction in E_{TH} and prolonging exciton lifetime (τ_{rec}) [29, 39, 91]. Yet, from the device point of view, QY can be strongly limited by the bandgap of electron transport layer. However, the contribution of MEG to solar efficiency has seldom been evaluated, and under most circumstances their peak external quantum efficiency (EQE) is under 100% [12, 26, 34, 65, 92, 93].

2.3.1 *Quantum-Size Effect in CQDs*

Apart from bulk materials, CQDs have advantages such that they are distinguished by their discrete energy levels; thus, they are called sub-bandgap, and the onset of their optical absorption can be tuned by changing the size of the CQDs (see Fig. 2.5a). Here the discrete sub-energy levels, i.e., S_e (for electrons) and S_h (for holes) can actually provide the opportunity to reverse the Auger recombination leading to MEG effect in CQDs upon energetic irradiance. On the other hand, as demonstrated in Fig. 2.5b, the surface defects are perceived as dangling bonds which appear due to the large surface-to-volume ratio. These defects provide energy levels within the bandgap, i.e. close to the $1S_e$ and $1S_h$ (Fig. 2.5b). However, these energy levels can be pushed outside the bandgap with surface modification, by altering their location via manipulation of the interaction between the surface atoms of the CQD and the ligand molecules. Besides, Fig. 2.5b, c illustrates the process of solution-phase ligand exchange with metal halide precursors and ammonium acetate. Consequently, the CQDs are surrounded by iodide molecules for surface passivation.

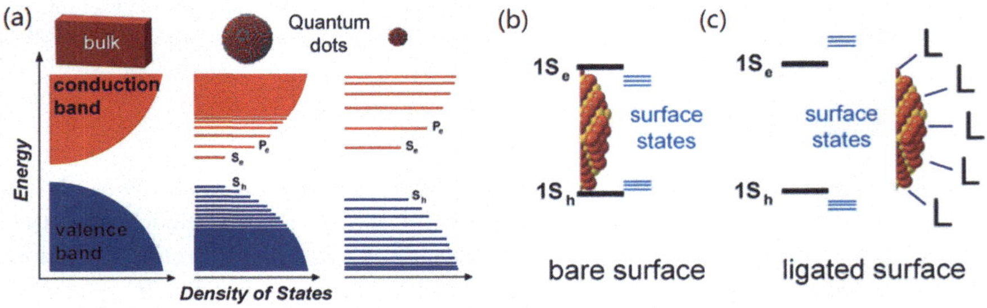

Fig. 2.5 (**a**) Evolution of the electronic structure of semiconductors from bulk to CQDs at different sizes. (**b**) Dangling bonds at the CQD surface and their energy states within the bandgap. (**c**) Altering the energy level of the surface states, i.e. pushing them outside the bandgap, by manipulating the interaction between the CQD surface atoms and ligand molecules. (Reprinted with permission from Ref. 10)

In particular, the bandgap of PbS CQDs can be tuned with in VIS and IR spectral regions (between 1–2 eV) simply by controlling their size during solution processing [17, 68, 73–75]. An interesting application of this effect is to build a cascaded-junction solar cell from several layers of QDs with multiple bandgaps [94]. Hence, a mono-disperse PbS CQD-based cascaded-junction solar cell with three different bandgap values of absorber, i.e., 1.03/1.23/1.37 eV, showed a higher PCE of 9.05% compared to 8.55% for the solar cell with a single bandgap of 1.23 eV [94]. This improvement is due mainly to the contribution of increased J_{SC}, i.e. from 28.86 to 32.51 mA/cm^2, entailed by the smooth charge-carrier transport.

2.3.2 QD Materials for PVs

In pursuit of the MEG effect, a strong quantum confinement regime is relevant with CQD radius being much smaller than Bohr radius, a_B. In this regard, lead chalcogenides (PbX, with X = S, Se or Te) appear to be ideal with a_B in the range 24–152 nm [13], estimated by the following Eq. 2.1: [40].

$$a_B = \varepsilon_\infty \frac{m}{m^*} a_0 \tag{2.1}$$

The relative high-frequency dielectric constant (ε_∞) of the material plays a prominent role in realizing a large a_B, and ε_∞ should be as high as possible. In general, PbX QDs have high ε_∞ experimental values, in the range of ~18–40 (unitless) [13, 95, 96]. In contrast, the other parameters in Eq. 2.1, namely, the effective electron mass (m^*), mass of the particle (m) and the Bohr radius of the hydrogen atom (a_0), are less important for calculation of the Bohr radius for different materials. The QD radius of a_B ~4–10 times gives rise to splitting of bands, as shown in Fig. 2.5. This hinged on the observation of the experimental QY determined from time-resolved photoluminescence (TRPL) spectroscopy being higher in CQDs than in bulk [97]. Moreover, the light effective mass of charge carriers ($0.09m_e$) promotes charge delocalization, resulting in a high mobility and diffusion length in the QD array film.

In Fig. 2.6a, a high resolution transmission electron microscopy (HRTEM) image is shown for monodispersed PbS CQDs. It can be seen that these nanocrystals are uniform in size and have monodispersed distribution. The distance between CQDs is at the scale of their diameter. The HRTEM image (in Fig. 2.6a inset) further magnifies the TEM image for lattice spacing of the compound, which is 3.00 Å in the <200> crystal direction. To characterize them, X-ray spectroscopy (XPS), absorbance spectra, and X-ray diffraction (XRD) patterns were employed to determine their energy band levels (E_C and E_V), light absorption coefficients, and characteristic exciton peak positions (in Fig. 2.6b, c). In particular, XRD pattern was used to study the crystal structure of the QD array film. Further, these PbS CQD array films exhibit carrier mobility ranging from 10^{-3} to 10^{-4} cm^2/V·s depending on surface ligands [103, 104], where the minority carrier (electrons) mobility is

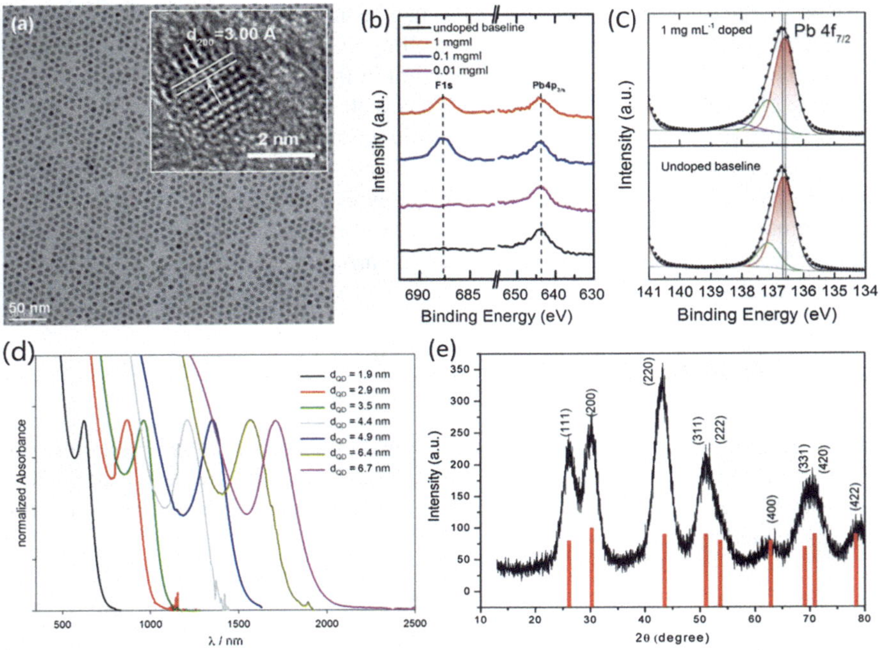

Fig. 2.6 (**a**) Transmission electron microscopy (TEM) image of PbS CQDs. The inset shows high-resolution (HR) TEM image of oleic acid (OA)-capped PbS CQDs. (Reprinted with permission from Refs. 98, 99) (**b**)–(**c**) X-ray photoelectron spectroscopy (XPS) for PbS CQDs with different solution concentrations [100]. (**d**) Absorption spectra for PbS CQDs with indicated diameters of 1.9–6.7 nm. (Reprinted with permission from Ref. 101). (**e**) X-ray diffraction (XRD) pattern for PbS CQDs. (Reprinted with permission from Ref. 102)

measured by a time of flight (TOF) method [105] and that of the majority carrier (holes) is measured by charge extraction by the linearly increasing voltage (CELIV) method [106].

As shown in Table 2.1, PbS CQD-based solar cells represented the highest PCE, with an experimental value of over 10% [26–28] after significant surface modification of the CQDs [11, 35, 92, 93, 102, 109]. In contrast, PbSe CQD solar cells have demonstrated EQEs higher than 100% at $h\nu$ of ~3.5 eV for smallest bandgap CQDs (0.72 eV), implying MEG participates in the light-harvesting process at an $h\nu/E_g$ value as high as ~$5E_g$ [49]. Further, the PbTe device shows lower PCE. This value is 1.5%, despite its high peak IQE of about 120% [51]. To a first approximation, this is correlated with a relatively higher Bohr radius in PbSe (46–66 nm) and PbTe (152 nm) QDs compared to that in PbS CQDs (20–23.5 nm) [13, 110]. Therefore, PbS devices show the highest efficiency among PbX CQD PVs, yet its EQE has not exceeded unity. By contrast, PbSe- and PbTe-based devices show suitability to fabricate CQD PVs with high efficiency, reflected by high EQE values of more than the unity, although their efficiency still remains at relatively low levels.

Table 2.1 Peak efficiencies and related parameters in CQD solar cells (CQD PVs) for different materials and device structures

Light absorber	Structure	PCE (%)	V_{OC} (V)	J_{SC} (mA/cm²)	*FF* (%)	Stability	Ref. (year)
PbS	ITO/ZnO/PbS-PbX_2-AA/PbS-EDT/Au	11.2	0.61	27.23	67.8	1000 h	[26] (2017)
PbSe	Au/PbSe-MPA/TiO_2/FTO	8.2	0.53	25.1	61.5	57 days	[107] (2017)
PbTe	Ag/MoO_x/PbTe QDs/ZnO/ITO	1.9	0.313	18.2	32	–	[51] (2017)
$CuInS_2$(CIS)	Cu_2S/polysulfide aqueous electrolyte solution/CIS QD/TiO_2	6.66	0.58	19.73	58	–	[108] (2014)
ZnCuInSe (ZCISe)	TiO_2/MPA capped ZCISe/ZnS/SiO_2/polysulfide aqueous electrolyte solution /MC/Ti	11.61	0.739	25.25	62.2	–	[52] (2016)
$AgBiS_2$	Ag/MoO_x/polymer/$AgBiS_2$/ZnO/ITO	6.31	0.45	22.1	63	–	[53] (2016)
Sb_2Se_3	Au/HTL/Sb_2Se_3/CdS/ITO	6.5	0.427	25.5	59.3	105 days	[54] (2017)

Apart from PbX, emerging CQD solar cells also exhibit satisfactory performance, as seen from Table 2.1. It lists currently active compounds and potentially promising materials for CQD device fabrication, including PbX, $CuInS_2$, $CuInSe_2$, $AgBiS_2$, α-Zn_3P_2, and SnS. Among these, I-III-VI_2 materials such as $CuInS_2$ (CIS) and $CuInSe_2$ (CISe) have been successful in harvesting solar photons with an ultimate high PCE of 7.01–11.6% [52, 111]. CIS has a high absorption coefficient (~10^5 cm^{-1}), and its CQDs show a suitable bandgap of 1.0–1.5 eV. Through surface modification and interface engineering, the efficiency leapt from 1.62% [112] to 7.01%. These included an In_2S_3 interfacial layer [113], $CuInS_2$/CdSe/ZnSe structures, passivation of surface defects using ZnSe [114], employment of $CuInSe_{1-x}S_x$ alloy QD absorber layer to extend the absorption range for higher J_{SC} [115], and CIS/ZnS core/shell structure [108] to improve the solar cell efficiency. On the other hand, the efficiency of the CISe device reached 8.1% after adjusting ZnS shell thickness, which was attributed to suppression of defect trapping [111]. At the same time, Zn doping of CISe formed alloyed QDs, which avoided the influence of the ZnS shell on electron injection from QDs; hence, the efficiency increased to a high value of 11.6% [52]. Besides, they are inexpensive and nontoxic in comparison to PbX. Furthermore, other new types of materials, including $AgBiS_2$ [116], Sb_2Se_3 [54], α-Zn_3P_2 [117], and SnS [118] and PbSe/PbS core/shell structures [119], have also shown promising results in the fabrication of CQD solar cells.

2.3.3 MEG in CQDs

The left panel of Fig. 2.7a depicts the relaxation pathways of photo-excited states in QDs. After absorption of solar photons (γ_{abs}), the QD goes to an excited state (n^*) from the relaxed state n_0. First, it undergoes the cooling process (γ_{cool}). The excess energy released can excite a second electron through reverse Auger recombination, resulting in a multiple-exciton QD (n_2) at rate γ_{MEG}. In the case of incident photons with high energy, i.e., $4E_g$, this may result in multiple EHPs. The more EHPs, the more the contribution to photocurrent in solar cells based on CQDs. Alternatively, the excess energy can be passed to the lattice by heat dissipation, forming a single exciton n_1. As a rather compelling pathway, on the contrary, γ_{cool} should be minimized to avoid energy loss. On the other hand, n^* may also undergo photo-charging at rate γ_g; see the right panel of Fig. 2.7a. After cooling, the hot trion state (n_{2T}) may also be created by excitation by the next high-energy laser. However, the photo-charging process is beyond the scope of this work, and thus irrelevant to the current

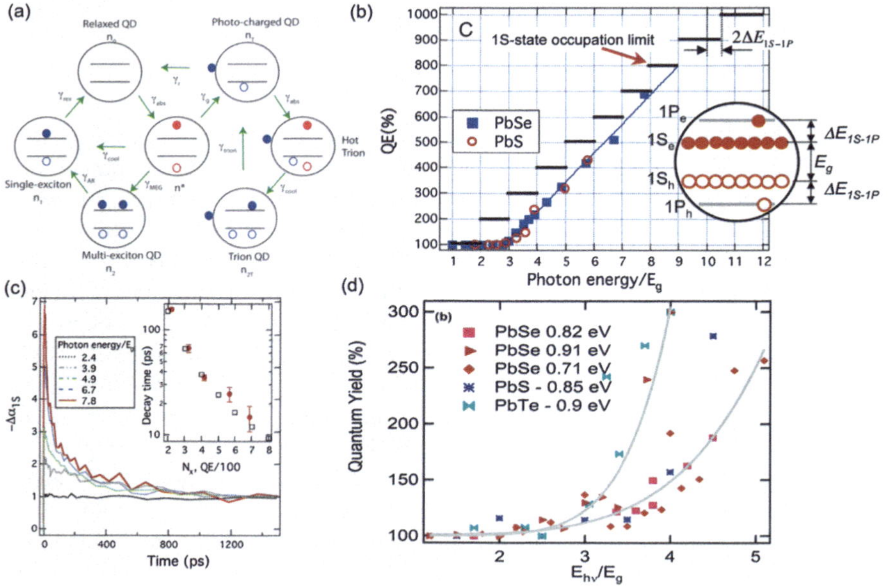

Fig. 2.7 (**a**) Relaxation pathways for excitons produced with high-energy photons in QDs. (**b**) Ideal staircase-like IQE derived from energy conservation and IQEs of PbSe, and PbS CQDs, measured as a function of $h\nu/E_g$, along with linear fits with a slope of $100\%/E_g$ and MEG thresholds of $2.5E_g$, $2.85E_g$, and $3E_g$, respectively [90]. (**c**) TA traces for PbS QDs (black), PbSe QDs (red), and PbTe QDs (blue). (**d**) Quantum efficiency for exciton formation versus normalized bandgap $h\nu/E_g$ for three PbSe QD sizes (diameter 3.9, 4.7, and 5.4 nm) one PbS and PbTe size (diameter 5.5 nm for both), with respective E_g of 0.91 eV, 0.82 eV, 0.73 eV, 0.85 eV, and 0.91 eV. (Reprinted with permission from Refs. 13, 38, 90)

discussion. For detailed information of the photo-charging process, the readers are referred to Refs. [15, 120].

In CQDs, QY indexes the ratio of the number of EHPs generated by a single photon at a certain energy. Ideally, QY of a nanocrystal QD follows the black staircase-like line (Fig. 2.7b), where the integer of normalized photon energy ($E_{h\nu}/E_g$) represents the ratio of photon energy over semiconductor bandgap energy. For the condition of $E_{h\nu}/E_g < 1$, the IQE value remains at 0, indicating that no photons are absorbed; whereas for $1 < E_{h\nu}/E_g < 3$, quantum efficiency (QE) remains at the level of 100%, meaning that one EHP is generated. Afterwards ($E_{h\nu}/E_g > 3$), QE starts to increase above 100%, meaning more than one pair of EHPs is generated, and so it continues. Hence, theoretically, $2E_g$ is the threshold energy E_{TH} for the MEG effect. In reality, experimental observation shows that the QE–$E_{h\nu}/E_g$ plot follows a linear relationship, with the same slope regardless of materials (PbS, PbSe, and CdSe). At the same time, E_{TH} is increased to ~2.9 E_g for those semiconductor QDs.

Figure 2.7c shows transient absorption (TA) traces collected for PbSe NCs with a probe pulse tuned to the lowest-energy (1S) exciton interband transition and with pump photons of energy ranging from $2.4E_g$ to $7.8E_g$. At low energies ($<2.9E_g$), single exciton is observed; whereas for high energies ($>2.9E_g$), a fast component associated with Auger recombination appears in the TA traces. This fast component indicates the generation of multi-excitons, but at the same time the early-time component of the TA dynamics becomes larger in amplitude, and the relaxation becomes faster for the highest energy of 7.8 E_g.

In a similar study, QY was measured for PbX QDs, which are displayed in Fig. 2.7d. It can be observed that QE is aligned at 100% for all QDs when the photon energy is less than $2E_g$. It becomes separated for PbSe QDs, at between 100 and 150% when the photon energy is between 2 and $3E_g$. Hence, again, the threshold energy is located at ~2–$3E_g$. QY begins to ramp after $3E_g$ and reaches a maximum of 300%. Among the three compounds, PbTe excelled in QY performance in comparison to other materials with similar bandgaps (0.8–0.91 eV). For PbSe in particular, QDs with wider E_g (0.91 eV) showed greatly advanced QY performance in comparison to the other two narrow bandgap QDs. From the QY spectra, it can be seen that QY is an intrinsic property, but it can also be affected by the bandgap of the materials involved. Indeed, highly efficient MEG by one photon has been observed in nanocrystals of lead chalcogenides [13].

The higher the QY in the nanocrystal QD, the higher the QE anticipated in devices. In particular, PbS CQDs exhibit QY with experimental value of 430% at photon energy $h\nu$ equal to ~$5.8E_g$; whereas, for PbSe QDs, this value reaches as high as 700% at even higher photon energy ($7.8E_g$). The threshold energy (E_{TH}), which defines the photon energy at which the QY breaks 100%, stays at ~$2.9E_g$, irrespective of materials [90]. Actually, it is advantageous to have as high QY as possible and at the same time a relatively low E_{TH}, e.g., $<2E_g$, in these low-dimensional CQDs. In fact, it is rather challenging for high-energy solar photons ($>5E_g$) to participate in the solar-to-power conversion because high-energy solar photons can be easily scattered or absorbed by the front electrode, which are typically TCOs, e.g. ZnO (E_g of ~3.4 eV). This results in a normalized photon energy of ~2–4E_g, depending on size-controllable bandgaps, i.e., 0.8–1.2 eV.

2.3.4 The MEG Effect on Device PCE

In the framework of CQD solar cells, the loss of solar energy is expected to be reduced to nearly zero by utilizing multiple EHPs, so that the ultimate efficiency is expected to reach 43% [121]. In the following section, this value is explained and the correlation between MEG and ultimate PCE of the device is uncovered. First, it is helpful to know that the enhancement of J_{ph} and V_{ph} through the MEG effect is beneficial to the device's PCE. According to Klimov, the maximum PCE (η*) of a device is expressed by the following Eq. 2.2: [11].

$$\eta^* = FF^* \eta_0^* eV_{OC}^* E_g^{-1} \tag{2.2}$$

where FF* is fill factor and η_0* is the MEG enhanced ultimate efficiency, where η_0* is expressed in Eq. 2.3a:

$$\eta_0^*(\omega_g) = \bar{q}(\omega_g) * \eta_0(\omega_g) \tag{2.3a}$$

where $\bar{q}(\omega_g)$ and $\eta_0(\omega_g)$ are the average IQE and ultimate efficiency, respectively. They are expressed as:

$$\bar{q}(\omega_g) = \int_{\omega_g}^{\infty} q(\omega) F_S(\omega) d\omega \left[\int_{\omega_g}^{\infty} F_S(\omega) d\omega \right]^{-1} \tag{2.3b}$$

$$\eta_0(\omega_g) = E_g \int_{\omega_g}^{\infty} F_S(\omega) d\omega \left[\int_{\omega_g}^{\infty} \hbar\omega F_S(\omega) d\omega \right]^{-1} \tag{2.3c}$$

where $F_s(\omega)$ is solar irradiance at angular frequency ω and $q(\omega)$ is the frequency-dependent QY of photon-to-exciton conversion. As a result, device efficiency is highly dependent on the solar radiance frequency. Because it is a result of the integral of the different frequencies. From this, the device efficiency is calculated as a function of bandgap of the absorber layer.

Figure 2.8a illustrates carrier generation in the nanocrystal CQD PVs under exposure to solar (F_s) and ambient (F_a) radiations. At the same time, the carrier recombination occurs at rate of R_r. Figure 2.8b displays detailed balanced PCE spectra with respect to the photon energy for different MEG threshold conditions. Without MEG, it gives a maximum efficiency of 32% at $h\nu$ of ~1.2 eV (see the black dashed line) and denoted as $h\nu_{max}$. With MEG, the efficiency is substantially increased to 36% for the 3E_g MEG condition. In particular, the threshold condition

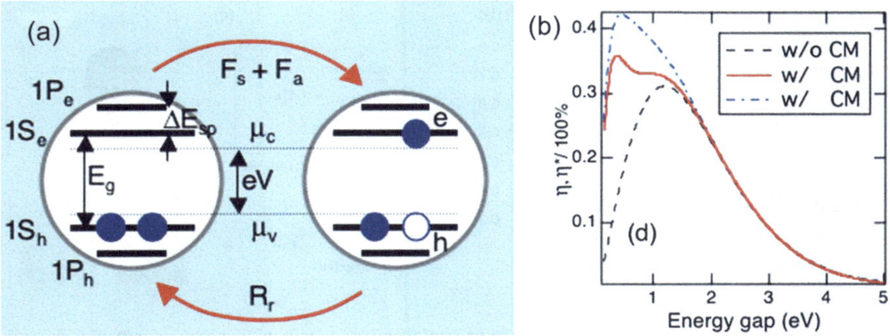

Fig. 2.8 (**a**) Diagram of carrier generation and recombination processes the nanocrystal CQDs in exposure to solar (F_s) and ambient (F_a) radiations. (**b**) Maximum PCE calculated as a function of absorber layer bandgap. (Reprinted with permission from Ref. 29)

of $2E_g$ can boost the efficiency to 43% at $0.5E_g$, which is the ultimate expectation in QDSCs. At the $2E_g$ condition, however, the maximum PCE can be anticipated to reach 60.3% without the decay of fast component in the TA trace; see Fig. 2.7c [122].

2.4 Recombination Loss Pathways

2.4.1 Sub-Bandgap Defects

Here, in CQD PVs, sub-bandgap defects (also known as midgap or intraband states) originate from unsaturated dangling bonds on CQDs, i.e., QD surface defects, because of their large surface-to-volume ratio or vacancies and interstitials [41, 123–125]. In the band diagram, they are located at specific energy levels, mainly along the photoactive layer. As mentioned earlier, these defects capture photo-generated carriers in the short timescale of ~1–100 ps. In order to reduce their density, surface modification through ligand exchange has been found helpful in passivation of these charge-active defects and thus improving J_{SC} in CQD PVs. Because these defects are reported to be responsible for reductions in photocurrent [126].

As can be seen in Fig. 2.9a, V_{OC} is much lower than its bandgap, estimated from its CQD absorption spectrum. Here, the V_{OC} deficit is closely related to the bandgap of the CQD layer, see Fig. 2.9b. For optimized CQD bandgap (~1.2–1.4 eV), the ideal V_{OC} should be in the range of ~0.6–0.8 eV, which is higher than currently obtained values from experiments. For oleate-capped PbS CQDs, for instance, hydroxyl group (-OH) bonded on the Pb-terminated polar (111) surface facet (in Fig. 2.9c) are reported to be the main cause of the sub-bandgap defect [124, 125]. For treatment, novel iodide-based ligands, 1-propyl-2,3-dimethylimidazolium iodide (PDMII), 1-ethyl-3-methylimidazolium iodide (EMII) and other surface pas-

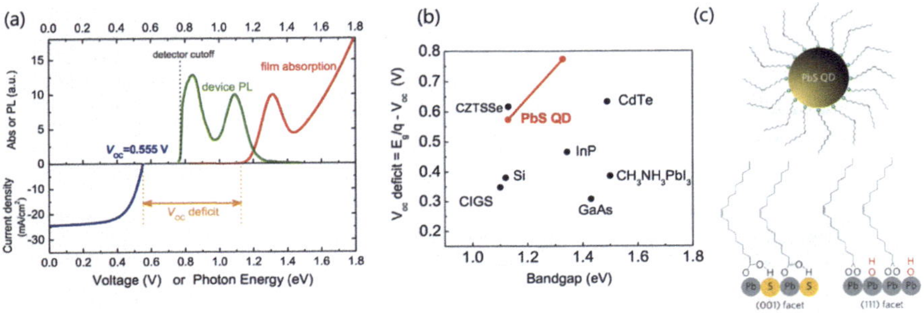

Fig. 2.9 (**a**) Absorption spectra of a PbS-TBAI CQD film and photoluminescence (PL) spectra of a PbS-TBAI QD/PbS-EDT CQD device; *J–V* characteristics of the certified the CQD solar cells. (**b**) The V_{OC} deficits in PbS CQD PV and other PV materials. (Reproduced from Ref. 36) (**c**) Schematics of PbS CQDs with oleic surface ligands showing the (001) face passivated by oleic acid and the (111) facet by deprotonated oleic ligands together with hydroxide groups. (Reprinted with permission from Ref. 124)

sivation treatment have substantially suppressed the activity of defects in the treated PbS CQDs, leading to an increase in V_{OC} and associated efficiency [125, 127, 128]. Moreover, the band tail of the trap distribution is becoming narrow as the QD size becomes larger [129]. This means a short capture time for the trap states. Even though some study shows that the maximum PCE is not severely depending on the band tailing [130].

Further, in order to elucidate bulk defects and their relation to device performance, an electro-optical technique, pump-push photocurrent (PPP) spectroscopy was set up to study the trapping dynamics in PbS devices. It was found that IR photo-induced absorption of CQD in the 0.2–0.5 eV region is partly associated with immobile charges [36]. Besides, it was observed that the early trapping dynamics rely strongly on surface-modifying ligands, rather than ETL, which are a more interface-defect-related recombination. This further confirmed that, in comparison to intraband trap states, the timescale of recombination due to interface defects of ETL/photoactive layer takes a longer time.

In a recent study of PbS by first-principle calculations, PbS QDs were classified into four types depending on whether they are stoichiometric/off-stoichiometric or cubic{111}-faceted; see Fig. 2.10a. On the other hand, stoichiometry (S/Pb ratio) in PbS QDs are crucial to their electronic structure and related properties. Figure 2.10b gives semiconducting properties with varied bandgap and vacuum levels in the limited ratio of 0.93–1.08. This opens the opportunity that band alignment can be realized by adjustment of the ratio in PbS CQDs. More importantly, at lower ratios, the midgap states are located closer to the conduction band, whereas these states are located closer to the valance band once the ratio is higher than 1. As we already know, midgap states (in green in Fig. 2.10b) are a result of surface defects. They can introduce charge recombination loss, so they should be reduced by passivation via surface modification, e.g., ligand exchange. The findings in this

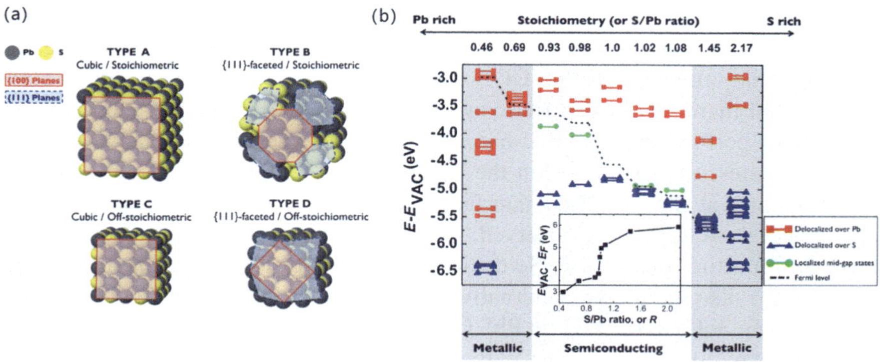

Fig. 2.10 (**a**) Classification of PbS CQDs into four different types (A, B, C, and D). (**b**) Kohn-Sham energy levels for PbS CQDs for a range of stoichiometry (S/Pb ratio). (Reprinted with permission from Ref. 131)

work provide a new route to high efficiency in PbS QD-based solar cells by adjusting the ratio to 1:1 so that PbS QDs stays in stoichiometry, where no midgap states should be observed.

2.4.2 *Interface Defects*

In addition, another recombination pathway by interface defects should also be considered in the device performance [9]. Instead of being located at fixed energy level, as are sub-bandgap defects (except for tail states with a continuous distribution), they have a continuous distribution along the energy levels [132]. The ultrafast interfacial charge separation can effectively compete with exciton-exciton annihilation considering their charge recombination timescale, and are occurring at a timescale faster than sub-bandgap defects [35]. Control over band offsets [9, 133, 134], doping [135], and interfacial trap state densities have been proved to have effective charge dissociation at the junction interface [109, 136]. Here, interface defects include both hetero-interface defects as well as defects located at the back-contact (PbS/Au) [9], both of which cause a reduction in diffusion length of carriers and influence J_{SC}. Also, it must be emphasized that the anodic interface should be Ohmic, meaning the Fermi level aligns with the work function of the metal or higher work function of the anode metal than the Fermi level of the QD, giving upwards band bending. Using scanning photocurrent and photo-voltage microscopies, lateral depletion width and related PV properties were interrogated in planar PbS CQD devices. From the results, the Au/PbS QD was proved to be Ohmic [137]. Even so, there might still be interface defects generated at the PbS QD layer/Au interface.

Currently, passivation of interface defects involves using an additional PbS-EDT layer and engineering the ETLs, e.g., by nitrogen doping of ZnO [35], replacement of ZnO by SnO_2 [55], thermal annealing at elevated temperature [128], or anode modification [129]. Upon nitrogen doping of the ZnO front electrode, V_{OC} and efficiency of PbS CQD devices was found to be improved by 50% as a result of increase in its carrier concentration. In contrast, this strategy did not improve the performance of PbSe-based cells. From experiments, a larger volume of defect states was found within the bandgap of the PbS QDs, which facilitate recombination at the PbS/ZnO interface; whereas, for the PbSe QD/ZnO solar cells, fewer such traps were observed, so the interfacial recombination channel is thus not a limiting factor in PbSe devices [35]. In addition, replacement of ZnO with SnO_2 is also reported as alternative for interface engineering. In comparison to ZnO, using SnO_2 as the ETL in PbS CQDSCs exhibits extended transmittance in the UV light region. The lifetimes in the devices was determined to follow a similar trend, with devices using SnO_2 as ETL showing improved lifetime. Hence, it indicates improved interfaces, at either photoactive layer/ETL or ETL/metallic contact. Further, in transient photocurrent (TPC) and photo-voltage (TPV) decay studies, it was found that the introduction of chlorine atoms (SnO_2-Cl) could achieve interface passivation in the low-temperature-processed SnO_2 film. In comparison with pure SnO_2 devices, the transport lifetime ($\tau_t = 2.7$ μs) is prolonged to $\tau_t = 3.8$ μs in SnO_2-Cl devices, and the charge recombination lifetime τ_{rec} is substantially prolonged from 60 μs to 140 μs in the Cl-doped SnO_2 device. This result agrees well with the alleviated charge recombination at the device interfaces and the high PCE performance in SnO_2-Cl devices (9.37%) [55]. Further, thermal annealing is reported as a convenient way to improve V_{OC} through reduction of interface defects. In detail, device optimization by modification of interfacial electron/hole extraction layers may provide additional performance enhancement [128]. At 140 °C, in particular, the device gives an optimized PCE value, reflected by an increase in J_{SC} rather than V_{OC} in comparison with low-temperature-treated devices. However, further understanding is necessary in regard to the electrical properties, and the effect of thermal annealing treatment on PbS CQD PVs could further increase performance of PCEs in air-stable [138]. Reference 128 reported that the CQD-ink photovoltaic device achieves much higher PCEs (certified 9.61%) than the control devices (7.85%) owing to improved charge drift and diffusion. This is reflected in the charge recombination lifetime τ_{rec} measurement through TPV measurement, from 16 to18 μs [128]. In PbS CQD solar cells, highly ordered wrinkled CP-$Mg_{1-x}Zn_xO$ thin films increase significantly the charge-carrier extraction from the CQD solid and therefore diminish the charge interfacial recombination at the CQD/ETL junction, leading to a 15.5% increase in PCE [139]. Furthermore, Cho et al. found that improved interface and reduced interface recombination loss through the piezoelectric effect lead to increase of 38% in J_{SC} for a final device of 5.01% [140].

Another example of interface engineering is to investigate the back-contact, where the PbS absorber is in contact with MoO_x/Ag in this specific case. Introduction of an additional HTL layer actually has no influence on the distribution of photogeneration; see Fig. 2.11. From this figure, it's clearly seen that Mo_xO/Ag layers are hole-driving. In the case of P3HT as HTL, holes are accumulated at the metal/

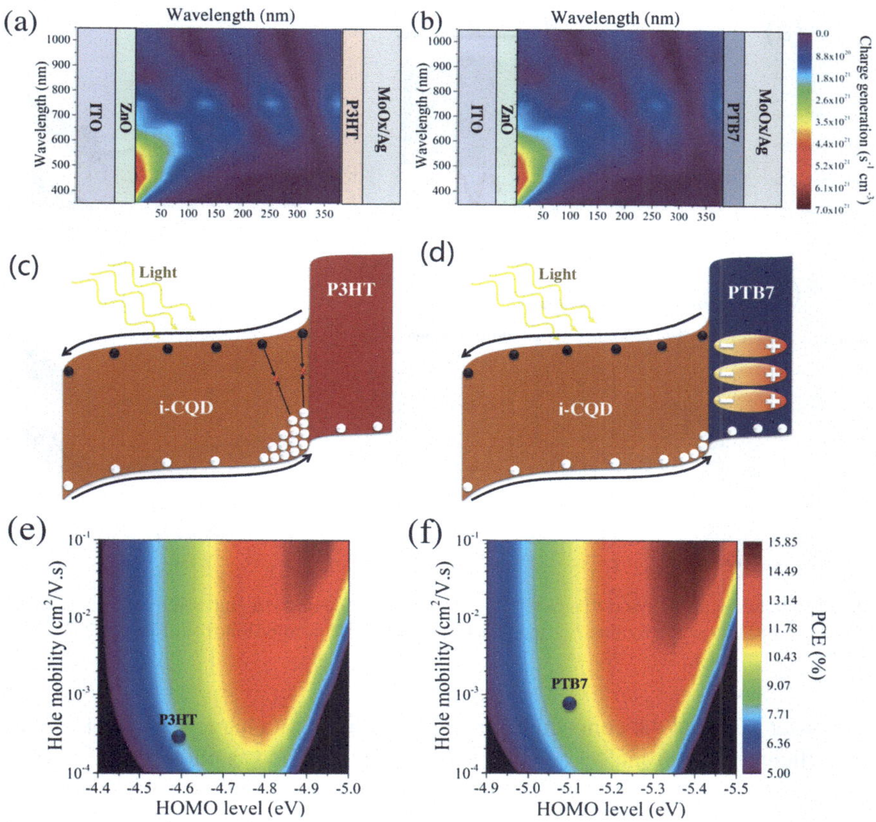

Fig. 2.11 Charge generation distribution in PbS-CQD device with (**a**) P3HT and (**b**) PTB7 HTLs. Schematics of interfaces for (**c**) i-CQD/P3HT and (**d**) i-CQD/PTB7. Simulated device performance dependence on various hole mobilities and HOMO levels of (**e**) P3HT and (**f**) PTB7. (Reprinted with permission from Ref. 64)

semiconductor interface, and cannot be transported efficiently because of the offset of band alignment here; whereas, in the case of PTB7 HTL, this situation is improved because of the electrical dipole effect in this HTL. Subsequently, hole mobility in the PTB7 HTL device is enhanced as well as its performance, reflected by increases in both J_{SC} and V_{OC}. Besides, with anode modification, V_{OC} is dramatically increased in the inverted-structure PbS CQD solar device due to enhanced hole extraction. This is realized with an additional layer of unique conjugated polymer PTB7 between ITO and the PbS-EDT layer. At the same time, the PTB7 layer can enable efficient charge separation, reducing interfacial charge recombination and reversing saturation current density. In comparison, V_{OC} is increased from 0.14 to 0.63 V, while J_{SC} is increased from 9.34 to 20.39 mA·cm^2, resulting in a PCE of 8.45% [141].

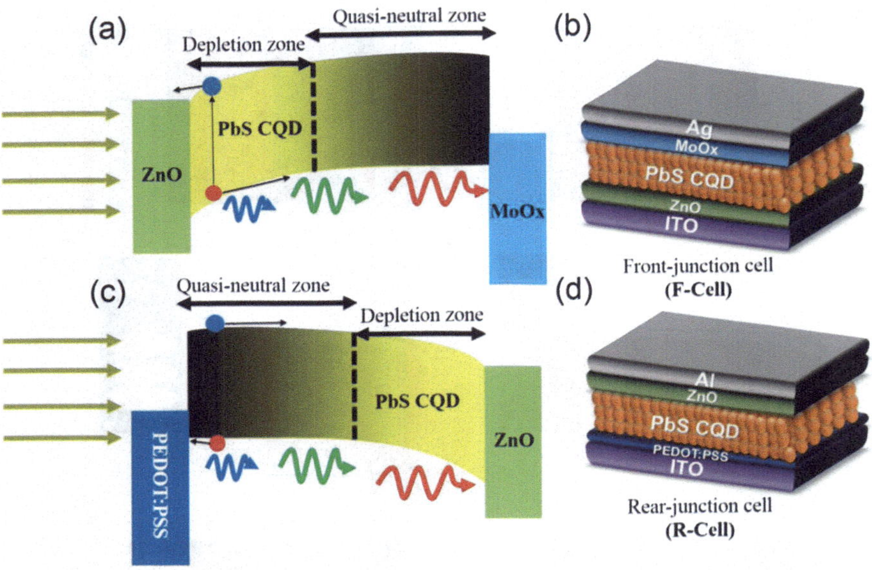

Fig. 2.12 Schematics of band diagram and corresponding device structures for regular structure (**a**) and (**b**) and inverted structure (**c**) and (**d**). (Reprinted with permission from Ref. 143)

In order to reveal the connection between the nature of charge separation and its solar efficiency, the front ZnO electrode in a CQD device was photo-doped, and its EQE spectra were compared at different doping densities [109]. The results showed that ZnO with the highest level of doping density gives the best overall EQE spectrum. The better performance with a high doping level is correlated with efficient charge extraction because a large portion of the depletion region is located in the PbS layer compared to in the low-doped samples. In this way, photo-generated carriers within region of the PbS/ZnO interface have a good probability of being harvested [142]. In addition, a regular structure (in Fig. 2.12a, b) is found to yield a PCE of 4.5%, far less than that in the inverted structure (in Fig. 2.12c, d) with a thin (60 nm) PbS layer (6.0%). This is realized by the improved EQE spectrum in the range of 400–800 nm because of efficient photogeneration/extraction at the interface of the PEDOT:PPS/PbS layers [143, 144].

For interface characterization, however, electrochemical IS (ECIS) was employed to investigate the junction properties of CQD PVs, e.g., recombination resistance, carrier lifetime, capacitance, and conductivity. The charge-carrier conduction mechanisms were revealed by comparison of devices with and without a PbS-EDT photo-anode in terms of extraction of photo-generated charge carriers, transportation, and collection of carriers. At zero bias, without PbS-EDT photo-anode, the device exhibits carriers with lifetime increasing from 4 to 14 μs under illumination, leading to a leap of PCE from 5.23 to 7.85% [144].

As mentioned earlier, engineering of the front electrode has proved successful in the improvement of the hetero-interface and the associated efficiency in CQD solar devices, because of its prolonged diffusion length as well as reduced trap-assisted recombinations [55, 139, 145]. Here, the front electrode is usually known as the ETL in many research articles [9, 64, 116], where its electron-transport properties are also concerned. Thus, in this work, we will not differentiate between those two terms. Moreover, nanowire structure and 2D materials (in particular graphene) can also be used as the front electrode to allow passage of solar photons for penetrating solar photons and at the same time to keep good electrical contact with the photoactive layers [146], like ZnO nanowires [133, 147, 148], SnO_2:Sb nanocrystals [56], $Mg_{1-x}Zn_xO$ buffer layers [9, 139, 149], and silver nanowires [72, 150].

On the other hand, HTLs have also been useful in carrier transport, band alignment, and improvement of PCE [14, 64, 77, 79, 133, 148]. For the HTL, *n*-type oxides consisting of molybdenum oxide (MoO_x) were used in PbS CQD solar cells, resulting in a 4.4% in PCE. This represents a more than 65% efficiency improvement in comparison with the case of the Au layer is directly vapored on the PbS CQD layer. It was found that the HTL involves a dipole formed at the hetero-interface to enhance band bending and allow efficient hole extraction [47]. Besides, Fermi level pining can be avoided by inserting a MoO_3 thin layer between the two layers, through which the Schottky junction and impedance of extraction of holes from PbS can be prevented. In this way, the performance of ZnO/PbS CQD PVs was significantly enhanced, reflected by an improvement in J_{SC}, V_{OC}, FF, and η [151].

2.5 Summary and Future Prospects

Despite the significant progress made with charge transport layers in CQD PVs, there still exist three fundamental limiting factors:

1. High QY of MEG at close to $2E_g$ in the QDs.
2. Surface traps, resulting in increase in surface activities, which remain the main pathway for carrier recombination.
3. With significant surface passivation remedies, the V_{OC} still remains at ~0.5–0.6 V in the type-I devices, with the optimized CQD bandgap ranging at ~1.3–1.4 eV [94].

This implies that the intra-bandgap defects [35] and consequential Fermi-level pinning [32, 36, 109, 152–154] may exist for such types of solar cells because such types of carrier/intrinsic defect interactions will definitely reduce the rate of injection of electrons to the ETL. As a result, defect recombination through different channels probably remains the largest limiting factor for improvement in PCE in PbS CQDSCs, together with limited QY in multiple EHPs. Because a high QY also requires high normalized photon energy $h\nu/E_g$.

2.5.1 Device Structure Engineering

In terms of interface engineering, nanostructured front electrode [155] (see Fig. 2.13a, b) could also be promising for the future application of highly efficient CQD PVs. Because they have a good penetration into the photoactive layer, so that there is an improvement of charge collection efficiency due to enlarged hetero-interface [147, 156, 157], reduction in Fermi level pinning caused by charge recombination centers [158] and associated V_{OC} [158, 159], and more sites for charge separation [160, 161], as long as the criteria of good interface must be meet for prevention of charge recombination [147]. Besides, the length of the nanowire should be optimized to avoid a poor hole collection efficiency [155]. Moreover, given the short diffusion lifetime of minority carriers in QDs, it is necessary to increase the hetero-interface area [147, 155, 158]. Actually, the insufficient L_D can lead to the problem of reduction in J_{SC}, similar to the consequence of inadequate thickness of CQD layer [147].

Further, buk heterojunction architecture has been successful in other types of solar cells, [109, 162–164] in Fig. 2.13c, and it is anticipated to improve the efficiency of CQD solar cells as well. For example, a CdI_2-treated CdSe QD buffer layer has been introduced at the interface between ZnO NPs and PbS QDs for the QDSCs. Through optimization of its carrier concentration and energy band alignment by surface- and size-tunable electronic properties, the CdSe QD buffer layer was proved to suppress the interface recombination and to contribute additional photogenerated carriers, increasing both the V_{OC} and J_{SC}, leading to a 25% increase in PCE of the PbS CQD PVs [165].

Apart from ZnO, TiO_2 nanowire electrodes grown by hydrothermal methods have been applied to improve the performance of N719 dye-sensitized solar cells with PCE increased from 2.34 to 7.16%, and the increase is mainly contributed by improvement in J_{SC} and FF due to enlarged effective surface area in the nanowire structured TiO_2 electrode [166]. A N-doped TiO_2 anatase nanowire can also alter the band alignment between rutile TiO_2 and fluorine doped tin oxide (FTO) so that electron transfer from TiO_2 to FTO is more efficient [167]. Besides, metal-

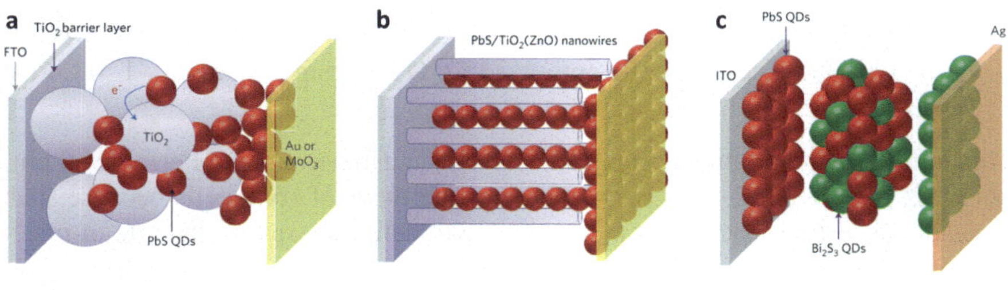

Fig. 2.13 Architectures for CQD photovoltaics: (**a**) porous structured electrode and (**b**) one-dimensional nanostructure electrode. (**c**) A bulk heterojunction of n-type Bi_2S_3 and p-type PbS CQDs. (Reprinted with permission from Ref. 109)

lic silver nanowires (AgNWs) are reported to be highly conductive and at the same time have good optical transmittance; hence, it is also available for PCE improvement [72, 150].

2.5.2 Emerging CQD Materials and Shapes

Despite the fact that solar cells based on PbS CQDs have achieved high efficiency, it is also sensible to ask the reason for the low efficiency in devices based on other lead chalcoginde nanocrystals in PbX family, where both of their peak QEs have exceeded 100%. In particular, the solar device based on PbSe nanorods (NRs) also have demonstrated peak efficiency of larger than 122% (see Fig. 2.14a, b), in the range of solar photons above threshold energy (~$3E_g$). This implies that the application of PbSe NRs in solar devices for their contribution of MEG to the device performance [168]. To be explicit, it can be noted that the

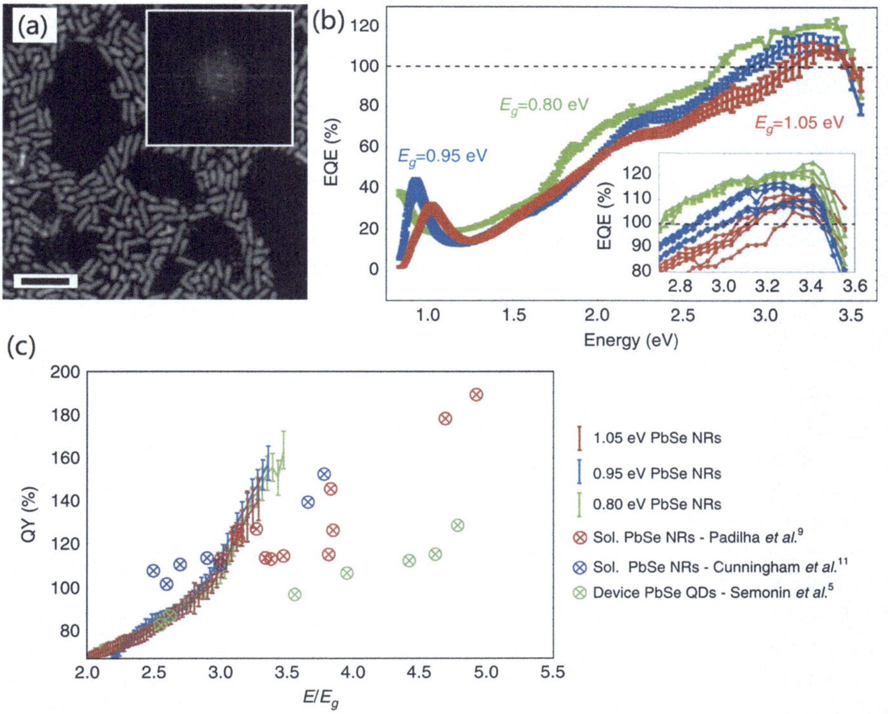

Fig. 2.14 (**a**) TEM images for PbSe nanorods. (**b**) EQEs for three types of bandgap PbSe nanorod photovoltaic devices. (**c**) Comparison of IQEs with literature. (Reprinted with permission from Ref. 168)

NRs display increased MEG QY in comparison to PbSe QDs for different normalized energy (see Fig. 2.14c) irrespective of bandgaps. A promising study found that η_{MEG} was enhanced in PbSe NRs with moderate aspect ratios compared to QDs. It was suggested that this increase resulted primarily from an increase in k_{MEG}, as recent measurements of intraband relaxation in PbSe NRs indicated that carrier cooling rates were not significantly affected by elongation of the nanocrystal [165].

In PbTe devices, charges produced by MEG were reported to be harvested efficiently with EQE exceeding 120%, and overall efficiency 1.9% achieved. Indeed, its J_{SC} (18.2 mA·cm^{-2}) is coherent with the EQE measurement and is comparable to that in highly efficient PbS solar devices, while V_{OC} (0.3 V) and *FF* (32%) remains only half of that seen in PbS devices. This is presumably due to the high susceptibility of PbTe to oxidation. Delayed early intraband relaxation of hot carriers was confirmed by employing ultrafast transient absorption spectroscopy (TAS) and two-pulse ultrafast photocurrent spectroscopy [51]. Further, the PbTe device showed lower photocurrent after 10 days, which was explained by the so-called "QD surface oxidation effect", meaning reduce the extractable photocurrent under short-circuit conditions [51]. Therefore, defect-related studies should be employed to further understand the role of PbTe in PV applications.

Instead of using bis(trimethylsilyl) sulfide (TMS_2) as sulfur source for PbS CQDs, a new method employing a CdS CQDs/ODE solution exploiting a cation exchange reaction has been discussed. The resulting CQDs show performance comparable with previous reports for solar cells. Through optimizing the architecture of the photoactive layer, excellent stability in air is achieved. An efficiency of 7.89% is retained in air after 110 days [169]. A combination of thiols and amines is very effective at reducing and then complexing with elemental selenium to form a highly reactive selenium precursor at room temperature; alloyed $CuInSe_xS_{2-x}$ CQDs can be synthesized by this new approach. Optimal sunlight harvesting can be finely tuned by controlling CQD size and composition [115].

Moreover, other material systems, novel nanostructures, synthesis methods, and associated defect passivation treatment have been investigated. It is also reported to be advantageous to use (1) a core/shell CQD structure, due to enhancement of the injection rate at the ETL and reduction of recombination. This includes PbS/CdS core/shell structures, CdS coatings, conformal coatings of ZnS on the ZnO ETL surface, fortification of graphene oxide, and (Zn)CuInS/ZnS core/shell coatings by in situ alloying approach; (2) different types of material synthesis methods, such as in situ growth of CdSe on TiO_2 nanotubes, increase of CQD loading and reduction of recombination rate using CBD; (3) chemical treatments by passivation agents, so that CQDs can aggregate, or their bandgap can be tuned, and multiple-exciton dissociation to methylene blue; and (4) modification of material properties, e.g., Mn doping in CdS QDs, and Mn-modified CdSe shells. Other practical methods include using mesoporous SnO_2 spheres as photoanode, graphene CQDs for CdSe/electrolyte interface, $TiCl_4$ treatment for porous SnO_2 in comparison to TiO_2, exciton quenching as the limiting factor. Sb_2Se_3 is also a new type of PV absorber with suitable bandgap (1.1–1.2 eV) and strong light absorption properties (absorption coefficient > 10^5 cm^{-1} at short wavelength). Currently, solar cells based on this materials have shown PCE up to 3.7% [170].

Silver sulfide (Ag_2S) is also considered as a promising PV absorber for its suitable bandgap (1.0 eV) and non-toxicity, but currently the device poor efficiency of 0.76% has only been demonstrated in CQD sensitized solar cells. Through functional theory (DFT) calculations, the authors have demonstrated that it is possible to tune the electronic properties of Ag_2S by surface modification [171].

Acknowledgments The research funding from the China Postdoctoral Science Foundation through the project (No. 2018M640906) is greatly acknowledged by the authors. L.Q. was supported by the National Natural Science Foundation of China (Grant No.:11774044).

References

1. Renewable Energy BP.
2. Renewable Energy Ren21.
3. Renewable Energy Capacity Growth IEA.
4. Photovoltaic Report IFS.
5. Jean, J., Brown, P. R., Jaffe, R. L., Buonassisi, T., & Bulović, V. (2015). Pathways for solar photovoltaics. *Energy & Environmental Science, 8*(4), 1200–1219.
6. Jean, J., Xiao, J., Nick, R., Moody, N., Nasilowski, M., Bawendi, M., & Bulović, V. (2018). Synthesis cost dictates the commercial viability of lead sulfide and perovskite quantum dot photovoltaics. *Energy & Environmental Science, 11*(9), 2295–2305.
7. Shockley, W., & Queisser, H. J. (1961). Detailed balance limit of efficiency of p-n junction solar cells. *Journal of Applied Physics, 32*(3), 510–519.
8. Zhang, X., Hägglund, C., Johansson, M. B., Sveinbjörnsson, K., & Johansson, E. M. J. (2016). Fine tuned nanolayered metal/metal oxide electrode for semitransparent colloidal quantum dot solar cells. *Advanced Functional Materials, 26*(12), 1921–1929.
9. Ding, C., Zhang, Y., Liu, F., Nakazawa, N., Huang, Q., Hayase, S., Ogomi, Y., Toyoda, T., Wang, R., & Shen, Q. (2017). Recombination suppression in PbS quantum dot heterojunction solar cells by energy-level alignment in the quantum dot active layers. *ACS Applied Materials & Interfaces, 10*(31), 26142–26152.
10. Kagan, C. R., Lifshitz, E., Sargent, E. H., & Talapin, D. V. (2016). Building devices from colloidal quantum dots. *Sciences, 353*(6302), aac5523.
11. Hu, L., Patterson, R. J., Hu, Y., Chen, W., Zhang, Z., Yuan, L., Chen, Z., Conibeer, G. J., Wang, G., & Huang, S. (2017). High performance PbS colloidal quantum dot solar cells by employing solution-processed CdS thin films from a single-source precursor as the electron transport layer. *Advanced Functional Materials, 27*(46), 1703687.
12. Tang, J., Kemp, K. W., Hoogland, S., Jeong, K. S., Liu, H., Levina, L., Furukawa, M., Wang, X., Debnath, R., Cha, D., Chou, K. W., Fischer, A., Amassian, A., Asbury, J. B., & Sargent, E. H. (2011). Colloidal-quantum-dot photovoltaics using atomic-ligand passivation. *Nature Materials, 10*(10), 765.
13. Murphy, J. E., Beard, M. C., Norman, A. G., Ahrenkiel, S. P., Johnson, J. C., Yu, P., Mićić, O. I., Ellingson, R. J., & Nozik, A. J. (2006). PbTe colloidal nanocrystals: Synthesis, characterization, and multiple exciton generation. *Journal of the American Chemical Society, 128*(10), 3241–3247.

14. Kirmani, A. R., Sheikh, A. D., Niazi, M. R., Haque, M. A., Liu, M., de Arquer, F. P. G., Xu, J., Sun, B., Voznyy, O., & Gasparini, N. (2018). Overcoming the ambient manufacturability-scalability-performance bottleneck in colloidal quantum dot photovoltaics. *Advanced Materials, 30*(35), 1801661.
15. Yuan, M., Liu, M., & Sargent, E. H. (2016). Colloidal quantum dot solids for solution-processed solar cells. *Nature Energy, 1*(3), 16016.
16. Carey, G. H., Abdelhady, A. L., Ning, Z., Thon, S. M., Bakr, O. M., & Sargent, E. H. (2015). Colloidal quantum dot solar cells. *Chemical Reviews, 115*(23), 12732–12763.
17. Zhao, H., & Rosei, F. (2017). Colloidal quantum dots for solar technologies. *Chem, 3*(2), 229–258.
18. Nozik, A. J., Beard, M. C., Luther, J. M., Law, M., Ellingson, R. J., & Johnson, J. C. (2010). Semiconductor quantum dots and quantum dot arrays and applications of multiple exciton generation to third-generation photovoltaic solar cells. *Chemical Reviews, 110*(11), 6873–6890.
19. Kim, J. Y., Voznyy, O., Zhitomirsky, D., & Sargent, E. H. (2013). 25th anniversary article: Colloidal quantum dot materials and devices: A quarter-century of advances. *Advanced Materials, 25*(36), 4986–5010.
20. Konstantatos, G. (2013). *Colloidal quantum dot optoelectronics and photovoltaics*. Cambridge: Cambridge University Press.
21. Murray, C., Norris, D. J., & Bawendi, M. G. (1993). Synthesis and characterization of nearly monodisperse CdE (E= sulfur, selenium, tellurium) semiconductor nanocrystallites. *Journal of the American Chemical Society, 115*(19), 8706–8715.
22. Hines, M. A., & Scholes, G. D. (2003). Colloidal PbS nanocrystals with size-tunable near-infrared emission: Observation of post-synthesis self-narrowing of the particle size distribution. *Advanced Materials, 15*(21), 1844–1849.
23. Qiao, K., Deng, H., Yang, X., Dong, D., Li, M., Hu, L., Liu, H., Song, H., & Tang, J. (2016). Spectra-selective PbS quantum dot infrared photodetectors. *Nanoscale, 8*(13), 7137–7143.
24. Ren, Z., Sun, J., Li, H., Mao, P., Wei, Y., Zhong, X., Hu, J., Yang, S., & Wang, J. (2017). Bilayer PbS quantum dots for high-performance photodetectors. *Advanced Materials, 29*(33), 1702055.
25. Lhuillier, E., Scarafagio, M., Hease, P., Nadal, B., Aubin, H., Xu, X. Z., Lequeux, N., Patriarche, G., Ithurria, S., & Dubertret, B. (2016). Infrared photodetection based on colloidal quantum-dot films with high mobility and optical absorption up to THz. *Nano Letters, 16*(2), 1282–1286.
26. Liu, M., Voznyy, O., Sabatini, R., de Arquer, F. P. G., Munir, R., Balawi, A. H., Lan, X., Fan, F., Walters, G., & Kirmani, A. R. (2017). Hybrid organic-inorganic inks flatten the energy landscape in colloidal quantum dot solids. *Nature Materials, 16*(2), 258.
27. Lu, K., Wang, Y., Liu, Z., Han, L., Shi, G., Fang, H., Chen, J., Ye, X., Chen, S., & Yang, F. (2018). High-efficiency PbS quantum-dot solar cells with greatly simplified fabrication processing via "solvent-curing". *Advanced Materials, 30*(25), 1707572.
28. Hou, Y., Chen, W., Baran, D., Stubhan, T., Luechinger, N. A., Hartmeier, B., Richter, M., Min, J., Chen, S., Quiroz, C. O. R., Li, N., Zhang, H., Heumueller, T., Matt, G. J. Osvet, A., Forberich, K., Zhang, Z.-G., Li, Y., Winter, B., Schweizer, P., Spiecker, E., & Brabec, C. J. (2016). Overcoming the interface losses in planar heterojunction perovskite-based solar cells. *Advanced Materials, 28*(25), 5112–5120.
29. Klimov, V. (2006). Detailed-balance power conversion limits of nanocrystal-quantum-dot solar cells in the presence of carrier multiplication. *Applied Physics Letters, 89*(12), 123118.
30. Scully, M. O. (2010). Quantum photocell: Using quantum coherence to reduce radiative recombination and increase efficiency. *Physical Review Letters, 104*(20), 207701.
31. McGuire, J. A., Joo, J., Pietryga, J. M., Schaller, R. D., & Klimov, V. I. (2008). New aspects of carrier multiplication in semiconductor nanocrystals. *Accounts of Chemical Research, 41*(12), 1810–1819.

32. Malgras, V., Nattestad, A., Kim, J. H., Dou, S. X., & Yamauchi, Y. (2017). Understanding chemically processed solar cells based on quantum dots. *Science and Technology of Advanced Materials, 18*(1), 334–350.
33. Polman, A., Knight, M., Garnett, E. C., Ehrler, B., & Sinke, W. C. (2016). Photovoltaic materials: Present efficiencies and future challenges. *Science, 352*(6283), aad4424.
34. Hoye, R. L., Ehrler, B., Böhm, M. L., Muñoz-Rojas, D., Altamimi, R. M., Alyamani, A. Y., Vaynzof, Y., Sadhanala, A., Ercolano, G., & Greenham, N. C. (2014). Improved open-circuit voltage in ZnO–PbSe quantum dot solar cells by understanding and reducing losses arising from the ZnO conduction band tail. *Advanced Energy Materials, 4*(8), 1301544.
35. Ehrler, B., Musselman, K. P., Böhm, M. L., Morgenstern, F. S., Vaynzof, Y., Walker, B. J., MacManus-Driscoll, J. L., & Greenham, N. C. (2013). Preventing interfacial recombination in colloidal quantum dot solar cells by doping the metal oxide. *ACS Nano, 7*(5), 4210–4220.
36. Chuang, C.-H. M., Maurano, A., Brandt, R. E., Hwang, G. W., Jean, J., Buonassisi, T., Bulović, V., & Bawendi, M. G. (2015). Open-circuit voltage deficit, radiative sub-bandgap states, and prospects in quantum dot solar cells. *Nano Letters, 15*(5), 3286–3294.
37. Nozik, A. J. (2002). Quantum dot solar cells. *Phyisca E, 14*(1-2), 115–120.
38. Beard, M. C. (2011). Multiple exciton generation in semiconductor quantum dots. *Journal of Physical Chemistry Letters, 2*(11), 1282–1288.
39. Beard, M. C., Knutsen, K. P., Yu, P., Luther, J. M., Song, Q., Metzger, W. K., Ellingson, R. J., & Nozik, A. J. (2007). Multiple exciton generation in colloidal silicon nanocrystals. *Nano Letters, 7*(8), 2506–2512.
40. Klimov, V. I. (2010). *Nanocrystal quantum dots*. Boca Raton, FL: CRC Press.
41. Klimov, V. I. (2000). *Optical nonlinearities and ultrafast carrier dynamics in semiconductor nanocrystals*. Washington: ACS Publications.
42. Midgett, A. G., Luther, J. M., Stewart, J. T., Smith, D. K., Padilha, L. A., Klimov, V. I., Nozik, A. J., & Beard, M. C. (2013). Size and composition dependent multiple exciton generation efficiency in PbS, PbSe, and PbS_x Se_{1-x} alloyed quantum dots. *Nano Letters, 13*(7), 3078–3085.
43. Fidler, A. F., Gao, J., & Klimov, V. (2017). Electron–hole exchange blockade and memoryless recombination in photoexcited films of colloidal quantum dots. *Nature Physics, 13*(6), 604.
44. Nishihara, T., Tahara, H., Okano, M., Ono, M., & Kanemitsu, Y. (2015). Fast dissociation and reduced auger recombination of multiple excitons in closely packed PbS nanocrystal thin films. *Journal of Physical Chemistry Letters, 6*(8), 1327–1332.
45. Emin, S., Singh, S. P., Han, L., Satoh, N., & Islam, A. (2011). Colloidal quantum dot solar cells. *Solar Energy, 85*(6), 1264–1282.
46. Luque, A., Martí, A., & Nozik, A. J. (2007). Solar cells based on quantum dots: Multiple exciton generation and intermediate bands. *MRS Bulletin, 32*(3), 236–241.
47. Gao, J., Perkins, C. L., Luther, J. M., Hanna, M. C., Chen, H.-Y., Semonin, O. E., Nozik, A. J., Ellingson, R. J., & Beard, M. C. (2011). N-type transition metal oxide as a hole extraction layer in PbS quantum dot solar cells. *Nano Letters, 11*(8), 3263–3266.
48. Zhang, X., Jia, D., Hägglund, C., Öberg, V. A., Du, J., Liu, J., & Johansson, E. M. J. (2018). Highly photostable and efficient semitransparent quantum dot solar cells by using solution-phase ligand exchange. *Nano Energy, 53*, 373–382.
49. Semonin, O. E., Luther, J. M., Choi, S., Chen, H.-Y., Gao, J., Nozik, A. J., & Beard, M. C. (2011). Peak external photocurrent quantum efficiency exceeding 100% via MEG in a quantum dot solar cell. *Sciences, 334*(6062), 1530–1533.
50. Zhang, J., Gao, J., Church, C. P., Miller, E. M., Luther, J. M., Klimov, V. I., & Beard, M. C. (2014). PbSe quantum dot solar cells with more than 6% efficiency fabricated in ambient atmosphere. *Nano Letters, 14*(10), 6010–6015.
51. Böhm, M. L., Jellicoe, T. C., Tabachnyk, M., Davis, N. J., Wisnivesky-Rocca-Rivarola, F., Ducati, C., Ehrler, B., Bakulin, A. A., & Greenham, N. C. (2015). Lead telluride quantum dot solar cells displaying external quantum efficiencies exceeding 120%. *Nano Letters, 15*(12), 7987–7993.

52. Du, J., Du, Z., Hu, J.-S., Pan, Z., Shen, Q., Sun, J., Long, D., Dong, H., Sun, L., Zhong, X., & Wan, L.-J. (2016). Zn–Cu–In–Se quantum dot solar cells with a certified power conversion efficiency of 11.6%. *Journal of the American Chemical Society, 138*(12), 4201–4209.
53. Bernechea, M., Miller, N. C., Xercavins, G., So, D., Stavrinadis, A., & Konstantatos, G. (2016). Solution-processed solar cells based on environmentally friendly $AgBiS_2$ nanocrystals. *Nature Photonics, 10*(8), 521.
54. Chen, C., Wang, L., Gao, L., Nam, D., Li, D., Li, K., Zhao, Y., Ge, C., Cheong, H., & Liu, H. (2017). 6.5% certified efficiency Sb_2Se_3 solar cells using PbS colloidal quantum dot film as hole-transporting layer. *ACS Energy Letters, 2*(9), 2125–2132.
55. Khan, J., Yang, X., Qiao, K., Deng, H., Zhang, J., Liu, Z., Ahmad, W., Zhang, J., Li, D., & Liu, H. (2017). Low-temperature-processed SnO_2-Cl for efficient PbS quantum-dot solar cells via defect passivation. *Journal of Materials Chemistry A, 5*(33), 17240–17247.
56. Bai, Y., Fang, Y., Deng, Y., Wang, Q., Zhao, J., Zheng, X., Zhang, Y., & Huang, J. (2016). Low temperature solution-processed Sb: SnO_2 nanocrystals for efficient planar perovskite solar cells. *ChemSusChem, 9*(18), 2686–2691.
57. Santra, P. K., Palmstrom, A. F., Tanskanen, J. T., Yang, N., & Bent, S. F. (2015). Improving performance in colloidal quantum dot solar cells by tuning band alignment through surface dipole moments. *Journal of Physical Chemistry C, 119*(6), 2996–3005.
58. Carey, G. H., Levina, L., Comin, R., Voznyy, O., & Sargent, E. H. (2015). Record charge carrier diffusion length in colloidal quantum dot solids via mutual dot-to-dot surface passivation. *Advanced Materials, 27*(21), 3325–3330.
59. Kagan, C. R., & Murray, C. B. (2015). Charge transport in strongly coupled quantum dot solids. *Nature Nanotechnology, 10*(12), 1013.
60. Santra, P. K., Palmstrom, A. F., Tassone, C. J., & Bent, S. F. (2016). Molecular ligands control superlattice structure and crystallite orientation in colloidal quantum dot solids. *Chemistry Materials, 28*(19), 7072–7081.
61. Stolle, C. J., Lu, X., Yu, Y., Schaller, R. D., & Korgel, B. A. (2017). Efficient carrier multiplication in colloidal silicon nanorods. *Nano Letters, 17*(9), 5580–5586.
62. Tischler, J., Kennedy, T., Glaser, E., Efros, A. L., Foos, E., Boercker, J., Zega, T., Stroud, R., & Erwin, S. (2010). Band-edge excitons in PbSe nanocrystals and nanorods. *Physical Review B–Condensed Matter and Materials Physics, 82*(24), 245303.
63. Song, J. H., Choi, H., Kim, Y. H., & Jeong, S. (2017). High performance colloidal quantum dot photovoltaics by controlling protic solvents in ligand exchange. *Advanced Energy Materials, 7*(15), 1700301.
64. Aqoma, H., Mubarok, M. A., Lee, W., Hadmojo, W. T., Park, C., Ahn, T. K., Ryu, D. Y., & Jang, S. Y. (2018). Improved processability and efficiency of colloidal quantum dot solar cells based on organic hole transport layers. *Advanced Energy Materials, 8*, 1800572.
65. Wang, R., Wu, X., Xu, K., Zhou, W., Shang, Y., Tang, H., Chen, H., & Ning, Z. (2018). Highly efficient inverted structural quantum dot solar cells. *Advanced Materials, 30*(7), 1704882.
66. Rekemeyer, P. H., Chuang, C.-H. M., Bawendi, M. G., & Gradecak, S. (2017). Minority carrier transport in lead sulfide quantum dot photovoltaics. *Nano Letters, 17*(10), 6221–6227.
67. Thanh, N. T., Maclean, N., & Mahiddine, S. (2014). Mechanisms of nucleation and growth of nanoparticles in solution. *Chemical Reviews, 114*(15), 7610–7630.
68. Pu, Y., Cai, F., Wang, D., Wang, J.-X., & Chen, J.-F. (2018). Colloidal synthesis of semiconductor quantum dots toward large-scale production: A review. *Industrial and Engineering Chemistry Research, 57*(6), 1790–1802.
69. Park, J., Joo, J., Kwon, S. G., Jang, Y., & Hyeon, T. (2007). Synthesis of monodisperse spherical nanocrystals. *Angewandte Chemie International Edition, 46*(25), 4630–4660.
70. LaMer, V. K., & Dinegar, R. H. (1950). Theory, production and mechanism of formation of monodispersed hydrosols. *Journal of the American Chemical Society, 72*(11), 4847–4854.
71. Mer, V. K. L. (1952). Nucleation in phase transitions. *Industrial and Engineering Chemistry Research, 44*(6), 1270–1277.
72. Watzky, M. A., & Finke, R. G. (1997). Transition metal nanocluster formation kinetic and mechanistic studies. A new mechanism when hydrogen is the reductant: Slow, continuous

nucleation and fast autocatalytic surface growth. *Journal of the American Chemical Society, 119*(43), 10382–10400.
73. Ostwald, W. (1900). Über die vermeintliche Isomerie des roten und gelben Quecksilberoxyds und die Oberflächenspannung fester Körper. *Zeitschrift für Physikalische Chemie, 34U*(1), 495–503.
74. Lifshitz, I. M., & Slyozov, V. V. (1961). The kinetics of precipitation from supersaturated solid solutions. *Journal of Physics and Chemistry of Solids, 19*(1-2), 35–50.
75. Wagner, C. D. (1961). Polymerization of solid ethylene by ionizing radiation: Evidence for ion-molecule condensation. *Journal of Physical Chemistry Letters, 65*(12), 2276–2277.
76. Reiss, H. (1951). The growth of uniform colloidal dispersions. *The Journal of Chemical Physics, 19*(4), 482–487.
77. Burda, C., Chen, X., Narayanan, R., & El-Sayed, M. A. (2005). Chemistry and properties of nanocrystals of different shapes. *Chemical Reviews, 105*(4), 1025–1102.
78. Kwon, S. G., Piao, Y., Park, J., Angappane, S., Jo, Y., Hwang, N.-M., Park, J.-G., & Hyeon, T. (2007). Kinetics of monodisperse iron oxide nanocrystal formation by "heating-up" process. *Journal of the American Chemical Society, 129*(41), 12571–12584.
79. Pan, J., El-Ballouli, A., Rollny, L., Voznyy, O., Burlakov, V. M., Goriely, A., Sargent, E. H., & Bakr, O. M. (2013). Automated synthesis of photovoltaic-quality colloidal quantum dots using separate nucleation and growth stages. *ACS Nano, 7*(11), 10158–10166.
80. Calió, L., Kazim, S., Grätzel, M., & Ahmad, S. (2016). Hole-transport materials for perovskite solar cells. *ACS Energy Letters, 55*(47), 14522–14545.
81. Kamat, P. V. (2013). Quantum dot solar cells. The next big thing in photovoltaics. *Journal of Physical Chemistry Letters, 4*(6), 908–918.
82. Chernomordik, B. D., Marshall, A. R., Pach, G. F., Luther, J. M., & Beard, M. C. (2016). Quantum dot solar cell fabrication protocols. *Chem Matter, 29*(1), 189–198.
83. Suárez, J. A., Plata, J. J., Márquez, A. M., & Sanz, J. F. (2017). Effects of the capping ligands, linkers and oxide surface on the electron injection mechanism of copper sulfide quantum dot-sensitized solar cells. *Physical Chemistry Chemical Physics, 19*(22), 14580–14587.
84. Watson, D. F. (2010). Linker-assisted assembly and interfacial electron-transfer reactivity of quantum dot– substrate architectures. *Journal of Physical Chemistry Letters, 1*(15), 2299–2309.
85. Islam, M. A., & Herman, I. P. (2002). Electrodeposition of patterned CdSe nanocrystal films using thermally charged nanocrystals. *Applied Physics Letters, 80*(20), 3823–3825.
86. Brown, P., & Kamat, P. V. (2008). Quantum dot solar cells. Electrophoretic deposition of CdSe–C_{60} composite films and capture of photogenerated electrons with nC_{60} cluster shell. *Journal of the American Chemical Society, 130*(28), 8890–8891.
87. Deepa, K., & Nagaraju, J. (2014). Development of SnS quantum dot solar cells by SILAR method. *Materials Science in Semiconductor Processing, 27*, 649–653.
88. Woo Jung, S., Kim, J.-H., Kim, H., Choi, C.-J., & Ahn, K.-S. (2011). CdS quantum dots grown by in situ chemical bath deposition for quantum dot-sensitized solar cells. *Journal of Applied Physics, 110*(4), 044313.
89. Zhou, R., Niu, H., Zhang, Q., Uchaker, E., Guo, Z., Wan, L., Miao, S., Xu, J., & Cao, G. (2015). Influence of deposition strategies on CdSe quantum dot-sensitized solar cells: A comparison between successive ionic layer adsorption and reaction and chemical bath deposition. *Journal of Materials Chemistry A, 3*(23), 12539–12549.
90. Schaller, R. D., Sykora, M., Pietryga, J. M., & Klimov, V. I. (2006). Seven excitons at a cost of one: Redefining the limits for conversion efficiency of photons into charge carriers. *Nano Letters, 6*(3), 424–429.
91. Stolle, C. J., Schaller, R. D., & Korgel, B. A. (2014). Efficient carrier multiplication in colloidal CuInSe2 nanocrystals. *Journal of Physical Chemistry Letters, 5*(18), 3169–3174.
92. Luther, J. M., Gao, J., Lloyd, M. T., Semonin, O. E., Beard, M. C., & Nozik, A. J. (2010). Stability assessment on a 3% bilayer PbS/ZnO quantum dot heterojunction solar cell. *Advanced Materials, 22*(33), 3704–3707.

93. Ip, A. H., Thon, S. M., Hoogland, S., Voznyy, O., Zhitomirsky, D., Debnath, R., Levina, L., Rollny, L. R., Carey, G. H., & Fischer, A. (2012). Hybrid passivated colloidal quantum dot solids. *Nature Nanotechnology, 7*(9), 577.
94. Hou, B., Cho, Y., Kim, B. S., Hong, J., Park, J. B., Ahn, S. J., Sohn, J. I., Cha, S., & Kim, J. M. (2016). Highly monodispersed PbS quantum dots for outstanding cascaded-junction solar cells. *ACS Energy Letters, 1*(4), 834–839.
95. Hall, R., & Racette, J. (1961). Band structure parameters deduced from tunneling experiments. *Journal of Applied Physics, 32*(10), 2078–2081.
96. Burstein, E., Perkowitz, S., & Brodsky, M. (1968). The dielectric properties of the cubic IV-VI compound semiconductors. *Journal de Physique Colloques, 29*(C4), C4-78–C4-83.
97. Allan, G., & Delerue, C. (2006). Role of impact ionization in multiple exciton generation in PbSe nanocrystals. *Physical Review B, 73*(20), 205423.
98. Wang, D., Baral, J. K., Zhao, H., Gonfa, B. A., Truong, V. V., El Khakani, M. A., Izquierdo, R., & Ma, D. (2011). Controlled fabrication of PbS quantum-dot/carbon-nanotube nanoarchitecture and its significant contribution to near-infrared photon-to-current conversion. *Advanced Functional Materials, 21*(21), 4010–4018.
99. Wei, Y., Ren, Z., Zhang, A., Mao, P., Li, H., Zhong, X., Li, W., Yang, S., & Wang, J. (2018). Hybrid organic/PbS quantum dot bilayer photodetector with low dark current and high detectivity. *Advanced Functional Materials, 28*(11), 1706690.
100. Kirmani, A. R., García de Arquer, F. P., Fan, J. Z., Khan, J. I., Walters, G., Hoogland, S., Wehbe, N., Said, M. M., Barlow, S., Laquai, F., Marder, S. R., Sargent, E. H., & Amassian, A. (2017). Molecular doping of the hole-transporting layer for efficient, single-step-deposited colloidal quantum dot photovoltaics. *ACS Energy Letters, 2*(9), 1952–1959.
101. Debellis, D., Gigli, G., Ten Brinck, S., Infante, I., & Giansante, C. (2017). Quantum-confined and enhanced optical absorption of colloidal PbS quantum dots at wavelengths with expected bulk behavior. *Nano Letters, 17*(2), 1248–1254.
102. Tang, J., Brzozowski, L., Barkhouse, D. A. R., Wang, X., Debnath, R., Wolowiec, R., Palmiano, E., Levina, L., Pattantyus-Abraham, A. G., Jamakosmanovic, D., & Sargent, E. H. (2010). Quantum dot photovoltaics in the extreme quantum confinement regime: The surface-chemical origins of exceptional air-and light-stability. *ACS Nano, 4*(2), 869–878.
103. Jeong, K. S., Tang, J., Liu, H., Kim, J., Schaefer, A. W., Kemp, K., Levina, L., Wang, X., Hoogland, S., & Debnath, R. (2011). Enhanced mobility-lifetime products in PbS colloidal quantum dot photovoltaics. *ACS Nano, 6*(1), 89–99.
104. Koleilat, G. I., Levina, L., Shukla, H., Myrskog, S. H., Hinds, S., Pattantyus-Abraham, A. G., & Sargent, E. H. (2008). Efficient, stable infrared photovoltaics based on solution-cast colloidal quantum dots. *ACS Nano, 2*(5), 833–840.
105. Naka, S., Okada, H., Onnagawa, H., & Tsutsui, T. (2000). High electron mobility in bathophenanthroline. *Applied Physics Letters, 76*(2), 197–199.
106. Juška, G., Arlauskas, K., Viliūnas, M., & Kočka, J. (2000). Extraction current transients: New method of study of charge transport in microcrystalline silicon. *Physical Review Letters, 84*(21), 4946.
107. Zhang, Z., Chen, Z., Yuan, L., Chen, W., Yang, J., Wang, B., Wen, X., Zhang, J., Hu, L., & Stride, J. A. (2017). A new passivation route leading to over 8% efficient PbSe quantum-dot solar cells via direct ion exchange with perovskite nanocrystals. *Advanced Materials, 29*(41), 1703214.
108. Pan, Z., Mora-Seró, I., Shen, Q., Zhang, H., Li, Y., Zhao, K., Wang, J., Zhong, X., & Bisquert, J. (2014). High-efficiency "green" quantum dot solar cells. *Journal of the American Chemical Society, 136*(25), 9203–9210.
109. Lan, X., Masala, S., & Sargent, E. H. (2014). Charge-extraction strategies for colloidal quantum dot photovoltaics. *Nature Materials, 13*(3), 233.
110. Etgar, L., Yanover, D., Čapek, R. K., Vaxenburg, R., Xue, Z., Liu, B., Nazeeruddin, M. K., Lifshitz, E., & Grätzel, M. (2013). Core/Shell PbSe/PbS QDs TiO_2 heterojunction solar cell. *Advanced Functional Materials, 23*(21), 2736–2741.

111. Kim, J.-Y., Yang, J., Yu, J. H., Baek, W., Lee, C.-H., Son, H. J., Hyeon, T., & Ko, M. J. J. A. (2015). Highly efficient copper–indium–selenide quantum dot solar cells: Suppression of carrier recombination by controlled ZnS overlayers. *ACS Nano, 9*(11), 11286–11295.
112. Kuo, K.-T., Liu, D.-M., Chen, S.-Y., & Lin, C.-C. (2009). Core-shell $CuInS_2$/ZnS quantum dots assembled on short ZnO nanowires with enhanced photo-conversion efficiency. *Journal of Materials Chemistry A, 19*(37), 6780–6788.
113. Wang, Y.-Q., Rui, Y.-C., Zhang, Q.-H., Li, Y.-G., & Wang, H.-Z. (2013). A facile in situ synthesis route for $CuInS_2$ quantum-dots/In_2S_3 co-sensitized Photoanodes with high photoelectric performance. *ACS Applied Materials & Interfaces, 5*(22), 11858–11864.
114. Chang, J.-Y., Lin, J.-M., Su, L.-F., & Chang, C.-F. (2013). Improved performance of $CuInS_2$ quantum dot-sensitized solar cells based on a multilayered architecture. *ACS Applied Materials & Interfaces, 5*(17), 8740–8752.
115. McDaniel, H., Koposov, A. Y., Draguta, S., Makarov, N. S., Pietryga, J. M., & Klimov, V. I. (2014). Simple yet versatile synthesis of $CuInSe_x$ S_{2-x} quantum dots for sunlight harvesting. *Journal of Physical Chemistry C, 118*(30), 16987–16994.
116. Bernechea, M., Miller, N. C., Xercavins, G., So, D., Stavrinadis, A., & Konstantatos, G. (2016). Solution-processed solar cells based on environmentally friendly $AgBiS_2$ nanocrystals. *Nature Photonics, 10*(8), 521.
117. Luber, E. J., Mobarok, M. H., & Buriak, J. M. (2013). Solution-processed zinc phosphide (α-Zn_3P_2) colloidal semiconducting nanocrystals for thin film photovoltaic applications. *ACS Nano, 7*(9), 8136–8146.
118. Xu, Y., Al-Salim, N., Bumby, C. W., & Tilley, R. D. (2009). Synthesis of SnS quantum dots. *Journal of the American Chemical Society, 131*(44), 15990–15991.
119. Choi, H., Song, J. H., Jang, J., Mai, X. D., Kim, S., & Jeong, S. (2015). High performance of PbSe/PbS core/shell quantum dot heterojunction solar cells: Short circuit current enhancement without the loss of open circuit voltage by shell thickness control. *Nanoscale, 7*(41), 17473–17481.
120. Xiao, J., Wang, Y., Hua, Z., Wang, X., Zhang, C., & Xiao, M. (2012). Carrier multiplication in semiconductor nanocrystals detected by energy transfer to organic dye molecules. *Nature Communications, 3*, 1170.
121. Schaller, R. D., Petruska, M. A., & Klimov, V. I. (2005). Effect of electronic structure on carrier multiplication efficiency: Comparative study of PbSe and CdSe nanocrystals. *Applied Physics Letters, 87*(25), 253102.
122. Schaller, R. D., & Klimov, V. I. (2004). High efficiency carrier multiplication in PbSe nanocrystals: Implications for solar energy conversion. *Physical Review Letters, 92*(18), 186601.
123. Bakulin, A. A., Neutzner, S., Bakker, H. J., Ottaviani, L., Barakel, D., & Chen, Z. (2013). Charge trapping dynamics in PbS colloidal quantum dot photovoltaic devices. *ACS Nano, 7*(10), 8771–8779.
124. Cao, Y., Stavrinadis, A., Lasanta, T., So, D., & Konstantatos, G. (2016). The role of surface passivation for efficient and photostable PbS quantum dot solar cells. *Nature Energy, 1*(4), 16035.
125. Azmi, R., Sinaga, S., Aqoma, H., Seo, G., Ahn, T. K., Park, M., Ju, S.-Y., Lee, J.-W., Kim, T.-W., & Oh, S.-H. (2017). Highly efficient air-stable colloidal quantum dot solar cells by improved surface trap passivation. *Nano Energy, 39*, 86–94.
126. Kemp, K., Wong, C., Hoogland, S., & Sargent, E. H. (2013). Photocurrent extraction efficiency in colloidal quantum dot photovoltaics. *Applied Physics Letters, 103*(21), 211101.
127. Martín-García, B., Bi, Y., Prato, M., Spirito, D., Krahne, R., Konstantatos, G., & Moreels, I. (2018). Reduction of moisture sensitivity of PbS quantum dot solar cells by incorporation of reduced graphene oxide. *Solar Energy Materials and Solar Cells, 183*, 1–7.
128. Aqoma, H., Al Mubarok, M., Hadmojo, W. T., Lee, E. H., Kim, T. W., Ahn, T. K., Oh, S. H., & Jang, S. Y. (2017). High-efficiency photovoltaic devices using trap-controlled quantum-dot ink prepared via phase-transfer exchange. *Advanced Materials, 29*(19), 1605756.

129. Gao, J., & Johnson, J. C. (2012). Charge trapping in bright and dark states of coupled PbS quantum dot films. *ACS Nano, 6*(4), 3292–3303.
130. Jean, J., Mahony, T. S., Bozyigit, D., Sponseller, M., Holovský, J., Bawendi, M. G., & Bulović, V. (2017). Radiative efficiency limit with band tailing exceeds 30% for quantum dot solar cells. *ACS Energy Letters, 2*(11), 2616–2624.
131. Kim, D., Kim, D.-H., Lee, J.-H., & Grossman, J. C. (2013). Impact of stoichiometry on the electronic structure of PbS quantum dots. *Physical Review Letters, 110*(19), 196802.
132. Kushnir, K., Chen, K., Zhou, L., Giri, B., Grimm, R. L., Rao, P. M., & Titova, L. V. (2018). Dynamics of photoexcited carriers in polycrystalline PbS and at PbS/ZnO heterojunctions: Influence of grain boundaries and interfaces. *Journal of Physical Chemistry C, 122*(22), 11682–11688.
133. Rekemeyer, P. H., Chang, S., Chuang, C. H. M., Hwang, G. W., Bawendi, M. G., & Gradečak, S. (2016). Enhanced photocurrent in pbs quantum dot photovoltaics via ZnO nanowires and band alignment engineering. *Advanced Energy Materials, 6*(24), 1600848.
134. Xia, P., Liang, Z., Mahboub, M., van Baren, J., Lui, C. H., Jiao, J., Graham, K. R., & Tang, M. L. (2018). Surface fluorination for controlling the PbS quantum dot bandgap and band offset. *Chem Matter, 30*(15), 4943–4948.
135. Hu, L., Zhang, Z., Patterson, R. J., Hu, Y., Chen, W., Chen, C., Li, D., Hu, C., Ge, C., & Chen, Z. (2018). Achieving high-performance PbS quantum dot solar cells by improving hole extraction through Ag doping. *Nano Energy, 46*, 212–219.
136. Huang, J., Huang, Z., Yang, Y., Zhu, H., & Lian, T. (2010). Multiple exciton dissociation in CdSe quantum dots by ultrafast electron transfer to adsorbed methylene blue. *Journal of the American Chemical Society, 132*(13), 4858–4864.
137. Wang, H., Yang, S., Wang, Y., Xu, J., Huang, Y., Li, W., He, B., Muhammad, S., Jiang, Y., & Tang, Y. (2017). Influence of post-synthesis annealing on PbS quantum dot solar cells. *Organic Electronics, 42*, 309–315.
138. Kurpiers, J., Balazs, D. M., Paulke, A., Albrecht, S., Lange, I., Protesescu, L., Kovalenko, M. V., Loi, M. A., & Neher, D. (2016). Free carrier generation and recombination in PbS quantum dot solar cells. *Applied Physics Letters, 108*(10), 103102.
139. Zhang, X., Welch, K., Tian, L., Johansson, M. B., Häggman, L., Liu, J., & Johansson, E. M. J. (2017). Enhanced charge carrier extraction by a highly ordered wrinkled MgZnO thin film for colloidal quantum dot solar cells. *Journal of Materials Chemistry C, 5*(42), 11111–11120.
140. Cho, Y., Giraud, P., Hou, B., Lee, Y. W., Hong, J., Lee, S., Pak, S., Lee, J., Jang, J. E., & Morris, S. M. (2018). Charge transport modulation of a flexible quantum dot solar cell using a piezoelectric effect. *Advanced Energy Materials, 8*(3), 1700809.
141. Lu, K., Wang, Y., Yuan, J., Cui, Z., Shi, G., Shi, S., Han, L., Chen, S., Zhang, Y., & Ling, X. (2017). Efficient PbS quantum dot solar cells employing a conventional structure. *Journal of Materials Chemistry A, 5*(45), 23960–23966.
142. Willis, S. M., Cheng, C., Assender, H. E., & Watt, A. A. R. (2012). The transitional heterojunction behavior of PbS/ZnO colloidal quantum dot solar cells. *Nano Letters, 12*(3), 1522–1526.
143. Aqoma, H., Barange, N., Ryu, I., Yim, S., Do, Y. R., Cho, S., Ko, D. H., & Jang, S. Y. (2015). Simultaneous improvement of charge generation and extraction in colloidal quantum dot photovoltaics through optical management. *Advanced Functional Materials, 25*(39), 6241–6249.
144. Wang, H., Wang, Y., He, B., Li, W., Sulaman, M., Xu, J., Yang, S., Tang, Y., & Zou, B. (2016). Charge carrier conduction mechanism in PbS quantum dot solar cells: Electrochemical impedance spectroscopy study. *ACS Applied Materials & Interfaces, 8*(28), 18526–18533.
145. Liu, H., Tang, J., Kramer, I. J., Debnath, R., Koleilat, G. I., Wang, X., Fisher, A., Li, R., Brzozowski, L., Levina, L., & Sargent, E. H. (2011). Electron acceptor materials engineering in colloidal quantum dot solar cells. *Advanced Materials, 23*(33), 3832–3837.
146. Morales-Masis, M., De Wolf, S., Woods-Robinson, R., Ager, J. W., & Ballif, C. (2017). Transparent electrodes for efficient optoelectronics. *Advanced Electronic Materials, 3*(5), 1600529.

147. Wang, H., Kubo, T., Nakazaki, J., Kinoshita, T., & Segawa, H. (2013). PbS-quantum-dot-based heterojunction solar cells utilizing ZnO nanowires for high external quantum efficiency in the near-infrared region. *Journal of Physical Chemistry Letters, 4*(15), 2455–2460.
148. Kawawaki, T., Wang, H., Kubo, T., Saito, K., Nakazaki, J., Segawa, H., & Tatsuma, T. (2015). Efficiency enhancement of PbS quantum dot/ZnO nanowire bulk-heterojunction solar cells by plasmonic silver nanocubes. *ACS Nano, 9*(4), 4165–4172.
149. Zhang, X., Santra, P. K., Tian, L., Johansson, M. B., Rensmo, H., & Johansson, E. M. J. (2017). Highly efficient flexible quantum dot solar cells with improved electron extraction using MgZnO nanocrystals. *ACS Nano, 11*(8), 8478–8487.
150. Zhang, X., Öberg, V. A., Du, J., Liu, J., & Johansson, E. M. J. (2018). Extremely lightweight and ultra-flexible infrared light-converting quantum dot solar cells with high power-per-weight output using a solution-processed bending durable silver nanowire-based electrode. *Energy & Environmental Science, 11*(2), 354–364.
151. Brown, P. R., Lunt, R. R., Zhao, N., Osedach, T. P., Wanger, D. D., Chang, L.-Y., Bawendi, M. G., & Bulovic, V. (2011). Improved current extraction from ZnO/PbS quantum dot heterojunction photovoltaics using a MoO_3 interfacial layer. *Nano Letters, 11*(7), 2955–2961.
152. Tang, J., & Sargent, E. H. (2011). Infrared colloidal quantum dots for photovoltaics: Fundamentals and recent progress. *Advanced Materials, 23*(1), 12–29.
153. So, D. (2016). *Copper indium sulfide colloidal quantum dot solar cells*. (Doctoral thesis, UPC, Institut de Ciències Fotòniques).
154. Martínez Montblanch, L. (2014). *N-type bismuth sulfide coloidal nanocrystals and their application to solution-processed photovoltaic devices*. (Doctoral thesis, UPC, Institut de Ciències Fotòniques).
155. Cheng, J. J., Chuang, C.-H. M., Hentz, O., Rekemeyer, P. H., Bawendi, M. G., & Gradečak, S. (2018). Dimension-and surface-tailored ZnO nanowires enhance charge collection in quantum dot photovoltaic devices. *ACS Applied Materials & Interfaces, 1*(5), 1815–1822.
156. Leschkies, K. S., Jacobs, A. G., Norris, D. J., & Aydil, E. S. (2009). Nanowire-quantum-dot solar cells and the influence of nanowire length on the charge collection efficiency. *Applied Physics Letters, 95*(19), 193103.
157. Ren, Z., Kuang, Z., Zhang, L., Sun, J., Yi, X., Pan, Z., Zhong, X., Hu, J., Xia, A., & Wang, J. (2017). Enhancing electron and hole extractions for efficient PbS quantum dot solar cells. *Solar RRL, 1*(12), 1700176.
158. Kramer, I. J., Zhitomirsky, D., Bass, J. D., Rice, P. M., Topuria, T., Krupp, L., Thon, S. M., Ip, A. H., Debnath, R., & Kim, H. C. (2012). Ordered nanopillar structured electrodes for depleted bulk heterojunction colloidal quantum dot solar cells. *Advanced Materials, 24*(17), 2315–2319.
159. Wang, H., Kubo, T., Nakazaki, J., & Segawa, H. (2017). Solution-processed short-wave infrared PbS colloidal quantum dot/ZnO nanowire solar cells giving high open-circuit voltage. *ACS Energy Letters, 2*(9), 2110–2117.
160. Barkhouse, D. A. R., Debnath, R., Kramer, I. J., Zhitomirsky, D., Pattantyus-Abraham, A. G., Levina, L., Etgar, L., Grätzel, M., & Sargent, E. H. (2011). Depleted bulk heterojunction colloidal quantum dot photovoltaics. *Advanced Materials, 23*(28), 3134–3138.
161. Hjerrild, N. (2013). *Silver nanowire transparent conductors for quantum dot photovoltaics*. Oxford: University of Oxford.
162. Yang, Y. M., Chen, W., Dou, L., Chang, W.-H., Duan, H.-S., Bob, B., Li, G., & Yang, Y. (2015). High-performance multiple-donor bulk heterojunction solar cells. *Nature Photonics, 9*(3), 190.
163. Lu, L., Zheng, T., Wu, Q., Schneider, A. M., Zhao, D., & Yu, L. (2015). Recent advances in bulk heterojunction polymer solar cells. *Chemical Reviews, 115*(23), 12666–12731.
164. Jung, J. W., Jo, J. W., Chueh, C. C., Liu, F., Jo, W. H., Russell, T. P., & Jen, A. K. Y. (2015). Fluoro substituted n type conjugated polymers for additive free all polymer bulk heterojunction solar cells with high power conversion efficiency of 6.71%. *Advanced Materials, 27*(21), 3310–3317.

165. Zhao, T., Goodwin, E. D., Guo, J., Wang, H., Diroll, B. T., Murray, C. B., & Kagan, C. R. (2016). Advanced architecture for colloidal PbS quantum dot solar cells exploiting a CdSe quantum dot buffer layer. *ACS Nano, 10*(10), 9267–9273.
166. Mali, S. S., Kim, H., Shim, C. S., Patil, P. S., Kim, J. H., & Hong, C. K. (2013). Surfactant free most probable TiO_2 nanostructures via hydrothermal and its dye sensitized solar cell properties. *Scientific Reports, 3*, 3004.
167. Qiu, Q., Li, S., Jiang, J., Wang, D., Lin, Y., & Xie, T. (2017). Improved electron transfer between TiO_2 and FTO interface by n-doped anatase TiO_2 nanowires and its applications in quantum dot-sensitized solar cells. *Journal of Physical Chemistry C, 121*(39), 21560–21570.
168. Davis, N. J., Böhm, M. L., Tabachnyk, M., Wisnivesky-Rocca-Rivarola, F., Jellicoe, T. C., Ducati, C., Ehrler, B., & Greenham, N. C. (2015). Multiple-exciton generation in lead selenide nanorod solar cells with external quantum efficiencies exceeding 120%. *Nature Communications, 6*, 8259.
169. Yao, X., Song, Z., Mi, L., Li, G., Wang, X., Wang, X., & Jiang, Y. (2017). Improved stability of depletion heterojunction solar cells employing cation-exchange PbS quantum dots. *Solar Energy Materials & Solar Cells, 164*, 122–127.
170. Leng, M., Luo, M., Chen, C., Qin, S., Chen, J., Zhong, J., & Tang, J. (2014). Selenization of Sb_2Se_3 absorber layer: An efficient step to improve device performance of CdS/Sb_2Se_3 solar cells. *Applied Physics Letters, 105*(8), 083905.
171. Amaya Suárez, J., Plata, J. J., Márquez, A. M., & Fernández Sanz, J. (2017). Ag_2S quantum dot-sensitized solar cells by first principles: The effect of capping ligands and linkers. *The Journal of Physical Chemistry. A, 121*(38), 7290–7296.

Chapter 3
The Development of Quantum Emitters Based on Semiconductor Quantum Dots

Hai-Zhi Song

Abstract Quantum emitters serve as the building blocks of quantum network, connecting quantum computing, quantum communication, and quantum metrology. Quantum dots (QDs) are widely considered as the best candidate for quantum emitters. This chapter reviews the scientific and technological development of quantum emitters based on QDs in recent years.

By a decade of effort, the controllability, purity, brightness, indistinguishability, and coherence of QD emitters are greatly improved so that they are much closer to the application level. The energy level of QD and its fine structure splitting can be routinely tuned simultaneously by post-growth technique, the single-photon purity has come up to 10^{-3}, and the brightness has been greatly improved. Excellent indistinguishability and coherence can even be realized in the frame of electrically driven bulky semiconductors. Photon entanglement becomes an easy job by precisely tuning the energy levels and the symmetry of QDs. The second big progress is the success of coupling QD emitters to photonic nanostructures. These couplings greatly enhance the excitation/emission rates and the control of radiative direction. The entanglement of independent emitters might be possible by controlling the nanostructures coupled to QD emitters. Nanoscale selective excitation is realized by surface plasmonic interference. QDs coupled to waveguides produce strongly correlated photon states. Thirdly, remarkable development has been made towards on-chip integration of QD emitters into planar circuits and nanophotonic systems. Successful integration with multimode interference beam splitters has shown good single-photon purity of QD emitters. Integration into nanophotonic structures makes the engineering-related scattering much less important in the characters of QD emitters. Site-controlled integration with semiconductor nano-waveguides significantly

H.-Z. Song (✉)
Institute of Fundamental and Frontier Sciences, University of Electronic Science and Technology of China, Chengdu, Sichuan, China

Southwest Institute of Technical Physics, Chengdu, Sichuan, China

P. Yu, Z. M. Wang (eds.), *Quantum Dot Optoelectronic Devices*,
Lecture Notes in Nanoscale Science and Technology 27,
https://doi.org/10.1007/978-3-030-35813-6_3

improves the collection efficiency of QD emitters. In short, QDs are further proved the best on-demand, entangled, on-chip integrated quantum emitters for quantum information processing.

3.1 Introduction

Quantum information processing is developing and becoming the decisive information technology in the future. Quantum information processing systems can be constructed by many physical features such as superconductors, photons, electron charge or spin, atom/ion traps, etc. [1]. In systems utilizing optical and photonic components, quantum photonic emitters are the fundamental building blocks for a variety of quantum technologies, including quantum communication [2], quantum metrology [3], and quantum computing [4, 5]. Photonics quantum technologies require efficient, on-demand, and even entangled sources of highly indistinguishable single photons [6]. Parametric down-conversion sources provide highly indistinguishable photons but are operated at very low brightness to maintain high single-photon purity and very limited possibility of scalable integration. Hence, the development of solid-state sources with nanostructures, such as semiconductor quantum dots (QDs) or color centers in diamond, has received a lot of attention [7]. Semiconductor QDs possess discrete quantum states similar to single atomic systems and, therefore, are also referred to as artificial atoms. They have great potential and have been widely studied because they have proven to be excellent quantum light sources [8]. Currently, semiconductor QDs are the best-performing photonic system possessing high characteristics in almost all the physical aspects including photon generation rate, quality of optical coherence, range of spectral tunability, and mature methods for their integration into nanostructures [9]. Most importantly, QDs serve as the most promising candidates as sources of on-demand entangled photon pairs, which are highly desired for the future quantum information processing, telecommunication, and metrology applications [7, 10–13]. For these reasons, semiconductor QDs-based quantum emitters were widely and deeply studied in the last two decades. As a result, single-photon sources based on QDs have even been readily used for efficient generation of spin-photon entanglement [14–16], quantum key distribution [17], quantum state transfer [18], and quantum relays [19]. Especially in recent years, the QD quantum emitters exhibited much greater progresses than before [20–22]. In this chapter, we review the scientific and technological development of semiconductor QDs as quantum emitters in recent years. The most important developments lie in the quality improvement of single-photon QD emitter, optimization of entangled photons from QDs, and the efforts to integrate QD emitters on a chip. In the following, therefore, we review the progresses in three aspects: QD-based single-photon emitters, QD-based entangled photon emitters, and on-chip integration of QD quantum emitters.

3.2 QD-Based Single-Photon Emitters

Semiconductor QD single-photon sources are suitable for on-demand photon generation. From the beginning of this century [23], QDs were deeply studied while serving as attractive single-photon sources. In the main aspects, qualities are continuously improved, and deep progresses have been recently taken.

Spontaneous emission enhancement is very important for advances in single-photon sources. For semiconductor QDs, the strong Purcell enhancement has been readily obtained [24–26] many years ago, but the quality of any of these devices is far from being ideal. In this weak coupling regime, although in 2015 a theoretical proposal suggested a specific way to achieve high indistinguishability with a broad spectral emitter using a high-Q cavity [27], and Sapienza et al. [28] presented a promising alternative approach based on circular Bragg grating "bull's-eye" cavities to get good single-photon purity, no bright source was generated with near-unity indistinguishability and pure single photons before 2016. Among many systems, microcavities and photonic crystals have been widely investigated to get Purcell-enhanced spontaneous emission [29, 30]. The most effective way to increase the source brightness is to embed the QDs into a planar λ-cavity either defined by DBRs or metal mirrors [31]. With these efforts, single photons with superior indistinguishability have been demonstrated in InGaAs self-assembled QDs. In 2016, Somaschi et al. [32] produced quantum emitters made of InGaAs QDs in electrically controlled cavities, as shown in Fig. 3.1. Their QD single-photon sources exhibit the best quality so far, including a near-unity indistinguishability of 0.9956 ± 0.0045, a

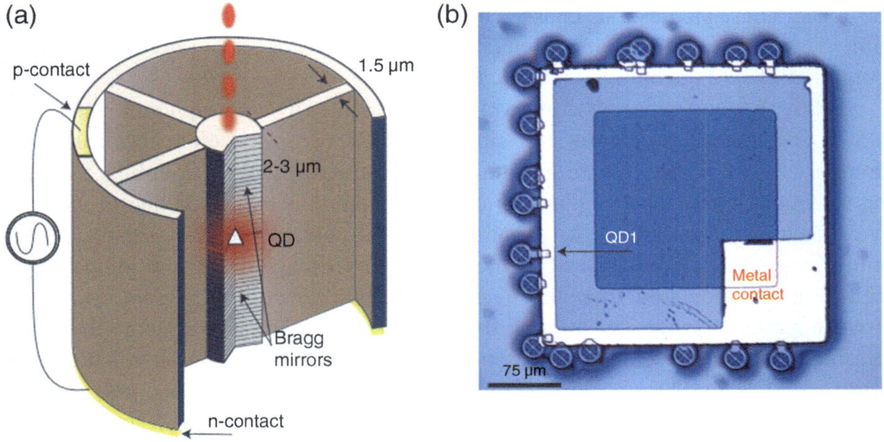

Fig. 3.1 Electrically controlled single-photon sources made by Somaschi et al. [32]. (**a**) Schematic of the devices under study: a micropillar coupled to a QD is connected to a surrounding circular frame by four one-dimensional wires. The top p-contact is defined on a large mesa adjacent to the frame. The n-contact is deposited on the back of the sample. (**b**) Optical microscope image showing 18 connected pillar sources electrically controlled through the metallic contact defined on the 300 × 300 μm² diode. Reuse with permission license 4702580294234, Copyright Clearance Center's RightsLink ® service

high purity of $g^{(2)}(0) = 0.0028 \pm 0.0012$, an extraction efficiency of 65%, and thus a brightness 100 times higher than before. Such QD single-photon sources have been well applied in scalable multiphoton experiments such as boson sampling [33]. To get higher extraction efficiency out of QD single-photon sources, one takes much effort to exact alignment technique. Niu and Wu groups [34] reported an extraction efficiency of 68% from single-photon sources based on QDs finely positioned in micropillar cavities. Single-photon emission with this degree of quality has been further confirmed by many works [35, 36] and thus opens the way to new levels of complexity and scalability in photonics quantum information processing. At the telecommunication band, however, there are not so successful realizations so far. The InAs/InP QD material system has been shown to emit efficiently at telecommunication wavelengths, and its single-photon emission at 1.5 μm was demonstrated in 2005 [37]. Song group designed novel micropillar and even monolithic micropillar cavities for weak and strong coupling of telecommunication-band photons with QDs to realize efficient, indistinguishable, and coherent single-photon sources [38–40]. More recently, bright single-photon emission was demonstrated at 1.3 and 1.5 μm using photonic crystal cavities [41, 42]. Furthermore, by molding the photonic environment of QDs at their native nanoscale, the emitted single photons can also be coupled to a guided mode with near-unity efficiency and made highly indistinguishable [43]. Thyrrestrup et al. [44] made an essential step towards a deterministic on-chip few-photon nonlinearity, which could form the basis of, e.g., a deterministic Bell-state analyzer [45] and is a prerequisite for coupling multiple QDs. They present near lifetime-limited linewidths for QDs embedded in nanophotonic waveguides through a resonant transmission experiment. It was found that the scattering of single photons from the QD can be obtained with an extinction of 66 ± 4%. These studies thus enable QD-based on-demand quantum emitters, quantum gates, quantum interference, etc., forming the basis of linear photonic quantum computation and quantum communication.

Although cavities are working well in improving QD single-photon sources, it was found that single-photon emission from QDs can also be enhanced in more compact ways such as with plasmonic nanostructures. The coupling strength between an emitter and the plasmonic resonant mode is enhanced due to the subwavelength mode volumes of plasmonic nanostructures [46–48]. Firstly, planar multilayer metamaterial nanostructures were proved capable of providing broadband Purcell effects on spontaneous emission. Pfeiffer et al. [49] demonstrated a novel approach to enhance emission from self-assembled semiconductor QDs using plasmonics. They observed enhancement and modification of the photon emission from a QD when the plasmonic nano-antenna array is tuned resonant to the QD emission energy, indicating that nano-antenna arrays are alternative to single optical antennas to enhance and control the emission of single quantum emitters with the potential of greater flexibility in design and possibly lower losses. In Li et al.'s recent work [50], Ag-SiO_2 multilayer metamaterial and nanostructured grating were used to enhance the 570–680 nm spontaneous emission of CdSe/ZnS QDs by a factor of 3–6, as shown in Fig. 3.2. Multilayer gratings help in positioning the QDs inside grating grooves and enhance the QDs' interaction with the surface plasmonic

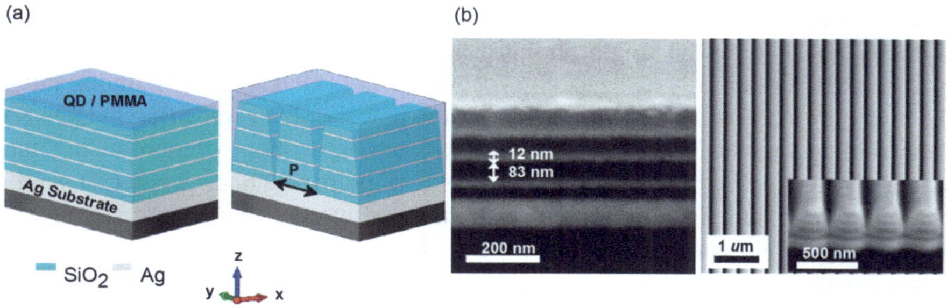

Fig. 3.2 Multilayer metamaterial nanostructures in [50]. (**a**, **b**) Schematic and SEM for a multilayer structure and multilayer gratings. The multilayer is composed of four pairs of Ag layers 12 nm thick and SiO_2 layers 83 nm thick. The multilayer is deposited on the 100 nm Ag substrate as a reflector. A thin layer of QDs in PMMA is spin-coated on the surface of the multilayer nanostructures. Adapted with permission from Ref. [50], Copyright (2017) American Chemical Society

modes. This work is helpful to understand the enhanced light-matter interactions between quantum emitters and metamaterial nanostructures and to apply metamaterial-based QD single-photon emitters in quantum information processing. In 2018, Leng et al. [51] reported straightforward observations of both the strong and intermediate coupling between single QDs and metal nanostructures. It is indicated that precise control over nanoscale structure will be required for high-yield fabrication of coupled plasmon-emitter systems. Their results represent an important step towards single-photon nonlinearities and quantum-photonic interfaces at the nanometer scale.

Applications in quantum information processing strongly require single-photon sources which can operate at room temperature, be electrically excited and controlled, and be compatible with other key elements of future quantum networks like quantum memories and single-photon detectors [52]. Important progresses happen just in the recent few years. Warburton group obtained lifetime-limited resonance linewidths by embedding QDs in electrically contacted bulk semiconductor structures [53]. Kirsanske et al. [54] presented a highly coherent single-photon source based on an electrically controlled QD integrated in a high-efficiency nanophotonic waveguide. Although much effort has been devoted to the self-assembled QDs, room-temperature single-photon emission can be more easily realized in colloidal core/shell QDs [55], which have recently been exploited for engineering the interaction between single photons and quantum emitters, single-photon transistors, and the long-range coupling of qubits [56]. In 2017, Lin et al. [57] explored a single-colloidal quantum dot as a quantum emitter in electrically driven single-photon source. Combined with an isolating layer in the device, they realized optimal antibunching single-photon generation at room temperature. This work highlights a pathway to the development of novel room-temperature quantum light sources for practical quantum information applications. To resolve the problem that such single-photons' purity is incomparable to their cryogenic temperature counterparts, Feng et al. [58] demonstrated the purification of photons to realize high-purity single photons sources based on the room-temperature colloidal QDs, as shown in Fig. 3.3. By properly shaping the single photons, they achieved a single-photon purity of $g^{(2)}(0) = 0.01$, a value only realized on cryogenic InGaAs QDs before [7]. Importantly, the high purity does not vary with the excitation power or between different QDs, so their finding enables increasing the generation rate without compromising the single-photon quality and provides a novel way of preparing high-purity QD single photons at room temperature. Furthermore, there arises another candidate for room-temperature single-photon emitter, III-nitride QD [59]. Large degree of inhomogeneous broadening has limited their use, but fortunately Arita et al. [60] recently realized ultraclean emission from single GaN QDs formed at macro-step edges in a GaN/AlGaN quantum well. They remarkably suppressed the spectral diffusion with high growth temperature and hence a reduced defect density and obtained a record

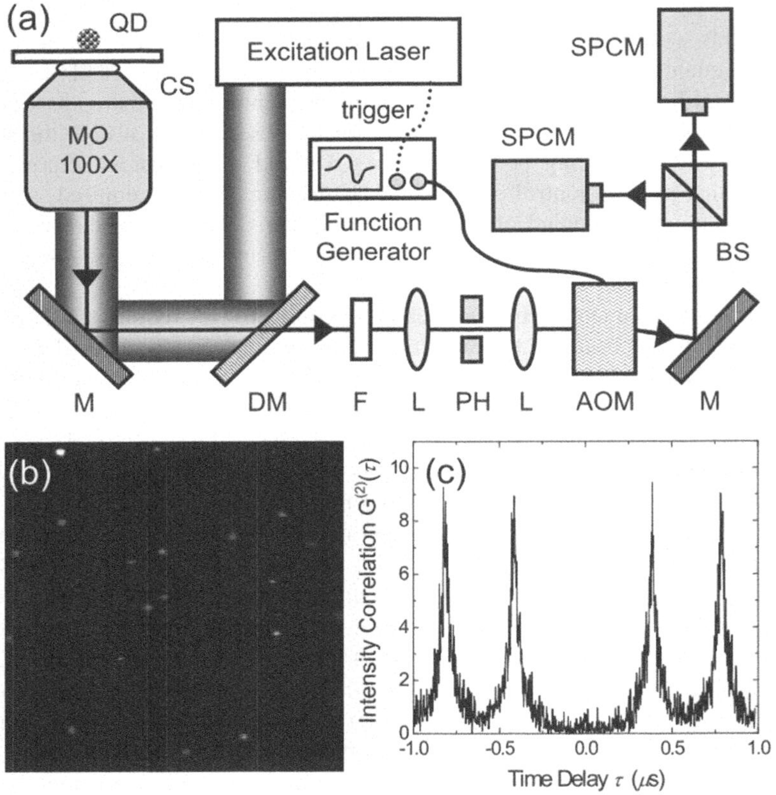

Fig. 3.3 Single photons from room-temperature colloidal QDs by Feng et al. [58]. (**a**) Single QDs are spin-coated on a cover glass (CS) and optically excited by focused laser pulses using a 100× microscopic objective (MO). (b) Photoluminescence image (20 × 20 μm) of single QDs. (c) Antibunching in the Hanbury–Brown–Twiss experiment. Adapted figure with permission from Ref. [58], Copyright (2017) by the American Physical Society

narrow linewidth of as small as 87 μeV and an upper limit of homogeneous broadening of 27 μeV in nitride QDs.

In order to build up quantum networks, it is essential to interface remote quantum emitters. This is an outstanding challenge, as the quantum states of dissimilar "artificial atoms" have to be prepared on demand with high fidelity and the generated photons have to be made indistinguishable in all possible degrees of freedom. In the QD single-photon emitter system, however, experiments have been performed using remote QDs [61] since the interaction between remote QDs is strongly required. In 2017, Trotta group [62] demonstrated an unprecedented two-photon interference from remote strain-tunable GaAs QDs embedded in a planar distributed Bragg reflector cavity. A novel phonon-assisted two-photon excitation scheme was exploited for the generation of highly indistinguishable entangled photon pairs, push-button biexciton state preparation, and robustness against environmental decoherence, so that the best remote visibility to date, 51%, was achieved. This result marks an important milestone for the practical realization of quantum repeaters and multiphoton quantum interaction within dissimilar artificial atoms. Delteil et al. [63] demonstrated heralded absorption of a single photonic qubit, generated by a single neutral QD, by a single-electron charged QD that is located 5 m away. The detection of this photon yields no information about the qubit. Although the current success rate is not high enough to implement a state transfer in a quantum network, it suggests a way for quantum information protocols based on cascaded quantum systems. If the success rate can be enhanced, a heralded quantum state transfer can be combined with a quantum error correction and it is possible to use erasure codes to achieve arbitrarily high efficiency in a quantum state transfer. Having demonstrated a technique to hinder the drifts of two separate QDs [64], Schmidt group further introduced an experimental setup to show the improved long-term two-photon interference visibility of frequency-stabilized QDs in a Hong–Ou–Mandel experiment in 2018 [65]. It is verified that stable two-photon interference from separate QDs is thus achieved by strain-mediated frequency stabilization. What is much more exciting, Pan group [66] reported quantum interference, entanglement, and nonlocality of two dissimilar photons from the Sun and a semiconductor QD on the Earth ~150 million km away from each other. It opens a route to QD-based quantum photonics at an astronomical scale. On the other hand, the combination of artificial and natural atoms allows saving QD single-photon states into atomic ones [67], being of particularly long storage times [68] and crucial ingredients to establish a quantum repeater [69]. Schmidt group [70] first reported the realization of a wavelength-tunable GaAs QD-based quantum light-emitting diode (see Fig. 3.4a), in which both the excitation and wavelength tunability are provided on the same chip while preserving the compatibility with the ^{87}Rb transitions. They sent the QD photons into an ^{87}Rb vapor cell, as shown in Fig. 3.4b, and observed slow single photons when the QD photon energy is fine-tuned through the hyperfine-split D_2 lines of Rb, which demonstrated the suitability of the device for future applications based on hybrid QD atomic systems.

In brief, QD-based single-photon sources have achieved indistinguishability of near unity, purity of the order of 10^{-3}, extraction efficiency of near 70%, and greatly

(a)

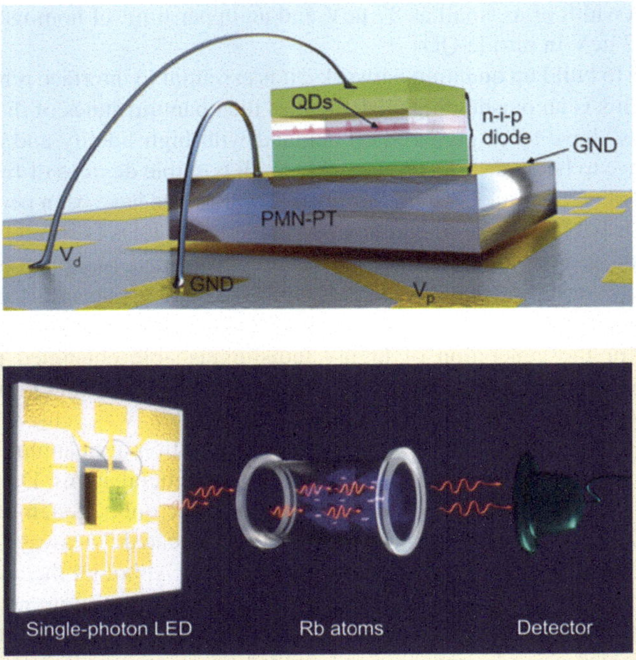

(b)

Fig. 3.4 (**a**) Sketch of the electrically pumped wavelength-tunable Q-LED in Huang et al.'s work [70]. (**b**) Sketch for the system where the LED light is fed into a ^{87}Rb-vapor cell and then slow single photons are observed at the detector when the QD transition energy is fine-tuned through the hyperfine-split D_2 lines of Rb

improved brightness even at telecommunication band; plasmonic and other nanostructures begin to take a significant role in improving QD-based single-photon emitters; room-temperature operating and electric driving are no longer hard tasks for QD-based single-photon source; quantum correlation is even realized between remote QD (and atom) quantum emitters. All these suggest a successful application of QD-based single-photon sources in quantum information processing in the near future.

3.3 QD-Based Entangled Photon Emitters

The entanglement concept has yet come out of experimental stage because of the lack of deterministic sources of entangled states. In this sense, solid-state sources of entangled photons are of special significance [71]. Among many systems under investigation, semiconductor QDs are particularly appealing emitters of on-demand, polarization-entangled photon pairs. About this aspect, Trotta group has recently given a good review on semiconductor QDs as ideal polarization-entangled photon source on demand [22].

While it was originally believed that QDs must exhibit a limited degree of entanglement, due to the typical decoherence effects of the solid state, recent studies have made thoughts change. The strongest limitation for QDs to entangle photons is the broken symmetry of as-grown dots, caused by the anisotropy of strain, composition, and shape irregularity, causing fine structure splitting (FSS) in the exciton emission [72]. To create the entangled states, it is preferable that the FSS is smaller than the radiative lifetime-limited linewidth of ~1 μeV [73]. A few post-growth tuning methods, such as uniaxial strain induced by piezoelectric materials [64], electric field-induced quantum-confined Stark effect [74, 75], magnetic field-induced Zeeman shifts [76], or laser annealing techniques [30], were advanced for semiconductor QDs emitting at <1 μm and at telecommunication range [77]. The full cancellation of the FSS, however, was yet achieved in spectral ranges beyond 1 μm. Another advantage of post-growth tuning is the simultaneous varying of the FSS and the emission energy, which is quite helpful to increase the yield of QD emitters [78]. As a result of these techniques, the highest rate of entanglement generation was achieved at nearly 10 kHz [79]. What is more important, by means of uniaxial strain tuning, Schmidt group [80] demonstrated the successful elimination of the FSS for telecom-wavelength InGaAs/GaAs QDs, which are integrated onto uniaxial piezoelectric actuators, as shown in Fig. 3.5. This technique enables tuning QD emission wavelengths and their FSS without degrading the performances of the QD emitters. Cancelling FSS within the experimental resolution limit, this work goes by an important step towards the high-yield generation of entangled photon pairs at telecommunication wavelength. As an outlook, the uniaxial strain in combination with Stark tuning [81] makes it possible to realize a QD source of entangled photons at telecom wavelengths in which the energy is tuned independent of FSS. In addition, as the strain-suppressed FSS in semiconductor QDs was realized on an on-chip platform [82], a quantum repeater network would be possibly achievable using similar architecture for telecom wavelengths. Besides optically excited QDs, electrically driven sources of entangled photons are of special importance because the system is much simpler than that with an excitation laser [83]. Embedded QDs into diode structures, several pulsed entangled light-emitting diodes have been presented [84]. The highest fidelity under pulsed electrical pumping was achieved by Chung et al. in 2016, which is 0.68 with FSS of 0.2 μeV without any post-selection. The results are significantly lower with respect to optical excitation [85] and suggest entanglement-degrading effects are dominant. The full potential of these devices might only be accessible via an electrically driven resonant population process. In 2019, Shields group [86] reported a week-long transmission of polarization-entangled photons from a single InAs/GaAs QD over a metropolitan network fiber 18 km long. It provides a reliable and stable technology for lowly complexed, robust quantum information network.

It is clear that the shape of the QDs and the in-plane anisotropy of the confining potential have a dominant role in determining the FSS. It is better to grow QDs of zero FSS rather than using post-selection or additional tuning. In 2013, Juska et al. [87] presented pyramidal site-controlled, highly symmetrical InGaAsN QDs on a patterned GaAs (111)B substrate. Up to 15% of the QDs can generate polarization-entangled photons with a fidelity up to 0.72 under non-resonant excitation. A differ-

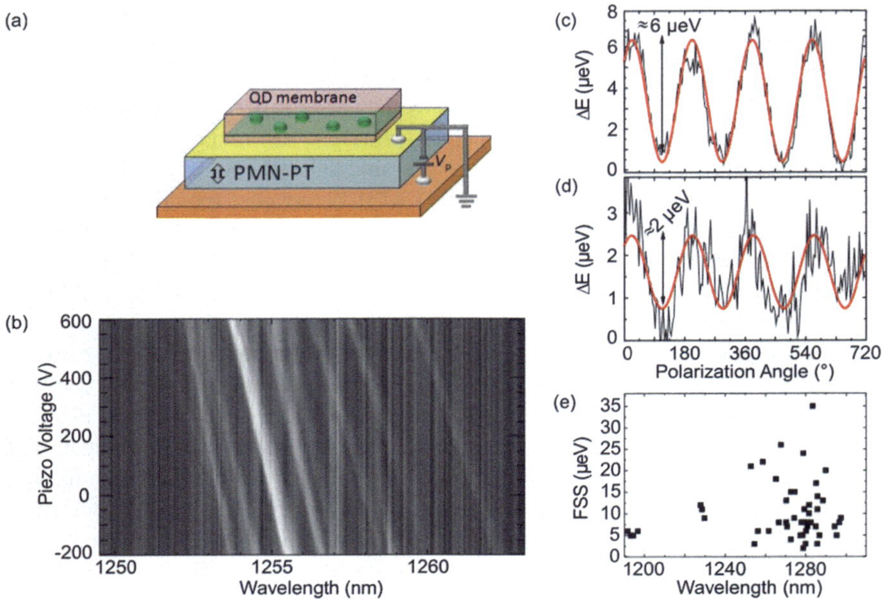

Fig. 3.5 Device structure and measured result in Hofer et al.'s work [80]. (**a**) Device structure showing the QD-containing nanomembrane integrated onto a piezoelectric substrate; (**b**) exemplary spectral tuning of a single QD; (**c**) polarization series of selected transitions revealing a finite FSS; (**d**) second polarization series revealing the experimental resolution limit of about 2 μeV in the determination of the FSS; (**e**) statistics of FSS distribution for the as-grown sample

ent way of symmetrically engineering excitonic wavefunctions of QDs is to grow large low-strain InGaAs QDs, exhibiting FSS well below 10 μeV despite anisotropic structure [88]. There are still some other approaches for growing QDs to emit entangled photons. The first method is a droplet epitaxy method [89]. Years ago, Wang et al. fabricated QDs by etching of nanoholes into an AlGaAs substrate via local droplet etching [90]. In 2013, Huo et al. [91] demonstrated that droplet QDs have ultra small FSS of 4 μeV in average. In addition to the in-plane symmetry, a tall morphology is useful to reduce the FSS, as the exchange interaction is determined by the electron-hole wavefunction overlap [92]. In 2017, Huber et al. [93] achieved an entanglement fidelity of 0.94, the best fidelity on a QD so far, without any post-growth techniques, as is shown in Fig. 3.6. The achieved fidelities are attractive for ready-to-use quantum relays, including error correction protocols. In 2018, Basset et al. offered an alternative solution by presenting a modified approach on GaAs (111) substrates, where the Ga droplets are crystallized and the subsequent barrier layer is deposited at a high substrate temperature of around 520 °C [94]. The authors claimed an improved crystalline quality of the QDs by suppressing the defect density normally high in low-temperature growth. With the exception of GaAs QDs, however, other material systems have yet met the demanding requirements of a high, and statistically frequent, excitonic confinement symmetry [95].

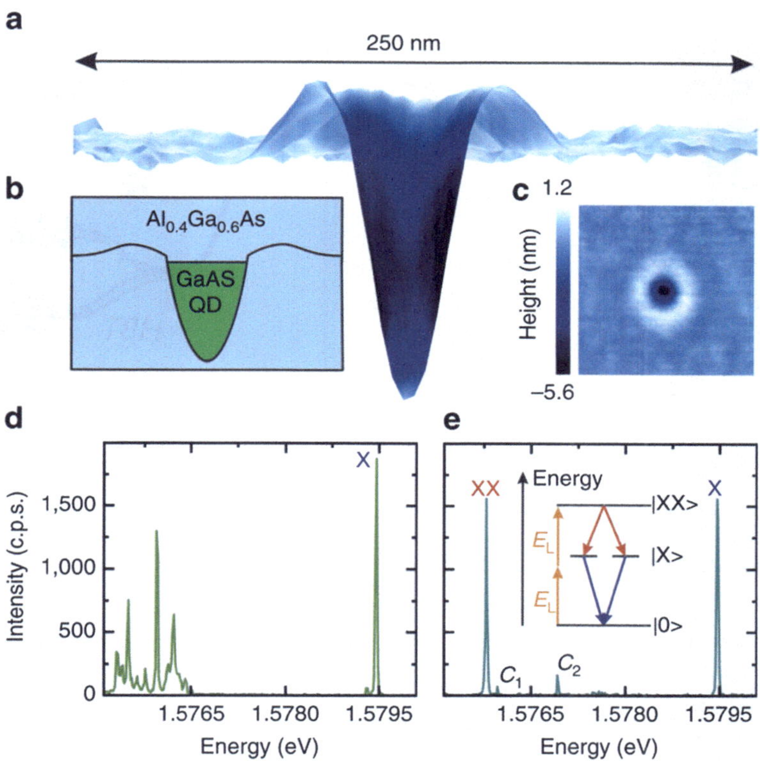

Fig. 3.6 Sample structure and spectral properties of the used QDs by Huber et al. [93]. (**a**) Cross-sectional 3D view of an atomic force microscopy (AFM) image of a nanohole in an AlGaAs layer. (**b**) Sketch of the sample structure. (**c**) Top view of the AFM measurement with color scale reflecting the local height. (**d**) Micro-photoluminescence spectrum of a representative QD under non-resonant excitation. (**e**) Spectrum of the same QD as in (**d**) under resonant two-photon excitation

Photonic states with multiple components are an essential resource for secure quantum relays, measurement-based quantum computers, and quantum-enhanced sensors [96]. There are a few ways for higher-dimensional photonic states, but even state-of-the-art technologies are probabilistic or have limited fidelities [97]. An alternative is to directly create a complex photonic state through a single quantum emitter, which in principle allows the deterministic multi-qubit photonic states. Lee et al. [98] used a cavity-enhanced Raman transition in a QD to show enhanced spin preparation and then to sequentially generate time-bin-encoded single-photon W-states, as shown in Fig. 3.7. They use cavity-stimulated Raman emission effect to perform spin-state preparation on a trapped hole spin. The scalability of their scheme for W-state was shown by producing photons superposed across up to four time bins. Their technique allows the deterministic generation of arbitrary single-photon time-bin-encoded states, and this capability will be useful for single-mode quantum computation and for maximizing the key rates of QKD protocols over long distances. Multiphoton entangled states such as “NOON states” have also attracted

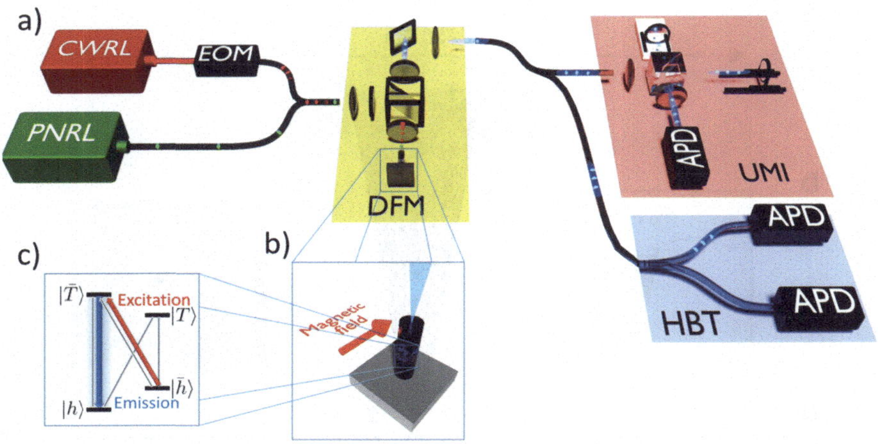

Fig. 3.7 (**a**) An illustration of the experimental setup of Lee et al. [98]. (**b**) The QD-micropillar system is cooled in a cryostat to 5 K and placed in a Voigt geometry magnetic field. (**c**) The energy level diagram shows the states and allowed transitions of a single-hole charged QD in a Voigt field. Used under the terms of the Creative Commons Attribution 3.0 licence

a lot of attention because of their possible application in high-precision, quantum-enhanced phase determination. NOON states have been generated in spontaneous parametric down-conversion processes and by mixing quantum and classical light on a beam splitter [99]. In contrast, Muller et al. [100] demonstrated super-resolving phase measurements based on two-photon NOON states generated by QD single-photon sources using Hong–Ou–Mandel effect. Using pulsed resonance fluorescence of a charged exciton state, they improved the accuracy of the phase uncertainty to be higher than the prediction of standard quantum limit. Their results point towards the realization of a real-world quantum sensor in the near future.

In brief, although originally considered limited, entanglement of photons from QDs has given a bit rate of 10 kHz, and strain tuning and electric driving are more powerful in realizing entangled QD photon sources; in situ growth technique improves the collective QD symmetry a lot so that a record entanglement fidelity of 0.94 was achieved; multiphoton entanglement of semiconductor QDs is getting into reality. These progresses pave the way for QDs to become the deterministic complex quantum emitters in quantum computing and quantum communication.

3.4 On-Chip Integration of QD Quantum Emitters

The lack of methods to simultaneously generate and control photons within the same chip has long been an obstacle to achieve efficient multi-qubit gates and chip-scale quantum photonics. Photonic integration is fortunately a recent method for building on-chip quantum information processing system possessing many functions including quantum emitting, quantum processing, quantum memorizing, and quantum measuring [101]. To obtain deterministic and scalable light-matter quantum interfaces for quantum information network, solid-state quantum emitters are required to be integrated into photonic chips [102]. This requirement includes deterministic single-photon sources, complex qubit manipulation circuits, on-chip detection, etc. Obviously, semiconductor QDs are one of the best building blocks [103], due to their integrability into microcavities and photonic nanostructures [104]. Together with the advantage that QDs enable on-chip electrical excitation [105] and wavelength-tunable entangled photon emission [106], many efforts have been devoted to integrate semiconductor QDs on a photonic chip, and much progress has been made.

In this field, there were challenges to realize III–V quantum photonic circuits, including deterministic integration of selected QDs into optical waveguides/cavities, efficient filtering of specific quantum transitions within the emission spectrum, on-chip pump suppression, and multiplexing of multiple QDs [106, 107]. Epitaxially grown InGaAs/GaAs QDs, however, are now routinely integrated into a variety of photonic structures [8]. In 2017, Midolo et al. [108] integrated electro-optic phase shifter with embedded QD emitters in a photonic waveguide. Photons emitted by single self-assembled QDs can be actively routed into the two outputs of the interferometer, and the phase shifter takes advantage of the anisotropic electro-optic response in GaAs. Davanco et al. [109] developed a heterogeneous photonic integration platform enabling direct integration of GaAs waveguides and cavities containing QDs with low-loss Si_3N_4 waveguides. On this platform, a highly efficient optical interface between single InAs/GaAs QDs and Si_3N_4 waveguides, QD emission enhancement in microcavities, and a path for strong coupling were realized. Kirsanske et al. [35] demonstrated how the emitted photons can efficiently be coupled off chip to an optical fiber by the implementation of a tapered waveguide section. Very recently, Thyrrestrup et al. [44] stepped further towards a deterministic on-chip photonic nonlinearity by embedding a QD in an electrically contacted nanophotonic membrane to present near lifetime-limited linewidths of quantum emitters. The incorporation of on-demand single-photon sources on a silicon photonic chip remained a difficult challenge [109]. However, Kim et al. [110] demonstrate the integration of silicon photonic devices with a solid-state single-photon emitter. These results put forward a significant progress to chip-scaled single-photon sources and on-chip optical quantum information processing. Furthermore, fabrication techniques that enable the deterministic integration of preselected quantum light emitters into nanophotonic elements play a key role when moving forward to circuits containing multiple emitters. Elshaari et al. [111] overcame almost all of the above challenges with a hybrid and scalable approach. In their scheme, as shown in Fig. 3.8, single III–V QD emitters are positioned and deterministically integrated in a CMOS-compatible photonic circuit. Even with emitters' excitation suppression more than 95 dB and single photons' routing bandwidth as much as 40 nm, the system size is 10^{-6} times the similar table-top approaches. It realizes in-plane pumping on integrated quantum emitters and multi-quantum emitter circuit, in which quantum emitters are independently selected and deterministically operated. These results put forward a significant step to chip-scaled single-photon sources and on-chip photonics quantum computing and harvest the full potential of photonics quantum technologies.

To realize in-plane manipulation of integrated quantum devices, there occurs actually a more fundamental technique. Epitaxial QDs usually possess flat morphologies so that they are mainly quantized along the z direction and only suited for efficient vertically emitting single-photon devices [32, 36, 68, 102, 112, 113]. For planar integrated quantum photonics applications [114–116], it is hopeful to have QDs with a quantization axis within the *x–y* plane. In spite of their importance, very little effort has been devoted to developing quantum sources optimized for planar photonic circuits, until Zadel et al.'s work [117], where nanowires-QDs were removed from the substrate and turned by 90° for efficient coupling to dielectric waveguides. Recently, Yuan et al. [118] more powerfully showed how to turn the quantization axis of a QD to be in the growth plane without rotating anything. Their technique preserves the compatibility between the QD hetero-structure and the planar photonic processing. Their scheme is shown in Fig. 3.9 as including high-quality, initially unstrained GaAs QDs with a relatively large height and micro-machined piezoelectric actuators supplying in-plane stress. Different from most previous experiments, in which stress was added after growth as a perturbation to fine-tune the emission properties of QDs [119], confinement can be seen here as a perturbation compared to the strain-induced effects in determining the orientation of the quantization axis and hence the optical selection rules and excitonic fine structure. It is suggested that uniaxial stress may be the tool to obtain quantum light sources with ideally oriented transition dipoles and enhanced oscillator strengths for integrated quantum photonics.

On-chip integration has another goal, interfacing all kinds of quantum elements such as quantum light sources, quantum gates, and quantum detectors on a single chip. It is desirable to enhance the quantum functionality and avoid losses between different technology platforms in a quantum network [5, 120, 121]. In fact, deterministically

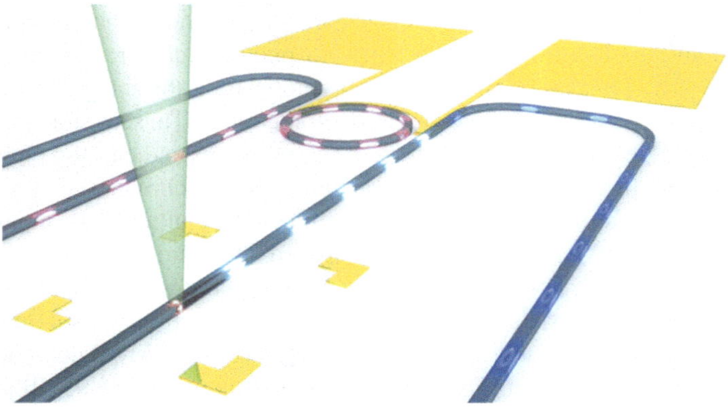

Fig. 3.8 Schematic view of the hybrid quantum photonic circuit fabricated by Elshaari et al. [111]. It consists of an InAsP QD in an InP nanowire (ruby colored) that is integrated with a SiN waveguide (blue) and on-chip tunable ring resonator filter. The ring resonator filter is tuned by applying voltage to the gold contacts (orange). The out-of-plane laser (green) excites the QD, which emits single photons (ellipsoids) into the waveguide

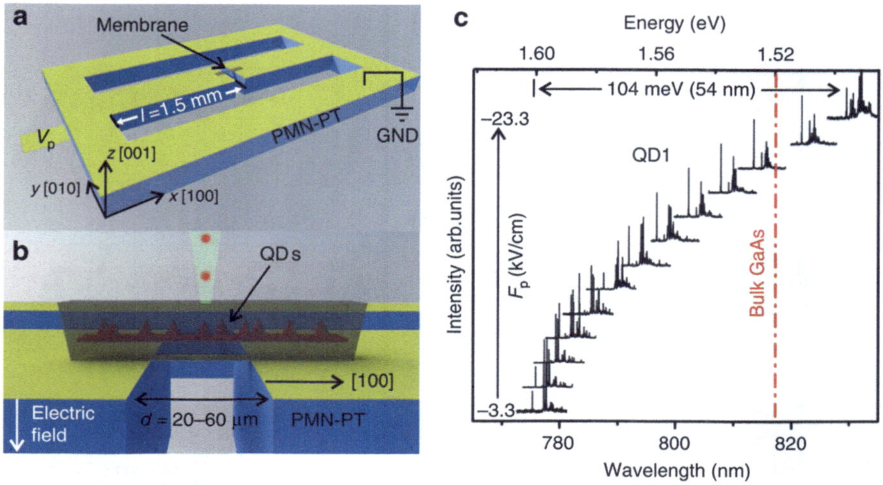

Fig. 3.9 Experimental configuration and emission energy tuning of a GaAs QD via uniaxial stress provided by a micro-machined PMN-PT actuator in Yuan et al.'s work [118]. (**a**) Sketch of the actuator featuring two fingers separated by a gap, where a semiconductor membrane with embedded QDs is bonded on the fingers and forms a bridge above the gap. (**b**) Side view of the device. (**c**) Normalized PL spectra of a GaAs QD measured for increasing uniaxial tensile stress

integrating quantum emitters into on-chip quantum nanophotonic elements has recently attracted much attention. These techniques include site-controlled QD growth [122–124], marker-aligning nanostructures to emitters [28], in situ lithography [125], and AFM-transferring quantum elements onto a chip [117]. Due to its high patterning resolution as well as spectral and spatial control, in situ electron beam lithography allows for integration of preselected quantum emitters into complex photonic systems. Schnauber et al. [126] presented deterministic integration of an InAs QD into a multimode interference beam splitter via in situ electron beam lithography, as shown in Fig. 3.10. The combined emitter-gate interface functionality is demonstrated with the on-chip single-photon purity of $g^{(2)}(0) = 0.13 \pm 0.02$. In combination with wavelength fine-tuning [127], the spectrally selective integration of multiple QDs in a single 2 × 2 beam splitter opens a window for conducting an on-chip Hong–Ou–Mandel experiment. Placing two resonant QDs in a single waveguide promises excitonic state transfer between the two quantum emitters. This work paves the way towards multi-node, fully integrated quantum photonic chips. The techniques that can super-resolve and selectively excite quantum emitters are of greater importance for manipulating integrated different quantum elements. Such a technique is often related to surface plasmons of metal nanostructures, which holds promise for a wide range of applications in the fields of integrated nanophotonic circuits [128–131] due to its subwavelength field confinement and microscale propagation length. The strong interaction between single quantum emitters and plasmonic nanowires has been exploited to make integrated quantum optical devices such as efficient single-photon sources and transistors [132–134]. To create

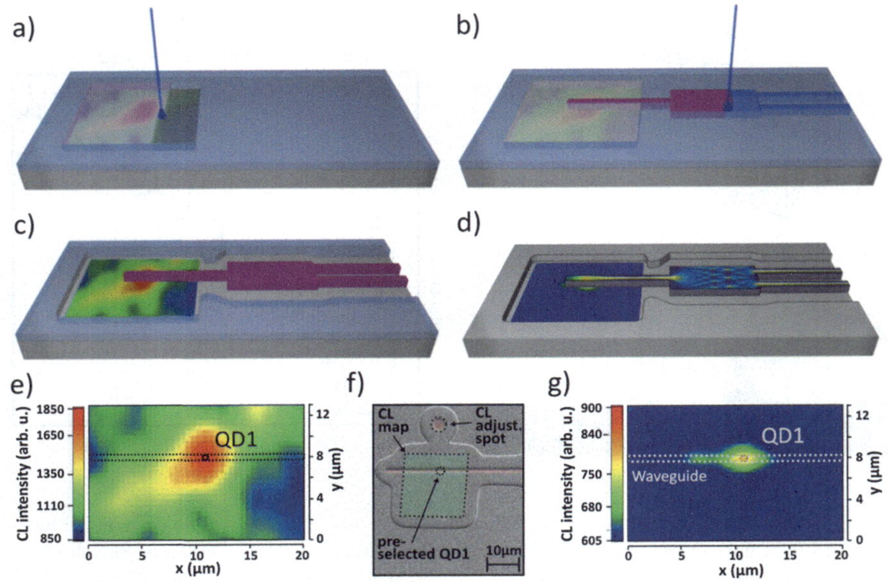

Fig. 3.10 The integrated structure and experimental data in Schnauber et al.'s work [126]. (**a–d**) Schematic visualization of the in situ EBL device manufacturing process; (**e**) CL map before EBL patterning; (**f**) false color SEM image of the fully processed WG containing QD; (**g**) CL map of QD1 after patterning and processing the sample. Adapted with permission from Ref. [126], Copyright (2018) American Chemical Society

a simpler, more flexible, and non-diffraction-limited optical resolving and excitation technique, Li et al. [135] controllably excited multiple QDs coupled with a silver nanowire by tuning the interference field of surface plasmons. Selective excitation of two QDs separated by 100 nm is realized because the interference size is much smaller than the diffraction limit. This work opens a new and effective way to manipulate the plasmon-mediated interaction between quantum emitters in the integrated quantum plasmonic circuit.

An outstanding challenge in quantum photonics is the scalability, which requires positioning of single quantum emitters in a deterministic fashion. To date, most of the QD quantum emitters were based on randomly nucleated quantum dots, which prevents them from effectively integrating into a chip. Site-selected QDs nucleated at specified positions on the substrate have been prepared for integrating single-photon emitters more than a decade ago [136] and are verified preferable to open up the possibility for deterministic integration of quantum dots with photonic structures [8, 137, 138]. Years ago, it was reported that site-controlled QD emitters in nanowire can generate single photons with excellent purity ($g^{(2)}(0) < 0.005$ [138] and polarization-entangled photon pairs with fidelities exceeding 80% [139–141]. These sources utilize $InAs_xP_{1-x}$QDs with $x \sim 30\%$ and operate at wavelengths of $\lambda \sim 950$ nm. To extend this approach to telecommunication band, in 2018, Haffouz et al. [142] realized the site-selected growth of bright single InAsP quantum dots embedded within InP photonic nanowire waveguides emitting at telecommunication wavelengths. With a properly designed waveguide, they increase the count rate of QD emitting in the telecom O-band from 0.4 to 35 kHz. This result paves the way towards integrating efficient single-photon sources at telecommunication wavelengths using site-controlling growth, but they did not show any single-photon or entangled emission. The precision of a QD position in a photonic nanostructure actually determines coupling strength and thus is crucial for controlling the quantum interaction [143, 144] and collective effects [145] in a quantum chip. Although all-optical imaging methods [146] have ever been reported, they cannot be well used to image QDs in photonic nanostructures. In 2017, de Assis et al. [147] demonstrated a nondestructive QD mapping technique by setting the QD-embedding structure into vibration. By using two cross-polarized mechanical modes, the 2D mapping of the QD position in the growth plane was carried out. In contrast to optical near-field methods and scanning tunneling microscopy [148], this technique can determine the location of QDs deeply buried in a microstructure with a higher spatial resolution. In 2018, Liu group [149] used nanofabrication and photoluminescence imaging to locate the sites of single InAs/GaAs QDs with precision better than 5 nm, allowing one to track the QD photon statistics as a function of the distance from a nanofabricated surface. Regarding integration with the interest of compactness and scalability, electrical carrier injection would be highly attractive because it avoids bulky external laser sources. However, simple electrically driven sources based on a PIN diode design do not allow for resonant excitation and suffer from charge noise introduced by excess carriers, both of which limit the indistinguishability of the emitted photons. This problem was resolved by Stock et al. [150] through integrating an electrically triggered optical excitation source and a resonantly excited QD emitter on the same chip. It is more desirable to tune the emitters in a localized way without interfering with other integrated elements. For this, monolithic integration of QDs with coherent, electrically controlled whispering-gallery-mode lasers has recently been carried out [151]. In this direction, Munnelly et al. [152] represent a major advancement by bringing multiple aspects of desirable functionality, including pure single-photon emitting, cavity enhancement, electrical control, and coherent excitation, together on a single chip. Their device exhibits low laser threshold currents of a few tens of μA, a small area of ~200 μm^2, and high single-photon purity with $g^{(2)}(0)$ of 0.07 ± 0.03. The electric field applied along the growth direction tunes the single-photon emission by 1.1 meV into resonance with the surrounding cavity mode for enhancing the emission via the Purcell effect.

In brief, QD-based quantum emitters have been successfully integrated with photonic structures into a single chip; QDs can be tuned to in-plane emitting style and can be well selected from an assemble so that they are more suitable and easier for on-chip integrating and quantum interacting with other quantum devices; development in site-controlling and on-chip electrical-driving QDs serves as a big step towards full quantum on-chip integration.

3.5 Summary

Semiconductor QDs are widely considered as the best candidate for quantum emitters. This chapter reviews the scientific and technological development of quantum emitters based on QDs in recent years. QD-based single-photon sources have achieved indistinguishability of near unity, purity of the order of 10^{-3}, extraction efficiency of near 70%, and greatly improved brightness even at telecommunication band; plasmonic and related nanostructures begin to take a significant role in improving QD-based single-photon emitters; room-temperature operating and electric driving are no longer hard tasks for QD-based single-photon source; quantum correlation even is realized between remote QD (and atom) quantum emitters. Entanglement of photons from semiconductor QDs has given a bit rate as high as 10 kHz, and strain tuning and electric driving are more powerful in realizing entangled QD photon sources; in situ growth technique improves the collective QD symmetry a lot so that a record entanglement fidelity of 0.94 was achieved; and multiphoton entanglement of semiconductor QDs is getting into reality. QDs are further proved the best on-demand, entangled, on-chip integrated quantum emitters for quantum information processing.

Acknowledgments This chapter was supported by Sichuan Science and Technology Program under Grant No. 2018JY0084, the Recruitment Program of Global Experts, and the Rongpiao Plan of Chengdu City, China.

References

1. Nielsen, M. A., & Chuang, I. L. (2000). *Quantum computation and quantum information*. Cambridge: Cambridge University Press.
2. Beveratos, A., et al. (2002). Single photon quantum cryptography. *Physical Review Letters, 89*, 187901.
3. Cheung, J., et al. (2007). The quantum candela: a re-definition of the standard units for optical radiation. *Journal of Modern Optics, 54*, 373.
4. Kok, P. W. J., et al. (2007). Linear optical quantum computing with photonic qubits. *Reviews of Modern Physics, 79*, 135.
5. Aspuru-Guzik, A., & Walther, P. (2012). Photonic quantum simulators. *Nature Physics, 8*, 285.
6. O'Brien, J. L., et al. (2009). Photonic quantum technologies. *Nature Photonics 3*, 687.
7. Aharonovich, I., et al. (2016). Solid-state single-photon emitters. *Nature Photonics 10*, 631.
8. Lodahl, P., et al. (2015). Interfacing single photons and single quantum dots with photonic nanostructures. *Reviews of Modern Physics, 87*, 347.
9. Atatüre, M., et al. (2018). Material platforms for spin-based photonic quantum technologies. *Nature Reviews Materials, 3*, 38.
10. Akopian, N., et al. (2006). Entangled photon pairs from semiconductor quantum dots. *Physical Review Letters, 96*, 130501.
11. Young, R. J., et al. (2006). Improved fidelity of triggered entangled photons from single quantum dots. *New Journal of Physics, 8*, 29.
12. Hafenbrak, R., et al. (2007). Triggered polarization-entangled photon pairs from a single quantum dot up to 30 K. *New Journal of Physics, 9*, 315.

13. Chunnilall, C. J., et al. (2014). Metrology of single-photon sources and detectors: a review. *Optical Engineering, 53*, 081910.
14. De Greve, K., et al. (2012). Quantum-dot spin-photon entanglement via frequency downconversion to telecom wavelength. *Nature, 491*, 421.
15. Gao, W. B. (2012). Observation of entanglement between a quantum dot spin and a single photon. *Nature, 491*, 426.
16. Schaibley, J. R. et al. (2013). Demonstration of quantum entanglement between a single electron spin confined to an InAs quantum dot and a photon. *Physical Review Letters, 110*, 167401.
17. Takemoto, K. et al. (2015). Quantum key distribution over 120 km using ultrahigh purity single-photon source and superconducting single-photon detectors. *Scientific Reports, 5*, 14383.
18. He, Y., et al. (2017). Quantum state transfer from a single photon to a distant quantum-dot electron spin. *Physical Review Letters, 119*, 60501.
19. Varnava, C., et al. (2016). An entangled-LED-driven quantum relay over 1 km. *Quantum Information, 2*, 16006.
20. Senellart, P., et al. (2017). High-performance semiconductor quantum-dot single-photon sources. *Nature Nanotechnology, 12*, 1026.
21. Lodahl, P., et al. (2018). Quantum-dot based photonic quantum networks. *Quantum Science and Technology, 3*, 013001.
22. Huber, D., et al. (2018). Semiconductor quantum dots as an ideal source of polarization-entangled photon pairs on-demand: a review. *Journal of Optics, 20*, 073002.
23. Benson, O., et al. (2000). Regulated and entangled photons from a single quantum dot. *Physical Review Letters, 84*, 2513.
24. Gérard, J. M., et al. (1998). Enhanced spontaneous emission by quantum boxes in a monolithic optical microcavity. *Physical Review Letters, 81*, 1110.
25. Lodahl, P., et al. (2004). Controlling the dynamics of spontaneous emission from quantum dots by photonic crystals. *Nature, 430*, 654.
26. Englund, D., et al. (2005). Controlling the spontaneous emission rate of single quantum dots in a two-dimensional photonic crystal. *Physical Review Letters, 95*, 13904.
27. Grange, T., et al. (2015). Cavity-funneled generation of indistinguishable single photons from strongly dissipative quantum emitters. *Physical Review Letters 114*, 193601.
28. Sapienza, L., et al. (2015). Nanoscale optical positioning of single quantum dots for bright and pure single-photon emission. *Nature Communications, 6*, 7833.
29. Tomas, M.-S., et al. (1999). Spontaneous-emission spectrum in an absorbing Fabry-Perot cavity. *Physical Review A: Atomic, Molecular, and Optical Physics, 60*, 2431.
30. Choy, J. T., et al. (2011). Enhanced single-photon emission from a diamond-silver aperture. *Nature Photonics, 5*, 738.
31. Barnes, W. L., et al. (2002). Solid-state single photon sources: light collection strategies. *European Physical Journal D: Atomic, Molecular, Optical and Plasma Physics, 18*, 197.
32. Somaschi, N., et al. (2016). Near-optimal single-photon sources in the solid state. *Nature Photonics, 10*, 340.
33. Loredo, J., et al. (2017). Boson sampling with single-photon fock states from a bright solid-state source. *Physical Review Letters, 118*, 130503.
34. Liu, S., et al. (2017). A deterministic quantum dot micropillar single photon source with >65% extraction efficiency based on fluorescence imaging method. *Scientific Reports 7*, 13986.
35. Mermillod, Q., et al. (2016). Harvesting, Coupling, and Control of Single-Exciton Coherences in Photonic Waveguide Antennas. *Physical Review Letters, 116*, 163903.
36. Ding, X., et al. (2016). On-demand single photons with high extraction efficiency and near-unity indistinguishability from a resonantly driven quantum dot in a micropillar. *Physical Review Letters, 116*, 20401.
37. Miyazawa, T., et al. (2005). Single-photon generation in the 1.55-μm optical-fiber band from an InAs/InP quantum dot. *Japanese Journal of Applied Physics, 44*, L620.
38. Song, H.-Z., et al. (2013). Design of Si/SiO_2 micropillar cavities for Purcell-enhanced single photon emission at 1.55 μm from InAs/InP quantum dots. *Optics Letters, 38*, 3241.

39. Song, H.-Z., et al. (2015). High quality-factor Si/SiO2-InP hybrid micropillar cavities with submicrometer diameter for 1.55-µm telecommunication band. *Optics Express, 23*, 16264.
40. Song, H.-Z., et al. (2017). In InGaAsP/InP Nanocavity for Single-Photon Source at 1.55-µm Telecommunication Band. *Nanoscale Research Letters, 12*, 128.
41. Kim, J., et al. (2016). Two-photon interference from a bright single-photon source at telecom wavelengths. *Optica, 3*, 577.
42. Birowosuto, M., et al. (2012). Fast Purcell-enhanced single photon source in 1,550-nm telecom band from a resonant quantum dot-cavity coupling. *Science Reports, 2*, 321.
43. Arcari, M. S., et al. (2014). Near-unity coupling efficiency of a quantum emitter to a photonic crystal waveguide. *Physical Review Letters, 113*, 093603.
44. Thyrrestrup, H., et al. (2018). Quantum optics with near lifetime-limited quantum-dot transitions in a nanophotonic waveguide. *Nano Letters, 18*, 1801.
45. Ralph, T. C., et al. (2015). Photon sorting, efficient bell measurements, and a deterministic controlled-gate using a passive two-level nonlinearity. *Physical Review Letters, 114*, 173603.
46. Ozel, T., et al. (2011). Anisotropic emission from multilayered plasmon resonator nanocomposites of isotropic semiconductor quantum dots. *ACS Nano, 5*, 1328.
47. Belacel, C., et al. (2013). Controlling spontaneous emission with plasmonic optical patch antennas. *Nano Letters, 13*, 1516.
48. Zhukovsky, S., et al. (2014). Hyperbolic metamaterials based on quantum-dot plasmon-resonator nanocomposites. *Optics Express, 22*, 18290.
49. Pfeiffer, M., et al. (2018). Coupling a single solid-state quantum emitter to an array of resonant plasmonic antennas. *Scientific Reports, 8*, 3415.
50. Li, L., et al. (2017). Enhanced quantum dot spontaneous emission with multilayer metamaterial nanostructures. *ACS Photonics, 4*, 501.
51. Leng, H., et al. (2018). Strong coupling and induced transparency at room temperature with single quantum dots and gap plasmons. *Nature Communications, 9*, 4012.
52. Boretti, A., et al. (2015). Electrically driven quantum light sources. *Advanced Optical Materials, 3*, 1012.
53. Kuhlmann, A. V., et al. (2015). Transform-limited single photons from a single quantum dot. *Nature Communications, 6*, 8204.
54. Kirsanske, G., et al. (2017). Indistinguishable and efficient single photons from a quantum dot in a planar nanobeam waveguide. *Physical Review B, 96*, 165306.
55. Michler, P., et al. (2000). Quantum correlation among photons from a single quantum dot at room temperature. *Nature, 406*, 968.
56. Reithmaier, G., et al. (2013). On-chip time resolved detection of quantum dot emission using integrated superconducting single photon detectors. *Scientific Reports, 3*, 6.
57. Lin, X., et al. (2017). Electrically-driven single-photon sources based on colloidal quantum dots with near-optimal antibunching at room temperature. *Nature Communications, 8*, 1132.
58. Feng, S.-W., et al. (2017). Purification of single photons from room-temperature quantum dots. *Physical Review Letters, 119*, 143601.
59. Lohrmann, A., et al. (2015). Single-photon emitting diode in silicon carbide. *Nature Communications, 6*, 7783.
60. Arita, M., et al. (2017). Ultraclean Single Photon Emission from a GaN Quantum Dot. *Nano Letters, 17*, 1902.
61. Gold, P., et al. (2014). Two-photon interference from remote quantum dots with inhomogeneously broadened linewidths. *Physical Review B, 89*, 035313.
62. Reindl, M., et al. (2017). Phonon-assisted two-photon interference from remote quantum emitters. *Nano Letters, 17*, 4090.
63. Delteil, A., et al. (2017). Realization of a cascaded quantum system: heralded absorption of a single photon qubit by a single-electron charged quantum dot. *Physical Review Letters, 118*, 177401.
64. Ding, F., et al. (2010). Tuning the exciton binding energies in single self-assembled quantum dots by piezoelectric-induced biaxial stress. *Physical Review Letters, 104*, 067405.

65. Zopf, M., et al. (2018). Frequency feedback for two-photon interference from separate quantum dots. *Physical Review B, 98*, 161302R.
66. Deng, Y.-H., et al. (2019, May 10). Quantum interference between light sources separated by 150 million kilometers. arXiv:1905.02868 [quant-ph].
67. Zhang, R., et al. (2009). Creation of long-term coherent optical memory via controlled nonlinear interactions in Bose-Einstein condensates. *Physical Review Letters, 103*, 233602.
68. Specht, H. P., et al. (2011). Single-atom quantum memory. *Nature, 473*, 190.
69. Shields, A. J. (2007). Semiconductor quantum light sources. *Nature Photonics, 1*, 215.
70. Huang, H., et al. (2017). Electrically-pumped wavelength-tunable GaAs quantum dots interfaced with rubidium atoms. *ACS Photonics, 4*, 868.
71. Orieux, A., et al. (2017). Semiconductor devices for entangled photon pair generation: a review. *Reports on Progress in Physics, 80*, 076001
72. Gong, M., et al. (2011). Exciton polarization, fine-structure splitting, and the asymmetry of quantum dots under uniaxial stress. *Physical Review Letters, 106*, 227401.
73. Rastelli, A., et al. (2012). Controlling quantum dot emission by integration of semiconductor nanomembranes onto piezoelectric actuators. *Physica Status Solidi B, 249*, 687.
74. Gerardot, B. D., et al. (2007). Manipulating exciton fine structure in quantum dots with a lateral electric field. *Applied Physics Letters, 90*, 041101.
75. Bennett, A. J., et al. (2010). Electric-field-induced coherent coupling of the exciton states in a single quantum dot. *Nature Physics, 6*, 947.
76. Hudson, A. J., et al. (2007). Coherence of an entangled exciton-photon state. *Physical Review Letters, 99*, 266802.
77. Sapienza, L., et al. (2013). Exciton fine-structure splitting of telecom-wavelength single quantum dots: Statistics and external strain tuning. *Physical Review B, 88*, 155330.
78. Patel, R. B., et al. (2010). Two-photon interference of the emission from electrically tunable remote quantum dots. *Nature Photonics, 4*, 632.
79. Delteil, A., et al. Generation of heralded entanglement between distant hole spins. (2016). *Nature Physics, 12*, 218.
80. Höfer, B., et al. (2019). Tuning emission energy and fine structure splitting in quantum dots emitting in the telecom O-band. AIP Advances 9, 085112.
81. Trotta, R., et al. (2012). Universal recovery of the energy-level degeneracy of bright excitons in InGaAs quantum dots without a structure symmetry. *Physical Review Letters, 109*, 147401.
82. Chen, Y., et al. (2016). Extended carrier lifetimes and diffusion in hybrid perovskites revealed by Hall effect and photoconductivity measurements. *Nature Communications, 7*, 7.
83. Salter, C. L., et al. (2010). An entangled-light-emitting diode. *Nature, 465*, 594.
84. Chung, T. H., et al. (2016). Selective carrier injection into patterned arrays of pyramidal quantum dots for entangled photon light-emitting diodes. *Nature Photonics, 10*, 782.
85. Keil, R., et al. (2017). Solid-state ensemble of highly entangled photon sources at rubidium atomic transitions. *Nature Communications, 8*, 15501.
86. Xiang, Z.-H., et al. (2019). Long-term transmission of entangled photons from a single quantum dot over deployed fiber. *Scientific Reports, 9*, 4111.
87. Juska, G., et al. (2013). Towards quantum-dot arrays of entangled photon emitters. *Nature Photonics, 7*, 527.
88. Musial, A., et al. (2014). Toward weak confinement regime in epitaxial nanostructures: Interdependence of spatial character of quantum confinement and wave function extension in large and elongated quantum dots. *Physical Review B, 90*, 045430.
89. Mano, T., et al. (2000). InAs quantum dots growth by modified droplet epitaxy using sulfur termination. *Japanese Journal of Applied Physics, 39*, 4580.
90. Wang, Z. M., et al. (2007). Nanoholes fabricated by self-assembled gallium nanodrill on GaAs (100). *Applied Physics Letters, 90*, 113120.
91. Huo, Y. H., et al. (2013). Ultra-small excitonic fine structure splitting in highly symmetric quantum dots on GaAs (001) substrate. *Applied Physics Letters, 102*, 152105.

92. Huo, Y. H., et al. (2014). Volume dependence of excitonic fine structure splitting in geometrically similar quantum dots. *Physical Review B, 90*, 041304.
93. Huber, D., et al. (2017). Highly indistinguishable and strongly entangled photons from symmetric GaAs quantum dots. *Nature Communications, 8*, 15506.
94. Basso Basset, F., et al. (2018). High-yield fabrication of entangled photon emitters for hybrid quantum networking using high-temperature droplet epitaxy. *Nano Letters 18*, 505.
95. Furis, M., et al. (2006). Bright-exciton fine structure and anisotropic exchange in nanocrystal quantum dots. *Physical Review B, 73*, 241313.
96. Giovannetti, V., et al. (2004). Quantum-enhanced measurements: beating the standard quantum limit. *Science, 306*, 1330.
97. Hacker, B., et al. (2016). A photon–photon quantum gate based on a single atom in an optical resonator. *Nature, 536*, 193.
98. Lee, J. P., et al. (2018). Multi-dimensional photonic states from a quantum dot. *Quantum Science and Technology, 3*, 024008.
99. Bennett, A. J., et al. (2016). Cavity-enhanced coherent light scattering from a quantum dot. *Science Advances, 2*, e1501256.
100. Muller, M., et al. (2017). Quantum dot single-photon sources for entanglement enhanced interferometry. *Physical Review Letters, 118*, 257402.
101. Pernice, W. H. P., et al. (2012). High-speed and high-efficiency travelling wave single-photon detectors embedded in nanophotonic circuits. *Nature Communications, 3*, 1325.
102. Claudon, J., et al. (2010). A highly efficient single-photon source based on a quantum dot in a photonic nanowire. *Nature Photonics, 4*, 174.
103. Gao, W. B., et al. (2013). Quantum teleportation from a propagating photon to a solid- state spin qubit. *Nature Communications, 4*, 2744.
104. Gazzano, O., et al. (2013). Bright solid-state sources of indistinguishable single photons. *Nature Communications, 4*, 1425.
105. Yuan, Z., et al. (2016). Electrically driven single-photon source. *Nature Communications, 7*, 10387.
106. Prtljaga, N., et al. (2014). Monolithic integration of a quantum emitter with a compact on-chip beam-splitter. *Applied Physics Letters, 104*, 231107.
107. Jöns, K. D., et al. (2015). Monolithic on-chip integration of semiconductor waveguides, beamsplitters and single-photon sources. *Journal of Physics D: Applied Physics, 48*, 085101.
108. Midolo, L., et al. (2017). Electro-optic routing of photons from single quantum dots in photonic integrated circuits. arXiv:1707.06522v1 [quant-ph], 20 July 2017.
109. Davanco, M., et al. (2017). Heterogeneous integration for on-chip quantum photonic circuits with single quantum dot devices. *Nature Communications, 8*, 889.
110. Kim, J.-H., et al. (2017). Hybrid Integration of Solid-State Quantum Emitters on a Silicon Photonic Chip. *Nano Letters, 17*, 7394.
111. Elshaari, A. W., et al. (2017). On-chip single photon filtering and multiplexing in hybrid quantum photonic circuits. *Nature Communications, 8*, 379.
112. Gschrey, M., et al. (2015). Highly indistinguishable photons from deterministic quantum-dot microlenses utilizing three-dimensional in situ electron-beam lithography. *Nature Communications, 6*, 7662.
113. Santori, C., et al. (2002). Indistinguishable photons from a single-photon device. *Nature, 419*, 594.
114. Kim, J.-H., et al. (2016). Two-photon interference from the far-field emission of chip-integrated cavity-coupled emitters. *Nano Letters, 16*, 7061.
115. Wu, X., et al. (2017). On-chip single-plasmon nanocircuit driven by a self-assembled quantum dot. *Nano Letters, 17*, 4291.
116. Keil, R., et al. (2016). Hybrid waveguide-bulk multi-path interferometer with switchable amplitude and phase. *APL Photonics, 1*, 81302.
117. Zadeh, I. E., et al. (2016). Deterministic integration of single photon sources in silicon based photonic circuits. *Nano Letters, 16*, 2289.

118. Yuan, X., et al. (2018). Uniaxial stress flips the natural quantization axis of a quantum dot for integrated quantum photonics. *Nature Communications, 9*, 3058.
119. Lin, H., et al. (2011). Stress tuning of strong and weak couplings between quantum dots and cavity modes in microdisk microcavities. *Physical Review B, 84*, 201301.
120. Silverstone, J. W., et al. (2016). Silicon quantum photonics. *IEEE Journal of Selected Topics in Quantum Electronics, 22*, 390.
121. Reithmaier, G., et al. (2015). On-chip generation, routing, and detection of resonance fluorescence. *Nano Letters, 15*, 5208.
122. Calic, M., et al. (2017). Deterministic radiative coupling of two semiconductor quantum dots to the optical mode of a photonic crystal nanocavity. *Scientific Reports, 7*, 4100.
123. Strauss, M., et al. (2017). Resonance fluorescence of a site-controlled quantum dot realized by the buried-stressor growth technique. *Applied Physics Letters, 110*, 111101.
124. Jöns, K. D., et al. (2013). Triggered indistinguishable single photons with narrow line widths from site-controlled quantum dots. *Nano Letters, 13*, 126.
125. Dousse, A., et al. (2008). Controlled light-matter coupling for a single quantum dot embedded in a pillar microcavity using far-field optical lithography. *Physical Review Letters, 101*, 267404.
126. Schnauber, P., et al. (2018). Deterministic integration of quantum dots into on-chip multimode interference beamsplitters using in-situ electron beam lithography. *Nano Letters, 18*, 2336.
127. Hallett, D., et al. (2018). Electrical control of nonlinear quantum optics in a nano-photonic waveguide. Optica, 5, 644.
128. Pyayt, A., et al. (2008). Integration of photonic and silver nanowire plasmonic waveguides. *Nature Nanotechnology, 3*, 660.
129. Guo, X., et al. (2009). Direct coupling of plasmonic and photonic nanowires for hybrid nanophotonic components and circuits. *Nano Letters, 9*, 4515.
130. Wei, H., et al. (2011). Cascaded logic gates in nanophotonic plasmon networks. *Nature Communications, 2*, 387.
131. Wei, H., et al. (2012). Nanowire-based plasmonic waveguides and devices for integrated nanophotonic circuits. *Nano Photonics, 1*, 155.
132. Chang, D. E., et al. (2007). A single-photon transistor using nanoscale surface plasmons. *Nature Physics, 3*, 807.
133. Kumar, S., et al. (2013). Efficient coupling of a single diamond color center to propagating plasmonic gap modes. *Nano Letters, 13*, 1221.
134. Li, Q., et al. (2015). Quantum yield of single surface plasmons generated by a quantum dot coupled with a silver nanowire. *Nano Letters, 15*, 8181.
135. Li, Q., et al. (2018). Plasmon-Assisted Selective and Super-Resolving Excitation of Individual Quantum Emitters on a Metal Nanowire. *Nano Letters, 18*, 2009.
136. Song, H. Z., et al. (2005). Site-controlled photoluminescence at telecommunication wavelength from InAs/InP quantum dots. *Applied Physics Letters, 86*, 113118.
137. Dalacu, D., et al. (2010). Deterministic emitter-cavity coupling using a single-site controlled quantum dot. *Physical Review B, 82*, 033381.
138. Dalacu, D., et al. (2012). Ultraclean emission from InAsP quantum dots in defect-free wurtzite InP nanowires. *Nano Letters, 12*, 5919.
139. Versteegh, M., et al. (2014). Observation of strongly entangled photon pairs from a nanowire quantum dot. *Nature Communications, 5*, 6298.
140. Jöns, K. D., et al. (2017). Bright nanoscale source of deterministic entangled photon pairs violating Bell's inequality. *Scientific Reports, 7*, 1700.
141. Huber, T., et al. (2014). Polarization entangled photons from quantum dots embedded in nanowires. *Nano Letters, 14*, 7107.
142. Haffouz, S., et al. (2018). Bright single InAsP quantum dots at telecom wavelengths in position-controlled InP nanowires: the role of the photonic waveguide. *Nano Letters, 18*, 3047.
143. Unold, T., et al. (2005). Optical control of excitons in a pair of quantum dots coupled by the dipole-dipole interaction. *Physical Review Letters, 94*, 137404.

144. Kasprzak, J., et al. (2011). Coherent coupling between distant excitons revealed by two-dimensional nonlinear hyperspectral imaging. *Nature Photonics, 5*, 123.
145. Temnov, V. V., et al. (2005). Superradiance and subradiance in an inhomogeneously broadened ensemble of two-level systems coupled to a low-Q cavity. *Physical Review Letters, 95*, 243602.
146. Betzig, E., et al. (1993). Single molecules observed by near-field scanning optical microscopy. *Science, 262*, 1422.
147. de Assis, P.-L., et al. (2017). Single molecules observed by near-field scanning optical microscopy. *Physical Review Letters, 118*, 117401.
148. Song, H. Z., et al. (2006). Scanning tunneling microscope study of capped quantum dots. *Applied Physics Letters, 85*, 2355.
149. Liu, J., et al. (2018). Single self-assembled quantum dots in photonic nanostructures: the role of nanofabrication. *Physical Review Applied, 9*, 064019.
150. Stock, E., et al. (2013). On-chip quantum optics with quantum dot microcavities. *Advanced Materials, 25*, 707.
151. Munnelly, P., et al. (2015). A pulsed nonclassical light source driven by an integrated electrically triggered quantum dot microlaser. *IEEE Journal of Selected Topics in Quantum Electronics, 21*, 681689.
152. Munnelly, P., et al. (2017). An electrically tunable single-photon source triggered by a monolithically integrated quantum dot microlaser. *ACS Photonics, 4*, 790.

Chapter 4
Gate-Defined Quantum Dots: Fundamentals and Applications

Guang-Wei Deng, Nan Xu, and Wei-Jie Li

Abstract Gate-defined quantum dots (GDQD) have gained rapid developments due to the progresses of modern nanofabrication technologies in recent years. Because of their high tunability, COMS compatibility, and long coherence time, gate-defined quantum dots are considered as one of the most likely candidates for quantum computation. This chapter reviews the fundamental concepts, recent developments, and applications of gate-defined quantum dots.

4.1 Introduction

Gate-defined quantum dots are man-made nanoscale devices in solid-state materials, typically containing a comparable number of electrons [1, 2]. Quantum dots can be made from superconducting [3] and semiconducting materials [4], together with their hybrid structures [5, 6]. The number of free electrons in a quantum dot can range from zero to several thousands, and these numbers can be tuned by gate voltages [1]. Both charge and spin states can be encoded in these free electrons, and these states can be measured and tuned all by splitting gates. Many efforts have been made to utilize such quantum dots for various of applications, including quantum computation [7, 8], quantum simulation [9, 10], and many-body physics [11, 12]. These applications require long coherence time, scalability, and high fidelity. Coherence time of GDQD spin qubits is considered to be affected by hyperfine interactions, and the dephasing time of typical GaAs qubits is below 1 μs [13]. Purified silicon spin qubits have coherence times up to 100 μs level [14]. Silicon-based donor qubits have shown a dephasing time as long as 600 ms, and the intrinsic dephasing time after dynamical decoupling

G.-W. Deng (✉) · N. Xu · W.-J. Li
Institute of Fundamental and Frontier Sciences, University of Electronic Science and Technology of China, Chengdu, China
e-mail: gwdeng@uestc.edu.cn

P. Yu, Z. M. Wang (eds.), *Quantum Dot Optoelectronic Devices*,
Lecture Notes in Nanoscale Science and Technology 27,
https://doi.org/10.1007/978-3-030-35813-6_4

manipulations can reach 30 s [15], showing an attractive prospects in quantum computation and quantum memory applications. For the scalability of GDQD, recent experimental efforts have shown that semiconductor quantum dots can be scaled by coupling with superconducting microwave resonators [16–21] and nano-electromechanical resonators [22–29]. Different from that of optical, ion-trap, or superconducting qubits, a high gate fidelity in semiconducting GDQD qubits is more difficult to reach. Yoneda et al. has reported a fidelity higher than 99.9% in silicon GDQD qubits [30], Muhonen et al. has achieved a fidelity as high as 99.9% in GDQD donor qubits [15]. In the following of this chapter, we introduce GDQD in three aspects: fundamental concepts, quantum bit applications, and future directions.

4.2 Fundamental Concepts of Gate-Defined Quantum Dots

4.2.1 Concepts of Quantum Dots

Quantum dots are artificial charge containers that can contain anything from a single electron to thousands of electrons. They range in size from a few nanometers to several microns, and their shape and electron interaction can be precisely controlled by advanced nanofabrication techniques. The physical properties of quantum dots have much in common with the quantum behavior that occurs naturally in atoms [1].

As one of the key quantum mechanical phenomena, electron tunneling of gate-defined quantum dots can be controlled by capacitance coupled splitting gates. Most properties of quantum dots are studied by electron tunneling. Typically electrons in a quantum dot can tunnel through a lateral geometry and a vertical geometry (see Fig. 4.1). The geometry of quantum dots is very similar to semiconductor transistors, where the source, dot, and drain can be thought of as an electronic ocean that houses many electrons. Electrons can tunnel between source, dot, and drain, under suitable source-drain bias voltage and electrode voltage conditions. There are tunnel barriers between the dot and source/drain. And when the barrier becomes large, the tunneling rates will be weak, so that the electrons in the dot can change one by one.

Figure 4.1a shows the vertical geometric structure, originally designed by Seigo Tarucha and colleagues in Japan and Leo Kouwenhoven and co-workers at Delft [31]. The dots are cylindrical, hundreds of nanometers in diameter, and 10 nm thick and can hold up to 100 electrons. The source and drain electrodes are electrically conductive doped semiconductor layers separated from the quantum dots by two non-conductive barrier layers. By applying a negative voltage to the metal side gate around the pillar, the diameter of the quantum dot is gradually squeezed, reducing from 500 nm to 0, leading to the reduction of the number of electrons one at a time. Figure 4.1b shows a lateral geometry of quantum dots, allowing the current through the device to be measured in response to a bias voltage V_{SD} and a gate voltage V_G.

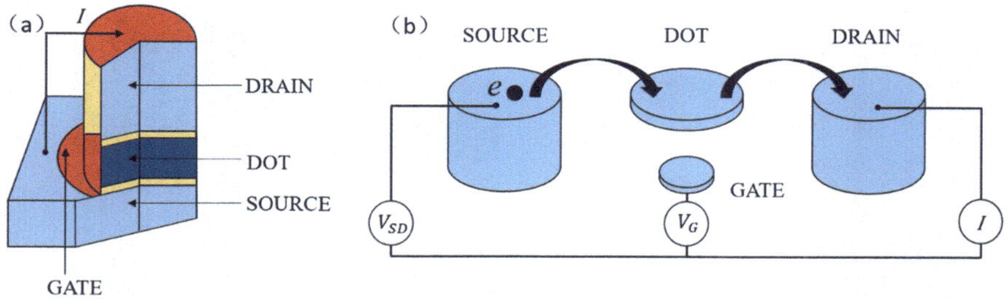

Fig. 4.1 Schematic picture of a quantum dot (**a**) in a vertical geometry and (**b**) in a lateral geometry. The quantum dot is connected to source and drain reservoirs via tunnel barriers. (Image (**a**) is reproduced from Kouwenhoven et al. [1], (**b**) is reproduced from Hanson et al. [32]) Copyright (2007) American Physical Society

When the size of the dot island is small enough, the capacitance C between it and the surrounding environment can be as small as $10^{-16}F$. Under such conditions, whenever a single electron tunnels from the outside into the conductive dot island (sometimes referred to as an isolated Coulomb island), it adds an additional charge energy to Coulomb Island e^2/C (e is an electronic charge). It is much larger than the thermal motion energy k_BT at low temperature (k_B is the Boltzmann constant and T is the absolute temperature). This will lead to a very interesting phenomenon: once an electron tunnels into the dot island, it will prevent the subsequent second electron from reentering the same island, because such a process will lead to an increase in the total energy of the system, which is not allowed to happen. This is the Coulomb blockade phenomenon. Obviously, it is only possible to enter another electron outside the island after waiting for an electron to leave Coulomb island. Thus, using the Coulomb blockade effect, it is possible to tunnel electrons into and out of Coulomb island one by one to achieve a single electron tunneling process. The concept of Coulomb blockade can be traced back to 1951 by Gorter et al. [33], and until 1987, the Coulomb blockade oscillation of conductance was observed directly on a system consisting of two micro-metal tunneling junctions [34].

Gate-defined quantum dots can be measured by all-electric technology which mainly includes two types: one is to measure the current of electrons passing through quantum dots (direct transport measurement) (see Fig. 4.2a), and the other is to use an electrometer usually called a quantum point contact (QPC) (see Fig. 4.2b). The latter is a kind of charge sensing method and can perform noninvasive detection without electrons passing through the quantum dot channel. It is known that spectrum measurements of quantum dots require coupling with macroscopic instruments, while any measurement causes backactions and brings decoherence [37]. In particular, charge sensing measurement techniques are often used to detect the excitatory spectra of a small number of electron quantum dots coupled to a reservoir [38] and used to measure the charge qubit coherence times [39].

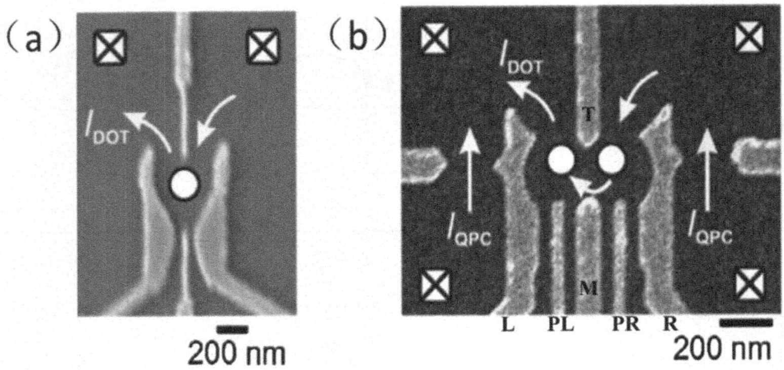

Fig. 4.2 Scanning electron micrographs of (**a**) a few-electron single-dot device and (**b**) a double dot device, showing the gate electrodes (light gray) on top of the surface (dark gray). White dots indicate the location of the quantum dots. Ohmic contacts are shown in the corners. White arrows outline the path of current I_{DOT} from one reservoir through the dots to the other reservoir. For the device in (**b**), the two gates on the side can be used to create two quantum point contacts, which can serve as electrometers by passing a current I_{QPC}. Note that this device can also be used to define a single dot. (Image (a) is reprinted with permission from [35]. Copyright (2000) American Physical Society and (b) is reprinted with permission from [36]. Copyright (2003) by the American Physical Society)

4.2.2 Material Systems for Gated-Defined Quantum Dots

In this part, we focus on lateral geometry and introduce different material systems, such as GaAs, silicon, carbon nanotube, InAs nanowire, and 2D materials.

4.2.2.1 Quantum Dots Based on GaAs

Lateral gated quantum dots are usually made of heterostructures of GaAs and AlGaAs (see Fig. 4.2). Free electrons are captured by doping algae layers with silicon at the two-dimensional interface. These materials gather at the two-dimensional interface to form a two-dimensional electron gas (2DEG), which can only move along the interface. A pattern is printed on the gate using electron beam lithography, and typically the chromium and gold layers are evaporated onto the gate surface to form a pattern [35]. Negative voltage is applied on the surface gate to limit electrons, and 2DEG with high mobility and low electron density will exhaust the electric field generated. Hence, electrons can pass in and out of the dot through two leads, but a depletion region of 100 nm around the gates prevents them from passing through the other gaps. Figure 4.2a, b describes the typical single quantum dot and double quantum dot scanning electron micrographs, respectively. The light gray gate electrode is placed on the dark gray surface. The white dots indicate the position of the quantum dots. The arrows indicate the path of the electrons in and out of the quantum dots, and there is an ohmic contact at each corner. For a double quantum dot device, two side gates can be used to generate two quantum point contacts (QPCs) that can be used as electrometers by measuring the transport current I_{QPC}.

Figure 4.2a shows a typical lateral single quantum dot device. In 2000, Ciorga et al. measured single quantum dot by using two specially designed gates [35]. One kind of gate can largely surround the quantum dot, and the voltage on the gate mainly determines the point potential. The other type of gate is thin and can be used to define the tunnel barrier. When the two types of gate are combined to measure electron transport through the quantum dots, it is possible to simultaneously change the point potential over a large range and maintain a sufficiently high tunnel rate.

As shown in Fig. 4.2b, Elzerman et al. demonstrated in 2003 how to control the number of electrons in double quantum dots while maintaining a tunable tunnel coupling to the reservoir [36]. The small circuit consists of a double quantum dot and two QPCs. The double quantum dot is defined by applying negative voltages to the six gates in the middle of the structure. Gate T in combination with the left (right) gate, L(R), defines the tunnel barrier from the left (right) dot to drain 1 (source 2). Gate T in combination with the middle, bottom gate, M, defines the tunnel barrier between the two dots. The narrow "plunger" gate, PL (PR), on the left (right) is used to change the electrostatic potential of the left (right) dot. A coaxial cable can be connected to one of the plunger gates (L or R), so that high-frequency signals can be applied to manipulate the electron states.

4.2.2.2 Quantum Dots Based on Silicon

Materials such as GaAs are limited by coherence time due to net nuclear spins [40]. As an emerging spin-based quantum computing material system, silicon has zero natural nuclear spins, which greatly reduces the dephasing of electron spins [41–43]. At the same time, silicon has a weak spin-orbit coupling, which can further improve the spin coherence time. Early work has focused on the partial oxidation of the underlying aluminum barrier gate on silicon, producing a very thin localized aluminum oxide layer [44]. As shown in Fig. 4.3a, high resistivity near-intrinsic silicon wafers were used in a typical device [44], with phosphorus-diffused regions as ohmic contacts. A 5-nm-thick SiO_2 gate oxide was thermally grown on the surface. The lower aluminum barrier gates were fabricated using electron-beam lithography (EBL), thermal evaporation, and lift off. These gates were then partially oxidized using a plasma oxidation technique. The aluminum gates were exposed to a low-pressure oxygen plasma (0.15 mbar) for 3 min at a temperature of $\sim$150 °C. This causes the aluminum to form a layer of aluminum oxide (a few nanometers thick) at the surface. The upper aluminum gate was aligned to the lower gates during a second EBL stage and again thermally evaporated and lifted off. The final step of processing was a low-temperature forming gas anneal. Electrical transport measurements were typically performed at a base temperature $\sim$50 mK in a dilution refrigerator, with an electron temperature of $\sim$100 mK. Standard low-frequency lock-in techniques were used to measure the two-terminal conductance and differential conduc-

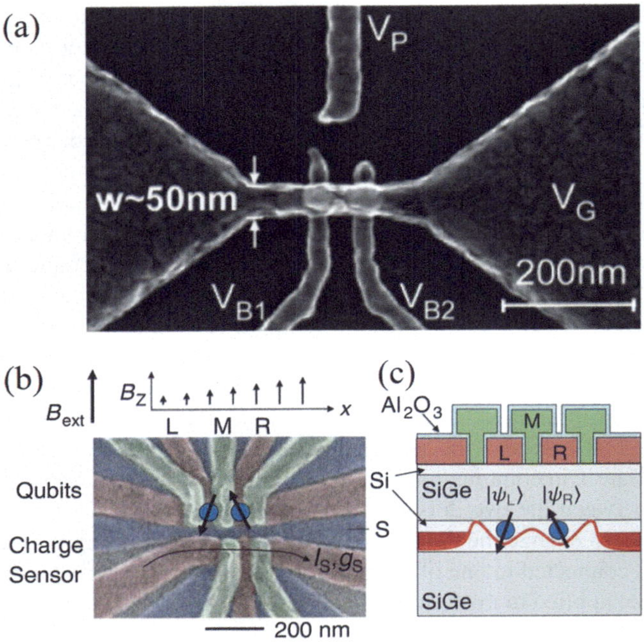

Fig. 4.3 Devices for silicon quantum dots. (**a**) Scanning electron microscopy (SEM) micrograph of a typical silicon single quantum dot made by a MOSFET architecture. (**b**) SEM image for a typical Si/SiGe double quantum dot. A Co micromagnet is used to generate a slanting Zeeman field for magnetic control. (**c**) Cross section of the double quantum dot, where the electron gas comes from the heterostructure between Si and SiGe. (Image (a) reprinted with permission from [44]. Copyright (2007) American Chemical Society. (b) and (c) are reprinted with permission from [45]. Copyright (2018) American Association for the Advancement of Science)

tance through the dot. Measurements were performed in zero applied magnetic field. To measure the turn-on characteristic, a voltage was applied to all three gates simultaneously, resulting in an approximately continuous field along the length of the nanowire. At 4 K, because of the high resistivity of the wafer, the source-drain conductance is zero until the applied gate voltage is equal to the threshold value. As the applied gate voltage is increased above the threshold value, the source (drain) current increases smoothly toward a maximum conductance. At millikelvin temperatures, conductance fluctuations occur in the MOSFET turn-on characteristic. The use of this natural alumina-based technology in silicon devices can electrostatically generate small, well-defined points that result in quantum confinements, as evidenced by the excited-state polarization spectra. The adjustable barrier of these quantum dots enables further study of many of the fundamental properties of silicon quantum dots, such as the Kondo effect and spin and valley degradation. A similar structure can form double quantum dots using Si/SiGe heterostructures; see Fig. 4.3b, c.

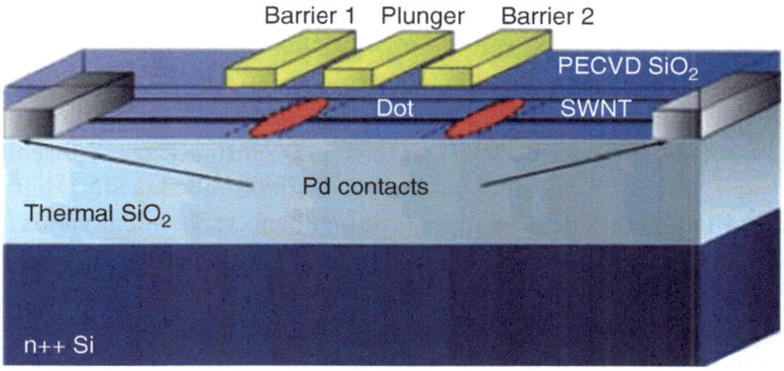

Fig. 4.4 Schematic of a gate-defined carbon nanotube quantum dot showing vertically integrated geometry and ohmic contacts. Pd provides high-conductance contacts at the metal-nanotube interface that do not form tunnel barriers at low temperatures. (Reprinted with permission from [48]. Copyright (2005) American Chemical Society)

4.2.2.3 Quantum Dots Based on Carbon Nanotubes

Carbon nanotubes have many attractive physical properties and various applications [46, 47]. Isolated quantum dots formed in carbon nanotubes can be defined by tunneling barriers along the metal-nanotube interface or by internal or induced defects along the nanotube. Figure 4.4 shows the vertically integrated geometry and ohmic contact of typical carbon nanotube quantum dots [48]. In an early work [48], Pd provides a high-conductivity contact on the metal nanotube interface and does not form a tunnel barrier at low temperatures. The formation of single quantum dot uses a three-gate configuration that locally depletes the carriers under outer gates, acting as tunnel barriers (barrier 1 and barrier 2) and defining points. The notional "plunger" removes the chemical potential from the point and from the segment. The gate response of tube conductance of a single dot device using PECVD SiO_2 exhibits p-type field-effect behavior. That is, when any top gate bias becomes positive enough, the tube runs out. The work does not rely on inherent defects in metal-nanotube interfaces or tunnel barriers but instead uses electrostatic gates to define quantum dots. The controlled gate at any point in the pipeline greatly enhances the function of nanotube devices and can be applied to various potential applications of quantum coherent logic devices.

4.2.2.4 Quantum Dots Based on Nanowires and 2D Materials

In addition to the nanotube materials mentioned above, nanowires in the systems of various materials also realize the exploration of quantum dots. Chemically synthesized semiconductor nanowires are often used as base materials for nanoscale devices and sensors in a bottom-up manner. The semiconductor material InAs has a small effective mass and a strong effect of spin orbit coupling, which can be useful for studying the quantum spin phenomenon. Local gate acting on uniform InAs nanowires provides a more flexible method to realize the quantum dot structure. Figure 4.5a shows the SEM micrograph of a typical device [49]. First, electron beam lithography (EBL) and metal evaporative deposition of gold electrode grids (vertical lines with both ends tilted to the left) can form the splitting gates. Subsequently, the grid is covered with a 25-nm-thick amorphous silicon nitride (SiN) dielectric film [light blue in Fig. 4.5a] deposited in a plasma-enhanced chemical vapor deposition process. By using reactive ion etching, the tilted ends of the electrodes are then selectively freed of the SiN insulating layer. InAs nanowires are deposited on the sample, and after individual wires are located, Ni/Au contacts (shown schematically by the yellow lines) are defined in a second EBL step. At the same time, five local gate electrodes [lg1 to lg5, red in Fig. 4.5a] are fabricated below the nanowire together with the remaining electrodes labeled "outer" [green in Fig. 4.5a]. Finally, these "outer" electrodes are connected to the Si substrate. Extensive efforts are made in investigating double quantum dots defined by electrostatic gates in other one-dimensional systems, such as Ge hut wire [50]; see Fig. 4.5c, d.

Graphene is a promising candidate for quantum dot applications due to the expected long spin coherence time and flexibility in device designs because of its two-dimensional nature. Graphene double dot devices have been realized by etching graphene into two small islands separated by a narrow constriction, where the inter-dot coupling was shown to be tunable by a side gate [20, 52–55]. Very recently, new kinds of 2D materials have attracted much attention for their applications in quantum dot study. MoS_2 quantum dots have been demonstrated to be highly tunable [51], showing future applications for quantum computing.

4.2.3 Electron States of GDQD

4.2.3.1 Charge and Spin States in GDQD

Gate-defined quantum dots are usually defined by nano-electrodes on semiconductor materials using nano-processing techniques. The two-dimensional electron gas (2DEG) formed at the semiconductor heterojunction can be confined in the potential well constructed by the voltage applied to these electrodes. According to the principles of quantum mechanics, quantized discrete energy levels are formed in the potential well, and electrons occupy different discrete energy levels. In semiconductor quantum dot systems, electron qubits are generally classified into spin-type states and charge-type states.

Charge-type qubits are encoded in different positional states of a single electron in a double quantum dot. A single electron can be placed on the left quantum dot (|L>) or the right quantum dot (|R>), or specially located at the superposition of |L> and |R>. Unlike spin-type qubits, the flipping speed of charge qubits between |L> and |R> is determined by tunneling coupling between two quantum dots, and the flipping frequency can be very fast (~5 GHz). However, because of the strong cou-

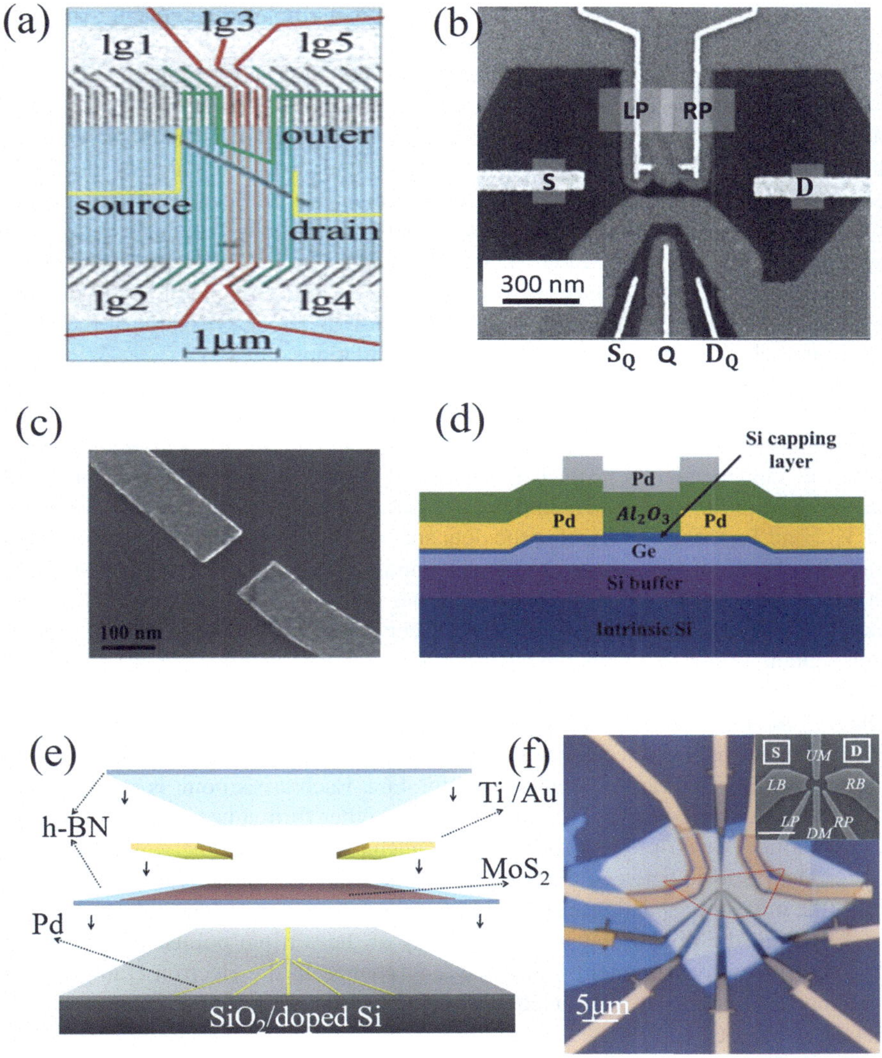

Fig. 4.5 Quantum dots based on other materials. (**a**) SEM micrograph of the InAs nanowire on the gold grid. The wire and its source and drain electrodes are isolated from the local gates (lg1-lg5 and "outer") by a 25-nm-thick silicon nitride film (bright blue area). (**b**) SEM micrograph of a typical graphene double quantum dot. (**c**) SEM micrograph of a Ge hut wire quantum dot. (**d**) Longitudinal section of the Ge hut wire sample in (**c**). (**e–f**) Double quantum dot structure made from MoS_2. (Image (**a**) is Reprinted with permission from Ref. [49] Copyright (2005) American Chemical Society, (**b**) is Reprinted with permission from Ref. [20] Copyright (2015) American Physical Society, (**c–d**) are Reprinted with permission from Ref. [50] with the permission of AIP Publishing, (**e–f**) are Reprinted with permission from Ref. [51] Copyright (2017)American Association for the Advancement of Science)

pling with environmental noise, the charge-type quantum bit coherence time is generally shorter (~2 ns). Completing a series of high-fidelity gate manipulations in such a short coherent time is a big challenge for GDQD charge qubits.

Spin-type qubits are defined by the spin characteristics (generally spin up and down) of electrons in GDQDs. For example, by applying an in-plane magnetic field, the electron spin undergoes a Zeeman splitting, and two-level states of the single electron can be formed by the spin directions (up or down). Because spin qubits are weakly coupled to ambient noise, they usually have a long coherence time (~μs order). To control the qubit and realize the electron spin flip, it is generally required to apply an oscillating electric field or a magnetic field to generate electron spin resonance. However, spin flip is usually very slow (~100 ns) because of the weak coupling, which limits the speed of qubit manipulations.

Here we introduce a simple capacitance theory for electron states in double quantum dots without going into the details of the derivations [32]. Consider two quantum dots (see Fig. 4.6a) labeled 1 and 2, whose electrochemical potentials are controlled independently by the gate voltages $V_{G,1}$ and $V_{G,2}$, respectively. Figure 4.6d shows the equilibrium electron numbers (N_1, N_2) of the quantum dots as a function of $V_{G,1}$ and $V_{G,2}$, for the case that the dots are completely uncoupled. Such a plot is called a charge stability diagram. The lines indicate the values of the gate voltages at which the number of electrons in the ground state changes. Note that the lines are exactly horizontal and vertical, since the electrochemical potential in either dot is independent of the charge on the other dot and each gate voltage only affects one of the dots. When the dots are capacitively coupled, addition of an electron on one dot changes the electrostatic energy of the other dot. Also, the gate voltages $V_{G,1}$ and $V_{G,2}$ generally have a direct capacitive coupling to quantum dot 2 [1]. The resulting charge stability diagram is sketched in Fig. 4.6e. Each cross point is splitting into two so-called triple points. The triple points together form a hexagonal or "honeycomb" lattice. At a triple point, three different charge states are energetically degenerate. The distance between the triple points is set by the capacitance between the dots (the inter-dot capacitance) C_m. At low source-drain bias voltage, electron transport through the double dot is possible only at these triple points. In contrast, a charge sensing measurement will detect any change in the electron configuration and therefore map out all transitions, including those where an electron moves between the dots.

Natural atoms are not accessible from the perspective of operability, while quantum dots are larger and more controllable. If we apply 1 T magnetic field to a quantum dot, the effect is equivalent to applying 10^6 T to a natural atom. However, such high magnetic field strength cannot be achieved experimentally by the state-of-the-art technology. Quantum dots in high magnetic fields can even realize the study of quantum Hall effect. For instance, Ray Ashoori from Massachusetts Institute of Technology in the United States reviewed how electronic states evolve when a magnetic field increases from zero to a quantum Hall state [56]. In 1991 Paul McEuen showed that simple models of constant Coulomb interactions no longer apply to high magnetic fields [57].

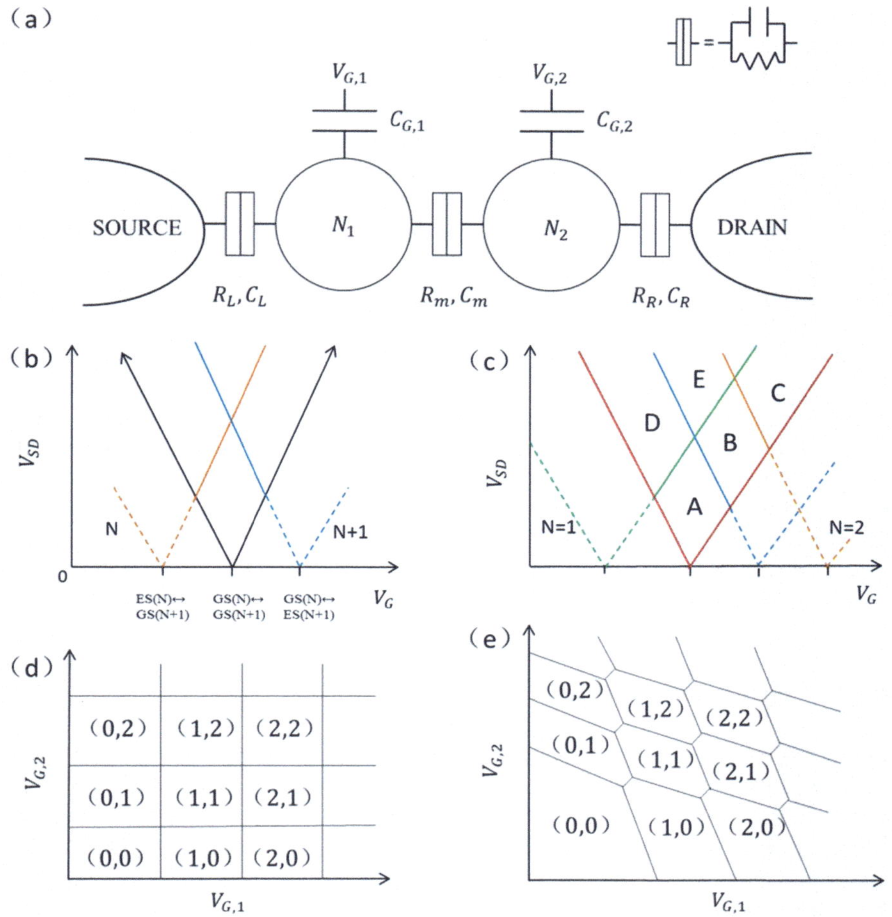

Fig. 4.6 Charge states of a typical double quantum dot. (**a**) Equivalent circuit model of a double quantum dot. (**b**) Schematic plot of the differential conductance dI_{DOT}/dV_{SD} as a function of V_{SD} and V_G. (**c**) Energetically allowed 1↔2 electron transitions as a function of V_{SD} and V_G. Charge stability diagrams for (**d**) uncoupled and (**e**) coupled double dots. The lines indicate the gate voltage values at which the electron number changes. All images are reprinted with permission from [32]. Copyright (2007) American Physical Society

Controlling the spin of a single electron and measuring the spin of a single electron are very difficult compared to charge control [32]. The study of electron spins in quantum dots is mainly about mesoscopic physics, the development of components for spins, and the construction of modules for quantum information processing. In the field of quantum information, the bound electron spin has the potential of long coherence time and is considered to be a potential form of quantum bit. However, it has been found in experiments that the detrimental effect of hyperfine coupling on the spin of the lattice core severely limits the phase coherence of the electron spin. Therefore, it is important to understand the dynamics of electron spin-nuclear coupling and to develop corresponding quantum control techniques to mitigate this coupling. A single electronic spin system is the simplest spin system. The electron has only two spin directions: spin up or down. One can observe a significant splitting of the orbital ground and the first excited state, which increases with increasing magnetic field. And a spin-up electron is added to the empty dot to form the one-electron ground state, as expected. The ground state of a two-electron dot in zero magnetic field is always a spin singlet (total spin quantum number $S = 0$), formed by the two electrons occupying the lowest orbital with their spins antiparallel: $|S\rangle=(|\uparrow\downarrow\rangle-|\downarrow\uparrow\rangle)/\sqrt{2}$. The first excited states are the spin triplets ($S = 1$), where the anti-symmetry of the total two-electron wave function requires one electron to occupy a higher orbital. Both the anti-symmetry of the orbital part of the wave function and the occupation of different orbitals reduce the Coulomb energy of the triplet states with respect to the singlet with two electrons in the same orbital. We include this change in Coulomb energy by the energy term E_k. The three triplet states are degenerate at zero magnetic field but acquire different Zeeman energy shifts E_z in finite magnetic fields. The lines in Fig. 4.6c corresponding to $\uparrow\leftrightarrow$S outline the region of transport. Indeed, a spin-down electron is added to the one electron spin-up ground state to form the two-electron singlet ground state.

4.2.3.2 Hybrid States of GDQD

In GDQD-based qubit systems, charge qubit can operate at high speed but short coherence time, while spin qubits have long coherence time but slow operating speed. In recent years, a novel qubit scheme of hybrid qubit encoding is generally adopted, which combines the characteristics of fast charge qubit manipulation and long spin coherence time. In 2012, a research group from the University of Wisconsin proposed the use of combining quantum dot spin and charge states to prepare hybrid qubits [58]. In this scheme, there are one or two electrons in each quantum dot. For example, in the schematic diagram of hybrid qubit energy levels, there is one electron in the quantum dot on the left and two electrons in the quantum dot on the right; see Fig. 4.7. The two states of the logical bit are, respectively,

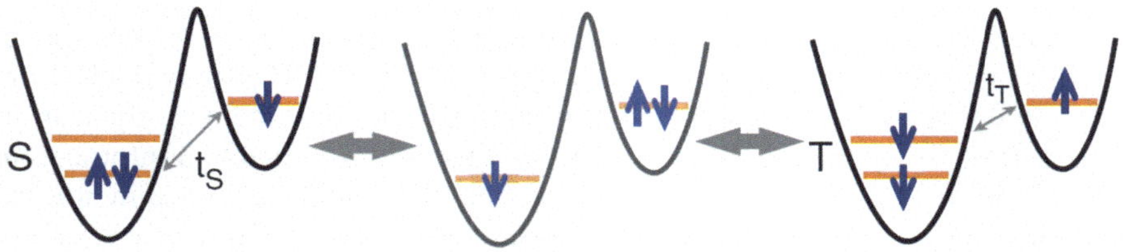

Fig. 4.7 Hybrid qubit level occupation diagram. (Reprinted with permission from [58]. Copyright (2012) American Physical Society)

$|0\rangle = |\downarrow\rangle|S\rangle$ and $|1\rangle = \sqrt{1/3}\,|\downarrow\rangle|T_0\rangle + \sqrt{2/3}\,|\uparrow\rangle|T_-\rangle$, where $|\downarrow\rangle$ and $|\uparrow\rangle$, respectively, represent the spins of electrons in the left quantum dot downward and upward and $|S\rangle$ and $|T\rangle$ are the spin singlet and triplet states of the right quantum dot, respectively. In this coding mode, the spin states of the different electrons of $|0\rangle$ and $|1\rangle$ ensure a longer relaxation time, and the parallel level structure of $|0\rangle$ and $|1\rangle$ effectively suppresses the charge noise so that the coherence time is much longer than typical charge qubit. On the other hand, unlike the traditional spin coding, the total spin state of the dual quantum dot system does not change, so the rapid evolution between quantum states can be achieved by gate electrode manipulation. The qubit operation speed can reach the order of 5 GHz [59], which is much faster than traditional spin qubits. At the same time, considering the scalability, the Wisconsin group also proposed a dual-hybrid qubit logic gate control scheme using capacitive coupling [58].

Triplet-singlet relaxation time T_1 of Si/SiGe double quantum dot was reported to be around 100 ms in 2012 [60]. It is expected that the coherence time T_2 can be up to the millisecond level. In 2014, the Wisconsin group reported an experiment about quantum control and process tomography of a semiconductor quantum dot hybrid qubit [61]. The group successfully characterized the hybrid qubit level structure and realized the coherent oscillation between different energy levels, which laid a foundation for the preparation of hybrid qubits. Also in 2014, they realized high-frequency pulse gate manipulation in the same hybrid qubit [59]. The fidelity of the 2π and z-axis (phase evolution) evolution of 2π around the Bloch ball x-axis (bit flip) reaches 85% and 94%, respectively, and the coherence time of the bit at the parallel energy level is about 10 ns. Figure 4.8a is a sample diagram used in their experiments, and Fig. 4.8c is a hybrid qubit level structure diagram and a pulse sequence diagram. Figure 4.8b, d shows coherent control of qubit states on a Bloch sphere. A schematic diagram of qubit flip and phase evolution is implemented. It should be noted that the flip of the hybrid qubit around the x-axis is the same as conventional charge qubit. The coherence time T_2^* extracted from experiment is about 2 ns, and the coherence time T_2 of the charge qubit is increased to about 150 ns by using the dynamic decoupling technique [59, 61].

Very recently, hybrid qubit has been realized in a GaAs double quantum dot [62] and GaAs triple quantum dot [63]; see Fig. 4.8e, f. In their double-dot experiments, the authors showed novel results that hybrid qubits can be realized in traditional GaAs quantum dot systems. They also demonstrated that such a hybrid qubit could be realized in multi-electron regime, for example, they showed their results in five-electron case. Larmor precession and Ramsey fringe experiments were performed, and they claimed that the qubit coherence time is significantly improved to the order of $T_2^* \sim 10$ ns level [62]. In their triple-dot experiment, the authors showed technique to tune the operation frequency of a hybrid qubit from 2 to 15 GHz by adjusting the gate voltages [63]. Such a large tunability could be very useful for future qubit control and for high-speed qubit operations.

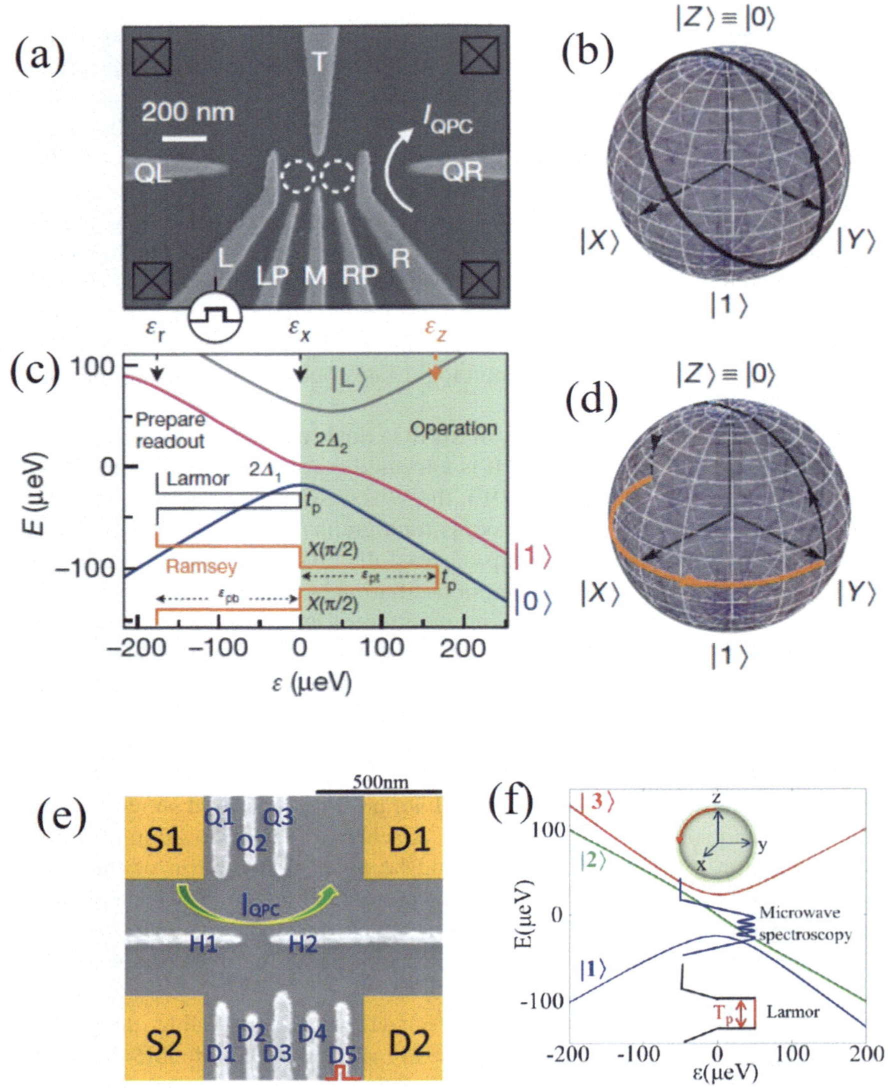

Fig. 4.8 Schematic diagrams of sample structures and quantum-state evolutions of hybrid qubit systems. (**a**) Scanning electron microscope picture of a typical Si/SiGe hybrid qubit. The pulse required for manipulation is applied to the electrode L. (**b**) Calculated level structure and pulse schematic of the hybrid qubit. (**c**) X rotations on the Bloch sphere correspond to qubit state flipping between |0> and |1>. (**d**) Schematic diagram of phase evolution of the bit on the Bloch sphere around the *Z*-axis under the driving of the Ramsey oscillation pulse shown in (**b**). (**e**) Scanning electron microscope picture of a GaAs hybrid qubit. The pulse required for manipulation is applied to the electrode D5. (**f**) The simulated energy levels for the GaAs hybrid qubit. (Images (**a**)–(**d**) is reprinted by permission from Ref. [61] Copyright (2014) Springer Nature, images (**e**)–(**f**) is reprinted by permission from Ref. [62] Copyright (2016) by the American Physical Society)

4.3 Quantum Bit Applications of Gate-Defined Quantum Dots

In this part, we will introduce some recent achievements of GDQD from aspects of coherence time, fidelity, and scalability. These achievements are important for quantum computing using GDQD.

4.3.1 Coherence Time

Both charge and spin states in a GDQD can be used as qubits [7, 32]. In general, a qubit can be considered as a two-level system with superposition and entanglement properties and can be typically defined using a sphere based on the ground state $|0\rangle$ and excited state $|1\rangle$ (see Fig. 4.9a) which is called Bloch sphere. Quantum states can be defined by the points on the sphere as

$$|\Psi = \cos(\theta/2)|0\rangle + e^{i\varphi}\sin(\theta/2)|1\rangle \tag{7.1}$$

An arbitrary qubit state can be described by a certain point on the surface of the Bloch sphere and determined by the two real numbers θ and φ. In Fig. 4.9b, the two blue lines show energy levels of a charge qubit using GDQD, where the energy space between the ground state $|0\rangle$ and excited state $|1\rangle$ can be tuned by the detuning ε of the two dots.

Coherence time of a qubit contains level relaxation time T_1 and dephasing time T_2. T_1 can be measured by measuring the Rabi oscillation of the two-level systems, while T_2 can be determined by Larmor precession or Ramsey fringe experiments. Because of noise and imperfections of the experimental instruments, the measured dephasing time (usually called T_2^*) will be somehow smaller than exact T_2 [65].

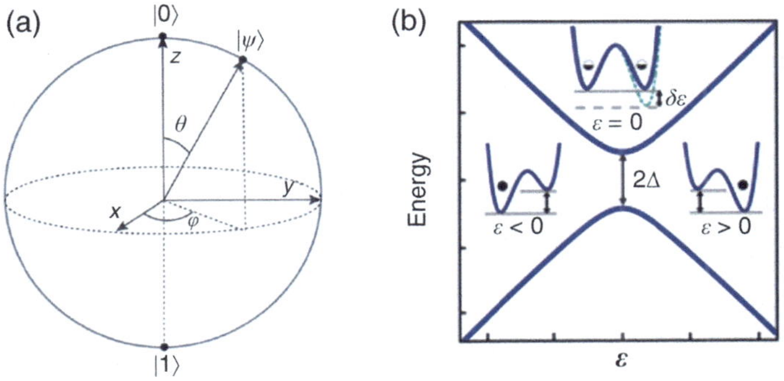

Fig. 4.9 Bloch sphere (**a**) and energy level (**b**) of a charge qubit. (Image (**b**) is reprinted with permission from [64] Copyright (2019) Springer Nature)

The physical meaning of the energy relaxation time is the decay time of the qubit from the excited state $|1\rangle$ to the ground state $|0\rangle$. In order to determine the energy relaxation, Rabi oscillation measurement under the same driving power is required to determine π pulse. By applying a π pulse, qubit state can be prepared to the excited state $|1\rangle$. Then one can read the probability $P(|1\rangle)$ of the excited state ($|1\rangle$) after a waiting time Δt. As the waiting time increases, the probability decays and finally falls back to the ground state $|0\rangle$, and the energy relaxation time T_1 can be extracted by fitting the formula $P(|1\rangle) = P_0 + A\exp(-t/T_1)$. For typical GaAs quantum dot-based charge qubit, the relaxation time T_1 is usually around several tens of nanosecond [39], while T_1 in Si/SiGe quantum dot-based qubit is as long as several tens of microsecond [43].

The phase decoherence time can be characterized by Ramsey fringe experiments [65]. When measuring Ramsey fringe, it is first necessary to measure the Rabi oscillation and determine the $\pi/2$ pulse. Applying a $\pi/2$ pulse with a detuning Δ to the qubit, the qubit state will be driven from the ground state $|0\rangle$to $(|0\rangle - i|1\rangle)/\sqrt{2}$ state. Because of the detuning, the qubit will be rotating along the XZ plane in the Bloch sphere. The final state as a function of the waiting time Δt will contribute to the dephasing time T_2^*.

4.3.2 *Fidelity*

Typically in quantum theory, fidelity is important to measure the distance of quantum states. Generally the fidelity is defined as [66].

$$F(\rho_1,\rho_2) = \left[Tr\left(\sqrt{\rho_1}\rho_2\sqrt{\rho_1}\right)^{1/2} \right]^2, \tag{7.2}$$

where $\rho_1$and$\rho_2$are the state density operators corresponding to input state (initial state) and the output state (end state), respectively. The value of $F(\rho_1,\rho_2)$ is between 0 and 1. When $F = 0$, it means that the information (quantum state) is completely distorted during transmission, indicating that the initial state and the final state are orthogonal to each other. When $F = 1$, it is expressed as ideal information transmission. The processes, that is, the initial state and the final state, are the same.

In GDQD qubit measuring experiments, the fidelity is usually determined by the coherence time and the manipulation time. A rough estimation of a fidelity ~99% means that a qubit can be manipulated by 100 times during its coherence time; also a fidelity above 99.9% means that 1000 operations can be finished during the coherence time.

In recent years, many efforts have been made by different groups to improve the coherence time and control fidelity, using various materials (see Table 4.1). For traditional GaAs spin qubit, T_2^* was reported to be about 66 ns in 2010 [70], while the authors did not mention the fidelity. In 2014, M. D. Shulman et al. reported a coherence time T_2^* ~2 μs with ~98% fidelity in GaAs spin qubit, using a real-time

Table 4.1 Coherence time and fidelity of GDQD spin qubit from different groups using various materials

References	Year	Materials	T_2^*	Fidelity
[30]	2018	Si/SiGe	20 μs	99.9%
[67]	2018	Si-MOS	33 μs	99.91%
[68]	2018	Si-MOS	–	99.86%
[69]	2014	Si/SiGe	1 μs	99%
[14]	2014	Si-MOS	120 μs	99.6%
[15]	2014	P in silicon	270 μs	99.99%
[70]	2010	GaAs	66 ns	–
[71]	2012	Si/SiGe	360 ns	–
[72]	2014	GaAs	2 μs	98%
[73]	2017	GaAs	700 ns	99%

Hamiltonian estimation method [72]. In 2017, J. M. Nichol et al. reported a coherence time T_2^* ~700 ns with ~99% fidelity in GaAs spin qubit and also measured an entangling gate fidelity of ~90% [73]. Silicon-based spin qubits are considered to have longer coherence time and have gained rapid development in recent years. In 2012, first coherent singlet-triplet oscillation experiment in an undoped Si/SiGe was published by B. M. Maune et al. demonstrating a coherence time T_2^* ~360 ns, which is two orders of magnitude over similar results in GaAs-based qubits before 2012 [71]. In 2014, E. Kawakami et al. reported a spin coherence time T_2^* ~1 μs with ~99% fidelity in a Si/SiGe qubit [69], while M. Veldhorst et al. reported a spin coherence time T_2^* ~120 μs with ~99.6% fidelity in a Si-MOS qubit [14]. J. T. Muhonen et al. reported a spin coherence time T_2^* ~270 μs with ~99.99% fidelity in a qubit with P atom doped in Si, and a benchmark coherence time as long as 30 s was achieved using Carr-Purcell-Meiboom-Gill (CPMG) sequence [15]. In 2018, J. Yoneda et al. reported a spin coherence time T_2^* ~20 μs with ~99.9% fidelity in a Si/SiGe qubit [30], K. W. Chan et al. reported a spin coherence time T_2^* ~33 μs with ~99.91% fidelity in a Si-MOS qubit using a noise spectroscopy method [67], and M. A. Fogarty et al. reported an integrated Silicon-MOS platform with ~99.86% fidelity in a Si-MOS qubit [68]. These experimental results demonstrate the superiority of semiconductor quantum dots in terms of coherence time and fidelity and have the potential to be further improved, laying a solid foundation for future semiconductor quantum computing platforms.

4.3.3 Scalability

In 2000, David DiVincenzo proposed his famous criteria to construct a quantum computer [75], where the first one is "a scalable physical system with well-characterized qubit." In a quantum computer architecture, the logic space on a quantum system of N qubits is described by a large group [known as $SU(2^N)$], which is much larger than the comparable group [SU [2]$^{\otimes N}$] of N unentangled bits, and cannot be emulated by N classical oscillators or N classical bits. Ultimately, it is the large Hilbert space of a quantum computer that allows its operations unavailable to classical computers. Declaring a technology "scalable" is a tricky business, because the resources used to define and control a qubit are diverse. They may include space on a microchip, classical microwave electronics, dedicated lasers, cryogenic refrigerators, and so on. Considering the very mature modern semiconductor integrated chip process, things might be easier when semiconducting quantum dot system compared with other qubit systems. Figure 4.10 shows a crossbar network using silicon quantum dot qubits, where more than 1000 qubits could be integrated using the method [74]. This method can be seen as an extension of quantum dots directly through the interaction of neighboring capacitors. So far, conditional rotation of two strongly coupled charge qubits [76] and quantum operations of three qubits [77] have been reported in GaAs systems (see Fig. 4.11). Although quantum process tomography measurements have not been realized in these experiments, the fast control speed and the high tunability show possible way to realize full quantum logic operations in semiconductor systems. Moreover, these results show a possible method for scalable quantum chip using GDQDs.

4.3.3.1 Superconducting Microwave Resonators

Superconducting microwave resonators are widely used for connecting distant qubits in superconducting quantum computing systems. Since A. Wallraff et al. reported the first experiment on strong coupling between superconducting qubits and microwave resonator [78], important progress has been made in this area. Two qubits coupling [79], three qubits entanglement [80], and 20 qubit cat states [81] have been realized in superconducting microwave resonator system. Similarly, gate-defined quantum dot can also be coupled to such superconducting microwave resonators. In 2004, L. Childress et al. proposed that mesoscopic cavity quantum electrodynamics with quantum dots can be realized using superconducting resonators [82]. In 2008 G. P. Guo et al. studied dispersive coupling between superconducting transmission line and quantum dots [83], and Z. R. Lin et al. reported cluster states generation using superconducting resonators and quantum dots [84]. However, these tasks are theoretical studied, and related experimental research is very difficult and slow. It was not until 2012 that a team from ETHs realized the first dispersive coupling of GaAs quantum dot with a transmission line superconducting microwave cavity [16] (see Fig. 4.12a); another team from Princeton University reported first circuit quantum electrodynamics with InAs spin qubit using GDQD [17]. In 2013, G. W. Deng et al. reported similar dispersive coupling between graphene quantum dots and a reflection line resonator [85] by a preprint online, and finally published there results in 2015 [20] (see Fig. 4.12b, c). In 2015, G. W. Deng et al. reported the first experiment on two GDQD qubits coupling using microwave resonator [21], and J. J. Viennot et al. reported coherent coupling of single spin to microwave cavity photons using CNT GDQD [86]. In 2017, X. Mi et al. reported the first strong cou-

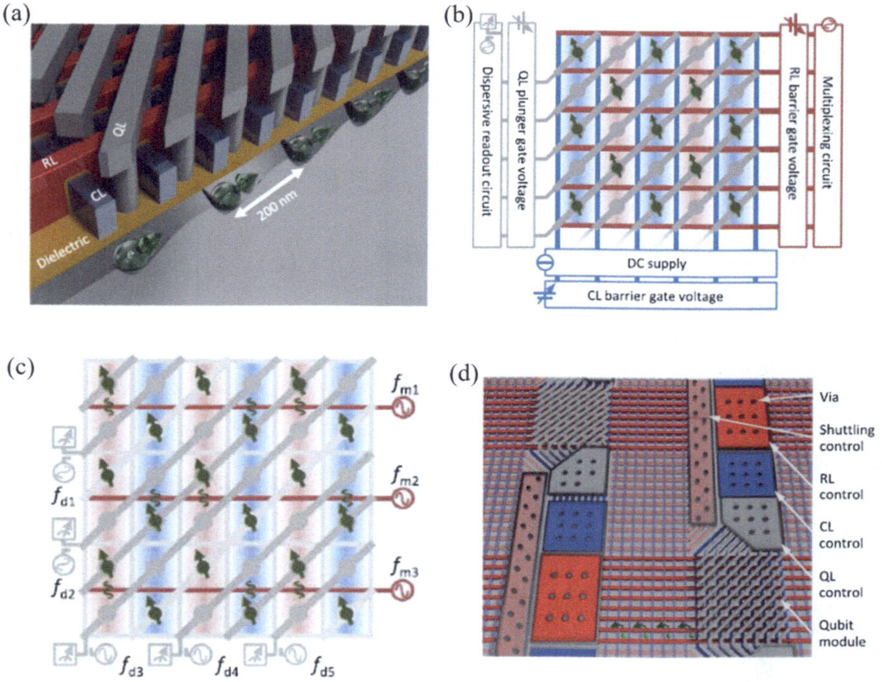

Fig. 4.10 A scalable architecture of silicon quantum dot crossbar array. (**a**) Model structure of the 3D gate design. (**b**) 2D quantum dot array can be realized by adjusting the cross gates in (**a**). (**c**) Electron shuttling and charge readout can be performed in a near-global manner. (**d**) Proposal for connecting different qubit modules. (Images are reprinted from Ref. [74]) Copyright (2018) American Association for the Advancement of Science)

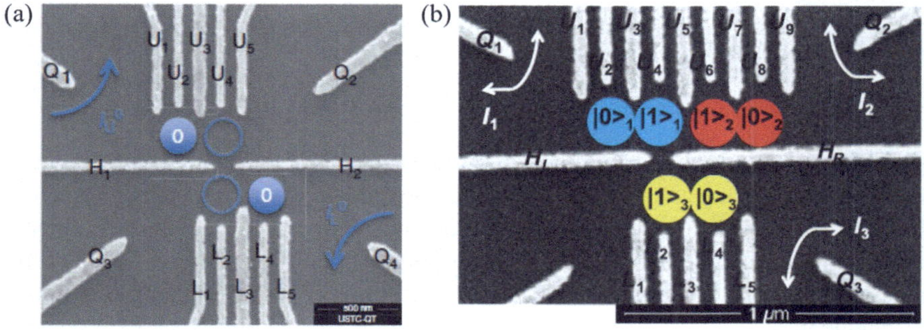

Fig. 4.11 SEM images of two (**a**) and three (**b**) strongly coupled GaAs charge qubits. (Image (**a**) is Reprinted with permission from Ref. [76] Copyright (2015) Springer Nature, (**b**) is Reprinted with permission from Ref. [77] Copyright (2018) American Physical Society)

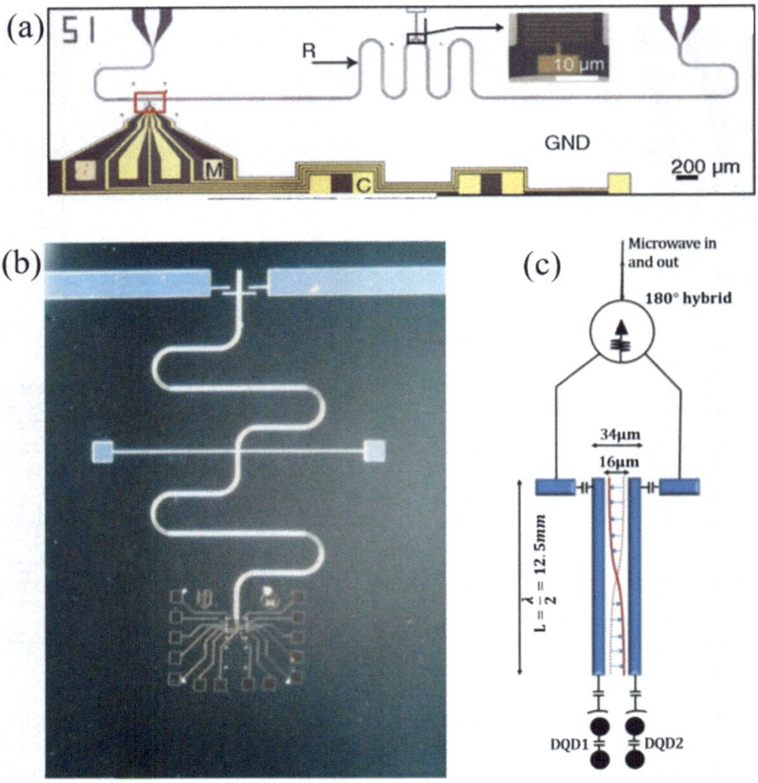

Fig. 4.12 Superconducting microwave resonators used for coupling quantum dots. (**a**) Typical transmission line resonator coupling with GaAs quantum dot. (**b**) Reflection line resonator used for multi-qubit coupling. (**c**) Circuit model for reflection line resonator coupling with quantum dots. (Image (**a**) is Reprinted with permission from Ref. [16] Copyright (2012) American Physical Society, (**b**) and (**c**) are reprinted with permission from Ref. [20] Copyright (2015) American Physical Society)

pling of a GDQD Si charge qubit to a microwave resonator [87], and A. Stockklauser et al. reported strong coupling of a GaAs charge qubit to a high impedance resonator [88]. In 2018, X. Mi et al. reported coherent spin-photon interface in silicon [89], N. Samkharadze et al. reported strong spin-photon coupling in silicon GDQD [90], and A. J. Landig et al. reported coherent spin-photon coupling using a resonant exchange qubit in GaAs GDQD system [91]. So far, many kinds of GDQD systems have been coupled to superconducting resonators, including GaAs [16, 19, 88, 91–93], InAs nanowire [17, 94], CNT [18, 86, 95], graphene [20, 21, 55], InSb nanowire [96], silicon [89, 90], and germanium hut wire [97] (see Fig. 4.13).

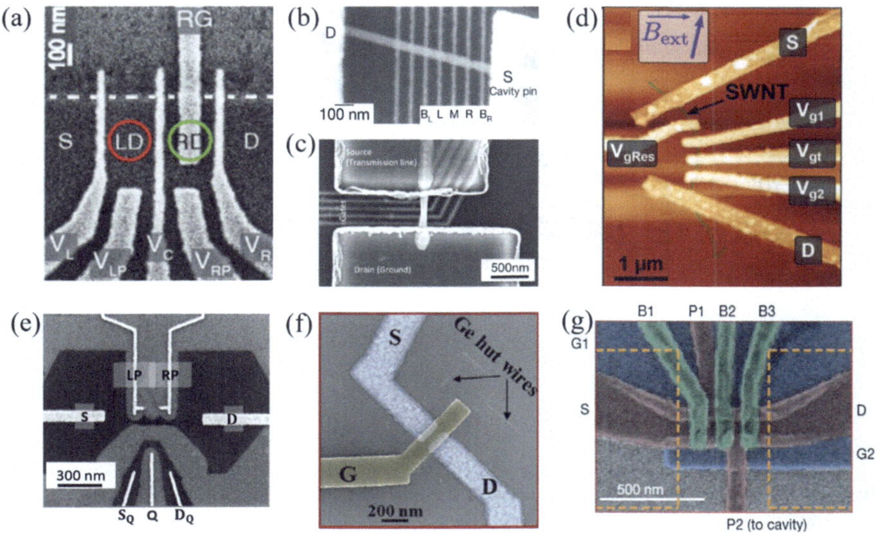

Fig. 4.13 Various GDQD systems coupled to microwave resonators. (**a**) GaAs double quantum dot. (**b**) InAs nanowire double quantum dot. (**c**) InSb nanowire double quantum dot. (**d**) Single-walled carbon nanotube double quantum dot. (**e**) Graphene double quantum dot. (**f**) Germanium hut wire single quantum dot. (**g**) Silicon double quantum dot. (Image (**a**) is Reprinted with permission from Ref. [16] Copyright (2012) American Physical Society, (**b**) is Reprinted with permission from Ref. [17] Copyright (2012) Springer Nature, (**c**) is Reprinted with permission from Ref. [96] with the permission of AIP Publishing, (**d**) is Reprinted with permission from Ref. [86] Copyright (2015) American Association for the Advancement of Science, (**e**) is Reprinted with permission from Ref. [20] Copyright (2015) American Physical Society, (**f**) is Reprinted with permission from Ref. [97] Copyright (2018) American Chemical Society, (**g**) is Reprinted with permission from Ref. [89] Copyright (2018) Springer Nature)

4.3.3.2 Nano-/Micromechanical Resonators

Scalable quantum computing methods based on artificial quantum systems can also be extensively realized by the nanomechanical resonator spin system [98, 99]. Spin has a longer decoherence time, and nanomechanical resonator can be manufactured on a large scale. The quantum motion of the nanomechanical resonator can strongly interact with the spin and induce strong coupling between the spins. As shown in Fig. 4.14, two switchable nanomechanical resonators are shown, and coupling strengths can be tuned by adjusting voltage gates between them.

Strong coupling between quantum dots and nanomechanical resonator has been reported in 2009 by two different groups [22, 23]. Charge dynamics of a quantum dot can be probed by a nanomechanical resonator [100, 101]. In 2014, A. Beny et al. reported real-space tailoring of the electron-phonon coupling in ultraclean nanotube mechanical resonators [24]. In 2015, P. Weber et al. reported switchable coupling of vibrations to two-electron CNT quantum dot states [102]. In 2016, by engineering

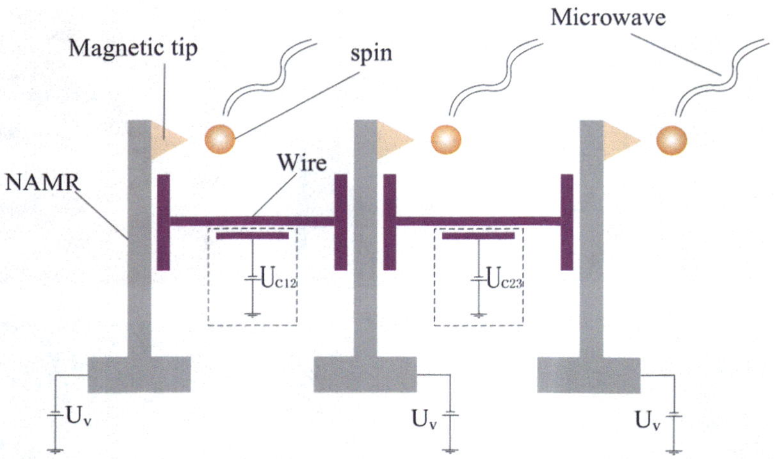

Fig. 4.14 Switchable coupling between two nanomechanical resonators. A switchable control gate (dashed box) is added between every two nearest nanomechanical resonators. (Image is Reprinted with permission Ref. [99] Copyright (2012) American Physical Society)

two interacted CNT nanomechanical resonators, A. Hamo et al. reported electron attraction mediated by Coulomb repulsion [103]. Using the coupling between quantum dots and nanomechanical resonators, S. X. Li et al. reported parametric strong mode coupling in CNT in 2016 [26], and D. Zhu et al. reported coherent phonon Rabi oscillations in 2017 [28]. In 2017, G. Luo et al. reported the first experiment on coupling graphene quantum dots with nanomechanical resonators [27], and later in 2018 they reported strong indirect coupling between graphene-based mechanical resonators via a phonon cavity [29]. In 2019, I. Khivrich et al. reported nanomechanical pump–probe measurements of insulating electronic states in a carbon nanotube [104]. These results lay a solid foundation for the realization of large-scale semiconductor quantum chips using nanomechanical resonators (Fig. 4.15).

4.4 Summary

Gate-defined quantum dots (GDQDs) are considered as one of the best candidates for quantum computing. This chapter reviewed the basic concepts and technological developments of GDQD in recent years. Quantum dots based on GaAs, silicon, carbon nanotubes, nanowires, and 2D materials were reviewed. Both spin and charge degree of freedoms in GDQD can be manipulated by adjusting gate voltages, and we introduced qubits based on spin, charge, and their hybrid states. From the perspective of constructing quantum bits, we discussed both coherence times and gate control fidelity of GDQD. Coherence time as long as several hundred microseconds and fidelity as large as 99.99% have been realized. Finally, we reviewed recent progresses on the scalability of GDQD from the perspective of superconduct-

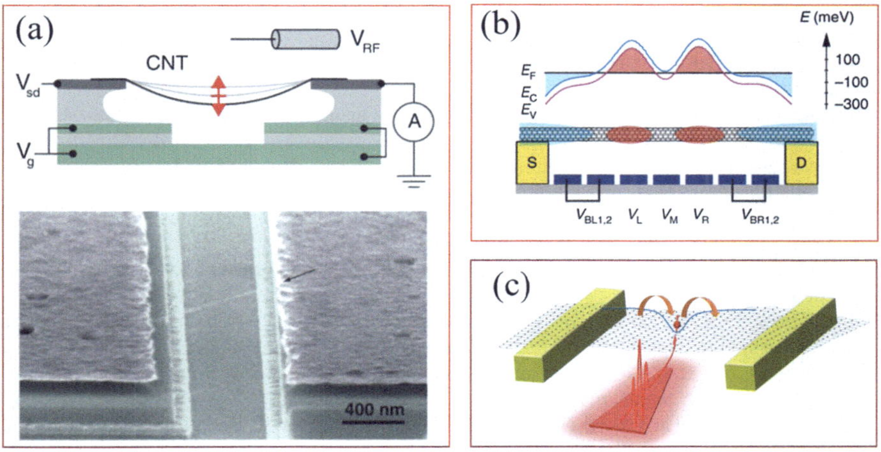

Fig. 4.15 Coupling between quantum dots and nanomechanical resonators. (**a**) Strong coupling between single CNT quantum dots and nanomechanical resonators. (**b**) Double quantum dots and electron-phonon interaction in suspended CNT nanomechanical resonators. (**c**) Coupling quantum dots and nanomechanical resonators in suspended graphene membrane. (Image (**a**) is Reprinted with permission from Ref. [22] Copyright (2009) American Association for the Advancement of Science, (**b**) is Reprinted with permission from Ref. [104] Copyright (2019) Springer Nature, (**c**) is Reprinted with permission from Ref. [27] Copyright (2017) Royal Society of Chemistry)

ing cavities and nanomechanical resonators. Although only two distant quantum dots have been integrated up to now, which is still far behind the superconducting quantum computing system, the strong coupling results, long coherence time, high fidelity, and CMOS compatibility have showed great application prospects in semiconducting quantum computing.

Acknowledgments This chapter was supported by the National Natural Science Foundation of China (Grants Nos. 61704164 and 91836102).

References

1. Kouwenhoven, L., & Marcus, C. (1998). Quantum dots. *Physics World, 11*, 35–40.
2. van der Wiel, W. G., et al. (2002). Electron transport through double quantum dots. *Reviews of Modern Physics, 75*, 1.
3. Ralph, D. C., Black, C. T., & Tinkham, M. (1995). Spectroscopic measurements of discrete electronic states in single metal particles. *Physical Review Letters, 74*, 3241–3244.
4. Alivisatos, A. P. (1996). Semiconductor clusters, nanocrystals, and quantum dots. *Science, 271*, 933.
5. Cleuziou, J. P., Wernsdorfer, W., Bouchiat, V., Ondarçuhu, T., & Monthioux, M. (2006). Carbon nanotube superconducting quantum interference device. *Nature Nanotechnology, 1*, 53–59.
6. De Franceschi, S., Kouwenhoven, L., Schönenberger, C., & Wernsdorfer, W. (2010). Hybrid superconductor–quantum dot devices. *Nature Nanotechnology, 5*, 703.

7. Loss, D., & DiVincenzo, D. P. (1998). Quantum computation with quantum dots. *Physical Review A, 57*, 120–126.
8. Zhang, X., et al. (2018). Semiconductor quantum computation. *National Science Review, 6*, 32–54.
9. Georgescu, I. M., Ashhab, S., & Nori, F. (2014). Quantum simulation. *Reviews of Modern Physics, 86*, 153–185.
10. Hensgens, T., et al. (2017). Quantum simulation of a Fermi–Hubbard model using a semiconductor quantum dot array. *Nature, 548*, 70.
11. Le Hur, K., Simon, P., & Loss, D. (2007). Transport through a quantum dot with SU(4) Kondo entanglement. *Physical Review B, 75*, 035332.
12. Le Hur, K., et al. (2016). Many-body quantum electrodynamics networks: Non-equilibrium condensed matter physics with light. *Comptes Rendus Physique, 17*, 808–835.
13. Koppens, F. H. L., Nowack, K. C., & Vandersypen, L. M. K. (2008). Spin echo of a single electron spin in a quantum dot. *Physical Review Letters, 100*, 236802.
14. Veldhorst, M., et al. (2014). An addressable quantum dot qubit with fault-tolerant control-fidelity. *Nature Nanotechnology, 9*, 981.
15. Muhonen, J. T., et al. (2014). Storing quantum information for 30 seconds in a nanoelectronic device. *Nature Nanotechnology, 9*, 986.
16. Frey, T., et al. (2012). Dipole coupling of a double quantum dot to a microwave resonator. *Physical Review Letters, 108*, 046807.
17. Petersson, K. D., et al. (2012). Circuit quantum electrodynamics with a spin qubit. *Nature, 490*, 380–383.
18. Delbecq, M. R., et al. (2011). Coupling a quantum dot, fermionic leads, and a microwave cavity on a chip. *Physical Review Letters, 107*, 256804.
19. Toida, H., Nakajima, T., & Komiyama, S. (2013). Vacuum Rabi splitting in a semiconductor circuit QED system. *Physical Review Letters, 110*, 066802.
20. Deng, G.-W., et al. (2015). Charge number dependence of the dephasing rates of a graphene double quantum dot in a circuit QED architecture. *Physical Review Letters, 115*, 126804.
21. Deng, G. W., et al. (2015). Coupling two distant double quantum dots with a microwave resonator. *Nano Letters, 15*, 6620.
22. Steele, G. A., et al. (2009). Strong coupling between single-electron tunneling and nanomechanical motion. *Science, 325*, 1103.
23. Lassagne, B., Tarakanov, Y., Kinaret, J., Garcia-Sanchez, D., & Bachtold, A. (2009). Coupling mechanics to charge transport in carbon nanotube mechanical resonators. *Science, 325*, 1107.
24. Benyamini, A., Hamo, A., Kusminskiy, S. V., Oppen, F. V., & Ilani, S. (2014). Real-space tailoring of the electron-phonon coupling in ultraclean nanotube mechanical resonators. *Nature Physics, 10*, 151–156.
25. Deng, G. W., et al. (2016). Strongly coupled nanotube electromechanical resonators. *Nano Letters, 16*, 5456–5462.
26. Li, S. X., et al. (2016). Parametric strong mode-coupling in carbon nanotube mechanical resonators. *Nanoscale, 8*, 14809–14813.
27. Luo, G., et al. (2017). Coupling graphene nanomechanical motion to a single-electron transistor. *Nanoscale, 9*, 5608–5614.
28. Zhu, D., et al. (2017). Coherent phonon Rabi oscillations with a high-frequency carbon nanotube phonon cavity. *Nano Letters, 17*, 915–921.
29. Luo, G., et al. (2018). Strong indirect coupling between graphene-based mechanical resonators via a phonon cavity. *Nature Communications, 9*, 383.
30. Yoneda, J., et al. (2018). A quantum-dot spin qubit with coherence limited by charge noise and fidelity higher than 99.9%. *Nature Nanotechnology, 13*, 102–106.
31. Tarucha, S., Austing, D. G., Honda, T., van der Hage, R. J., & Kouwenhoven, L. P. (1996). Shell filling and spin effects in a few electron quantum dot. *Physical Review Letters, 77*, 3613–3616.
32. Hanson, R., Kouwenhoven, L. P., Petta, J. R., Tarucha, S., & Vandersypen, L. M. K. (2007). Spins in few-electron quantum dots. *Reviews of Modern Physics, 79*, 1217–1265.

33. Gorter, C. J. (1951). A possible explanation of the increase of the electrical resistance of thin metal films at low temperatures and small field strengths. *Physica, 17*, 777–780.
34. Fulton, T. A., & Dolan, G. J. (1987). Observation of single-electron charging effects in small tunnel junctions. *Physical Review Letters, 59*, 109–112.
35. Ciorga, M., et al. (2000). Addition spectrum of a lateral dot from coulomb and spin-blockade spectroscopy. *Physical Review B, 61*, R16315–R16318.
36. Elzerman, J. M., et al. (2003). Few-electron quantum dot circuit with integrated charge read out. *Physical Review B, 67*, 161308.
37. Young, C. E., & Clerk, A. A. (2010). Inelastic backaction due to quantum point contact charge fluctuations. *Physical Review Letters, 104*, 186803.
38. Elzerman, J. M., Hanson, R., Willems van Beveren, L. H., Vandersypen, L. M. K., & Kouwenhoven, L. P. (2004). Excited-state spectroscopy on a nearly closed quantum dot via charge detection. *Applied Physics Letters, 84*, 4617–4619.
39. Petta, J. R., Johnson, A. C., Marcus, C. M., Hanson, M. P., & Gossard, A. C. (2004). Manipulation of a single charge in a double quantum dot. *Physical Review Letters, 93*, 186802.
40. Bluhm, H., et al. (2010). Dephasing time of GaAs electron-spin qubits coupled to a nuclear bath exceeding 200 μs. *Nature Physics, 7*, 109.
41. Zwanenburg, A. S. D. F. A., Morello, A., Simmons, M. Y., Hollenberg, L. C. L., Klimeck, G., Rogge, S., Coppersmith, S. N., & Eriksson, M. A. (2013). Silicon quantum electronics. *Reviews of Modern Physics, 85*, 961.
42. Saeedi, K., et al. (2013). Room-temperature quantum bit storage exceeding 39 minutes using ionized donors in silicon-28. *Science, 342*, 830–833.
43. Wang, K., Payette, C., Dovzhenko, Y., Deelman, P. W., & Petta, J. R. (2013). Charge relaxation in a single-electron Si/SiGe double quantum dot. *Physical Review Letters, 111*, 046801.
44. Angus, S. J., Ferguson, A. J., Dzurak, A. S., & Clark, R. G. (2007). Gate-defined quantum dots in intrinsic silicon. *Nano Letters, 7*, 2051–2055.
45. Zajac, D. M., et al. (2018). Resonantly driven CNOT gate for electron spins. *Science, 359*, 439.
46. Laird, E. A., et al. (2015). Quantum transport in carbon nanotubes. *Reviews of Modern Physics, 87*, 703–764.
47. Kuemmeth, F., Ilani, S., Ralph, D. C., & McEuen, P. L. (2008). Coupling of spin and orbital motion of electrons in carbon nanotubes. *Nature, 452*, 448–452.
48. Biercuk, M. J., Garaj, S., Mason, N., Chow, J. M., & Marcus, C. M. (2005). Gate-defined quantum dots on carbon nanotubes. *Nano Letters, 5*, 1267–1271.
49. Fasth, C., Fuhrer, A., Björk, M. T., & Samuelson, L. (2005). Tunable double quantum dots in InAs nanowires defined by local gate electrodes. *Nano Letters, 5*, 1487–1490.
50. Li, S.-X., et al. (2017). Measuring hole spin states of single quantum dot in germanium hut wire. *Applied Physics Letters, 110*, 133105.
51. Zhang, Z. Z., et al. (2017). Electrotunable artificial molecules based on van der Waals heterostructures. *Science Advances, 3*, e1701699.
52. Wang, L.-J., et al. (2010). A graphene quantum dot with a single electron transistor as an integrated charge sensor. *Applied Physics Letters, 97*, 262113.
53. Wei, D., et al. (2013). Tuning inter-dot tunnel coupling of an etched graphene double quantum dot by adjacent metal gates. *Scientific Reports, 3*, 3175.
54. Wang, L. J., et al. (2011). Gates controlled parallel-coupled double quantum dot on both single layer and bilayer graphene. *Applied Physics Letters, 99*, 112117.
55. Zhang, M.-L., et al. (2014). Measuring the complex admittance of a nearly isolated graphene quantum dot. *Applied Physics Letters, 105*, 073510.
56. Ashoori, R. C. (1996). Electrons in artificial atoms. *Nature, 379*, 413–419.
57. McEuen, P. L., et al. (1991). Transport spectroscopy of a coulomb island in the quantum hall regime. *Physical Review Letters, 66*, 1926–1929.
58. Shi, Z., et al. (2012). Fast hybrid silicon double-quantum-dot qubit. *Physical Review Letters, 108*, 140503.

59. Shi, Z., et al. (2014). Fast coherent manipulation of three-electron states in a double quantum dot. *Nature Communications, 5*, 3020.
60. Prance, J. R., et al. (2012). Single-shot measurement of triplet-singlet relaxation in a Si/SiGe double quantum dot. *Physical Review Letters, 108*, 046808.
61. Kim, D., et al. (2014). Quantum control and process tomography of a semiconductor quantum dot hybrid qubit. *Nature, 511*, 70.
62. Cao, G., et al. (2016). Tunable hybrid qubit in a GaAs double quantum dot. *Physical Review Letters, 116*, 086801.
63. Wang, B.-C., et al. (2017). Tunable hybrid qubit in a triple quantum dot. *Physical Review Applied, 064035*, 8.
64. Yang, Y.-C., Coppersmith, S. N., & Friesen, M. (2019). Achieving high-fidelity single-qubit gates in a strongly driven charge qubit with 1/f charge noise. *npj Quantum Information, 5*, 12.
65. Vandersypen, L. M. K., & Chuang, I. L. (2005). NMR techniques for quantum control and computation. *Reviews of Modern Physics, 76*, 1037–1069.
66. Jozsa, R. (1994). Fidelity for mixed quantum states. *Journal of Modern Optics, 41*, 2315–2323.
67. Chan, K. W., et al. (2018). Assessment of a silicon quantum dot spin qubit environment via noise spectroscopy. *Physical Review Applied, 044017*, 10.
68. Fogarty, M. A., et al. (2018). Integrated silicon qubit platform with single-spin addressability, exchange control and single-shot singlet-triplet readout. *Nature Communications, 9*, 4370.
69. Kawakami, E., et al. (2014). Electrical control of a long-lived spin qubit in a Si/SiGe quantum dot. *Nature Nanotechnology, 9*, 666.
70. Bluhm, H., Foletti, S., Mahalu, D., Umansky, V., & Yacoby, A. (2010). Enhancing the coherence of a spin qubit by operating it as a feedback loop that controls its nuclear spin bath. *Physical Review Letters, 105*, 216803.
71. Maune, B. M., et al. (2012). Coherent singlet-triplet oscillations in a silicon-based double quantum dot. *Nature, 481*, 344.
72. Shulman, M. D., et al. (2014). Suppressing qubit dephasing using real-time Hamiltonian estimation. *Nature Communications, 5*, 5156.
73. Nichol, J. M., et al. (2017). High-fidelity entangling gate for double-quantum-dot spin qubits. *npj Quantum Information, 3*, 3.
74. Li, R., et al. (2018). A crossbar network for silicon quantum dot qubits. *Science Advances, 4*, eaar3960.
75. DiVincenzo, D. P. (2000). The physical implementation of quantum computation. *Fortschritte der Physik, 48*, 771–783.
76. Li, H.-O., et al. (2015). Conditional rotation of two strongly coupled semiconductor charge qubits. *Nature Communications, 6*, 7681.
77. Li, H.-O., et al. (2018). Controlled quantum operations of a semiconductor three-qubit system. *Physical Review Applied, 024015*, 9.
78. Wallraff, A., et al. (2004). Strong coupling of a single photon to a superconducting qubit using circuit quantum electrodynamics. *Nature, 431*, 162–167.
79. Majer, J., et al. (2007). Coupling superconducting qubits via a cavity bus. *Nature, 449*, 443.
80. Neeley, M., et al. (2010). Generation of three-qubit entangled states using superconducting phase qubits. *Nature, 467*, 570.
81. Song, C., et al. (2019). Generation of multicomponent atomic Schrödinger cat states of up to 20 qubits. *Science, 365*, 574–577.
82. Childress, L., Sørensen, A., & Lukin, M. (2004). Mesoscopic cavity quantum electrodynamics with quantum dots. *Physical Review A, 69*, 042302.
83. Guo, G.-P., Zhang, H., Hu, Y., Tu, T., & Guo, G.-C. (2008). Dispersive coupling between the superconducting transmission line resonator and the double quantum dots. *Physical Review A, 78*, 020302.
84. Lin, Z.-R., Guo, G.-P., Tu, T., Zhu, F.-Y., & Guo, G.-C. (2008). Generation of quantum-dot cluster states with a superconducting transmission line resonator. *Physical Review Letters, 101*, 230501.

85. Deng, G.-W., et al. (2013). Circuit QED with a graphene double quantum dot and a reflection-line resonator. *arXiv, 1310*, 6118.
86. Viennot, J. J., Dartiailh, M. C., Cottet, A., & Kontos, T. (2015). Coherent coupling of a single spin to microwave cavity photons. *Science, 349*, 408.
87. Mi, X., Cady, J. V., Zajac, D. M., Deelman, P. W., & Petta, J. R. (2017). Strong coupling of a single electron in silicon to a microwave photon. *Science, 355*, 156.
88. Stockklauser, A., et al. (2017). Strong coupling cavity QED with gate-defined double quantum dots enabled by a high impedance resonator. *Physical Review X, 7*, 011030.
89. Mi, X., et al. (2018). A coherent spin–photon interface in silicon. *Nature, 555*, 599.
90. Samkharadze, N., et al. (2018). Strong spin-photon coupling in silicon. *Science, 359*, 1123.
91. Landig, A. J., et al. (2018). Coherent spin–photon coupling using a resonant exchange qubit. *Nature, 560*, 179–184.
92. Frey, T., et al. (2012). Quantum dot admittance probed at microwave frequencies with an on-chip resonator. *Physical Review B, 86*, 115303.
93. Basset, D. D. J. J., Stockklauser, A., Frey, T., Reichl, C., Wegscheider, W., Ihn, T. M., Ensslin, K., & Wallraff, A. (2013). Single-electron double quantum dot dipole-coupled to a single photonic mode. *Physical Review B, 88*, 125312.
94. Liu, Y. Y., et al. (2015). Semiconductor double quantum dot micromaser. *Science, 347*, 285–287.
95. Delbecq, M. R., et al. (2013). Photon-mediated interaction between distant quantum dot circuits. *Nature Communications, 4*, 1400.
96. Wang, R., Deacon, R. S., Car, D., Bakkers, E. P. A. M., & Ishibashi, K. (2016). InSb nanowire double quantum dots coupled to a superconducting microwave cavity. *Applied Physics Letters, 108*, 203502.
97. Li, Y., et al. (2018). Coupling a germanium hut wire hole quantum dot to a superconducting microwave resonator. *Nano Letters, 18*, 2091–2097.
98. Rabl, P., et al. (2010). A quantum spin transducer based on nanoelectromechanical resonator arrays. *Nature Physics, 6*, 602.
99. Zhou, L.-g., Gao, M., Peng, J.-l., & Wang, X.-b. (2012). Scalability of quantum computing based on nanomechanical resonators. *Physical Review A, 85*, 042326.
100. Ganzhorn, M., & Wernsdorfer, W. (2012). Dynamics and dissipation induced by single-electron tunneling in carbon nanotube nanoelectromechanical systems. *Physical Review Letters, 108*, 175502.
101. Meerwaldt, H. B., et al. (2012). Probing the charge of a quantum dot with a nanomechanical resonator. *Physical Review B, 86*, 115454.
102. Weber, P., et al. (2015). Switchable coupling of vibrations to two-electron carbon-nanotube quantum dot states. *Nano Letters, 15*, 4417–4422.
103. Hamo, A., et al. (2016). Electron attraction mediated by coulomb repulsion. *Nature, 535*, 395.
104. Khivrich, I., Clerk, A. A., & Ilani, S. (2019). Nanomechanical pump–probe measurements of insulating electronic states in a carbon nanotube. *Nature Nanotechnology, 14*, 161–167.

Chapter 5
Experimental Progress on Quantum Communication with Quantum Dot Based Devices

Chenzhi Yuan and Qiang Zhou

Abstract Quantum communication is a key branch in the field of quantum information. Quantum dot devices have very promising applications in quantum communication due to the advantage in emitting single photon with good purity and high generation rate. In this chapter, a brief introduction of quantum communication, i.e., quantum key distribution, quantum teleportation, and atom–photon entanglement, is given. The experimental progresses in applying quantum dot devices in these directions are summarized. In each direction, a typical example is introduced in details to show clearly the main principles and technology involved in these experimental progresses.

Keywords Quantum key distribution · Quantum teleportation · Spin–photon entanglement · Quantum network · Quantum dot-based devices

5.1 Brief Introduction of Quantum Communication

After the quantum physics was discovered, it has shifted our thinking of the fundamental mechanism of the nature. Besides their impact on the field of fundamental research, these insights have brought much new elements into information science, and a novel field, quantum information is created [1–5]. For several tasks in communication, computation, and information processing, quantum information science can be more powerful when compared with their classical counterparts [6–9]. As a branch in quantum information, quantum communication experiences a worldwide boom [10–13]. The extreme goal in this field is to establish quantum networks [14–17], rivaling the current classical Internet. For this goal, there are three important

---

C. Yuan · Q. Zhou (✉)
Institute of Fundamental and Frontier Sciences, University of Electronic Science and Technology of China, Chengdu, China
e-mail: zhouqiang@uestc.edu.cn

P. Yu, Z. M. Wang (eds.), *Quantum Dot Optoelectronic Devices*,
Lecture Notes in Nanoscale Science and Technology 27,
https://doi.org/10.1007/978-3-030-35813-6_5

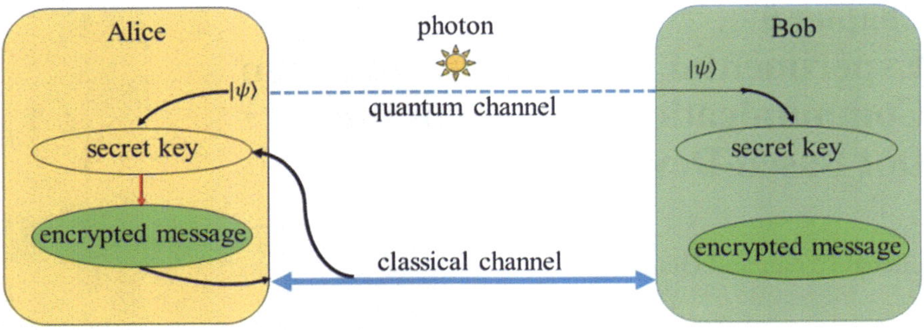

Fig. 5.1 The illustration of a general QKD process

topics on which significant progresses have been developed in the last decades. They are quantum cryptography distribution (QKD) [18–29], quantum teleportation [30–44], and photon–atom entanglement [44–59]. Quantum dot devices have considerable advantages in applications of these topics [60]. In the following paragraphs, we will alternatively introduce the basic background in the three topics and discuss why it is necessary to use quantum dot devices in these demonstrations.

5.1.1 Quantum Cryptography Distribution (QKD)

There are two routines for two spatially separated parties to share a secret message [10]. The first routine is to share a random key which can be used to encrypt the message. Specifically, the sender, Alice, gets a cipher by implementing a logical operator (an exclusive OR, XOR) of the message with the key. Only when the key is known, this cipher can be understood. The receiver, Bob, can retrieve the original message by using the key to decrypt the cipher. The intrinsic problem in route is that Alice and Bob must share the whole secret key. The second way is to use the public–private key cryptography [61], which is based on the fact that it is nearly impossible to find the prime factors of large numbers with the computing ability of the existing computers. It this scheme, the receiver, Bob creates a pair of keys. They are private and public keys, respectively. With the public keys, everybody can encrypt messages for Bob. However, only Bob himself can decrypt the messages since only he has the private key. This pledge got risky since Peter Shor proposed that a quantum computer could be much faster than classical computers in the task of prime numbers factoring [7]. According this proposal of Shor, an eavesdropper would understand the secret message with only the public key. Just as using the metal shields to defend against the shears also made of metal, we can use the principle of quantum physics to defend the risk pointed out by Peter Shor, and QKD is developed for this goal.

In contrast to classical cryptography, the generation of secret key in QKD is not based on the difficulty in prime factoring but on the quantum physical principles which are also the basis of quantum computer. Therefore, it can even defend the attack of a quantum computer. As illustrated in Fig. 5.1, two steps are usually involved in a QKD process. First, Alice and Bob exchange qubits via the quantum channel, measure the qubit, and get a raw key locally. The two raw keys obtained by Alice and Bob are strongly correlated, but not identical. In the second step, an interactive post-processing protocol is implemented by both Alice and Bob via communication in the classical channel. By this step, the two parties can distill two identical strings. The strings are only known by Alice and Bob themselves, and therefore they are two local copies of the secret key. It is notable that the classical channel in the second step has to be authenticated. This means that a third person cannot participate in the communication, though he or she can listen to the conversation. In contrast, the third person can perform any possible manipulation in the quantum channel. Therefore, the goal of QKD is to supply Alice and Bob schemes to guarantee the security of the exchanged messages against a potential adversarial eavesdropper, usually called Eve able to tap on the quantum channel and listen to the communication in the classical channel.

The first QKD scheme was proposed by Bennett and Brassard in 1984 [19], and it is known as the famous BB84 protocol. To explain how this scheme works, we can learn from the Fig. 27.2 in Ref. [62], which is shown in Fig. 5.2.

In BB84 protocol, four different qubit states are required, and the condition that they form two complementary bases should be satisfied. In Fig. 5.2, four linear polarization states of a photon are selected, and they form two complementary bases. The four states are labeled as horizontal (H), vertical (V), diagonal (D), and anti-diagonal (A). As illustrated in this figure, Alice first sends a single photon to Bob, which was prepared randomly in one of the four polarization states (H, V, D, and A), and records the prepared state at the same time. After receiving the photon, Bob analyzes it by a two-port polarization analyzer, the two ports of which form the

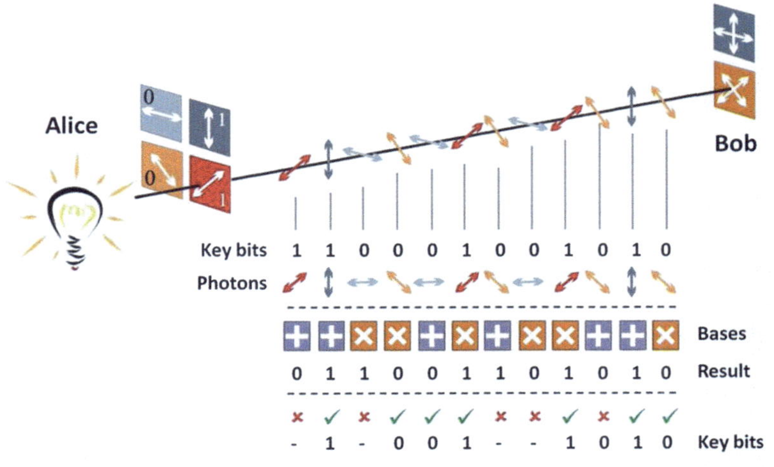

Fig. 5.2 The illustration showing the basic principle of BB84 protocol. (Adapted with permission from Ref. [62]. Copyright (2016) Springer Nature.)

basis H/V or D/A randomly. Then, he recorded the measurement results at the ports of the analyzer, as well as the corresponding basis. This process between Alice and Bob is repeated several times, until the photons which have been received by Bob are enough. At this time, Bob sends his measurement information to Alice via the classical channel. The information includes which photons have been detected successfully and which basis is used to detect the photon in one successful detection. In return, Alice sends Bob which photons she sent to him were prepared in the same bases, because only these photons can lead to the correct result in Bob's measurement. To generate key consistent with those in classical cryptograph, the states H and D (V and A) are mapped to 0 (1). By this mapping rule, Alice and Bob can obtain identical string composed of 0 s and 1 s, which is called the sifted key.

The principle guaranteeing the security of the BB84 protocol introduced above is that a measurement on a quantum state will make it collapse in most cases [10]. If the sifted keys obtained by Alice and Bob are identical, they can conclude that no eavesdropper exists and they can use the key to encode a confidential message. However, in practical experiment, the eavesdropping attacks are indistinguishable from the intrinsic noises originating from the dark counts in the detectors and transmission errors in quantum and classical channels, all of which can lead to the error in the sifted key. To solve this problem, an error threshold can be introduced [10, 19], with which Alice and Bob can distill a final secret key by the method of classical error correction and privacy amplification [10, 63]. If the error of a sifted key exceeds this threshold, the key is discarded and Alice and Bob should start a new key generation process, while they can use the key when its error is below the threshold.

It is clear that only one single photon is involved in one sending and receiving event in BB84 protocol. After the invention of this protocol and its several modified versions, the protocols based on distributed entangled photon pairs at Alice and Bob are proposed and realized. Since it is not within the core of this chapter, the readers interested in this topic can read the Ref. [18, 27, 28, 64].

5.1.2 Quantum Teleportation

Quantum teleportation is a process by which we can transfer the state of a quantum system to another quantum system, without bringing it along any routine between the two systems. Prior to the invention of this concept, the interesting idea of teleportation is from science fiction [65]. In 1931, C. H. Fort introduced a term "teleportation" to indicate a process in which objects can be transferred from one location to another, but does not need to experience the space between the two locations [65]. In 1993, C. H. Bennett, G. Brassard, C. Crépeau, R. Jozsa, A. Peres, and W. K. Wootters published a paper in which a quantum information protocol is proposed, having several characters of the teleportation in the fiction described above [30]. The term "quantum teleportation" was also first introduced in this paper. In this protocol, we can measure the unknown state of a physical system and subsequently reconstruct it at a remote location. The key source used in this process

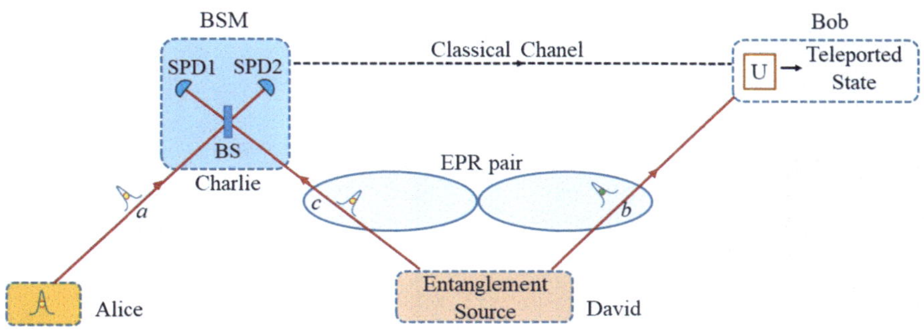

Fig. 5.3 The illustration of a general quantum teleportation process

is quantum entanglement. In Fig. 5.3, we illustrate how this protocol works by a case where all the particles involved are photons.

First, a pair of entangled photons, labeled as c and b, is distributed to C and B, respectively. Next, a third photon a is sent from a A to C. The state of this photon will be teleported, but is not known by C [10]. At the last step, a Bell state measurement (BSM) on the photons a and c is performed at C. This measurement is a joint projection of two-photon state, and it will destroy any information about the state of photon a. Because the photons c and b are entangled initially, the BSM will instantaneously project b onto the same state as a. It should be noted that the teleportation protocol is effective only when the result of BSM is an element in the set composed of four possible random outcomes. Therefore, A should tell B the result of the BSM via the classical channel. This communication can make B get a signal that the teleportation events are successful.

5.1.3 Quantum Cryptography Distribution (QKD)

A promising application of quantum teleportration is in the field of quantum repeater [66–68], which is essential in future large-scale quantum network. In a quantum repeater, BSM is implemented on two particles from independent entangled pairs, and then the entanglement transfers to other two photons. This process is called as entanglement swapping. Its implementation needs quantum memories [69–72] which can store the qubit in one location for a relative long duration.

The entanglement between photon and single atom or atomic ensemble is very appropriate to realize entanglement swapping, since the atom and atomic ensemble can be kept stationary for a long time and their decoherence time of their quantum state is long enough. In Fig. 5.4, the entanglement swapping between two pairs of entangled photon and atomic ensemble is introduced.

By excitation and emission process, entanglement between photon a (b) and atomic ensemble s_a (s_b) is probability created at both of the locations A (B) and B. Then, photons a and b are sent to a middle point between A and B, where a BSM is performed. The detection of a photon at the single-photon detector d or d will herald the entanglement between s_a and s_b.

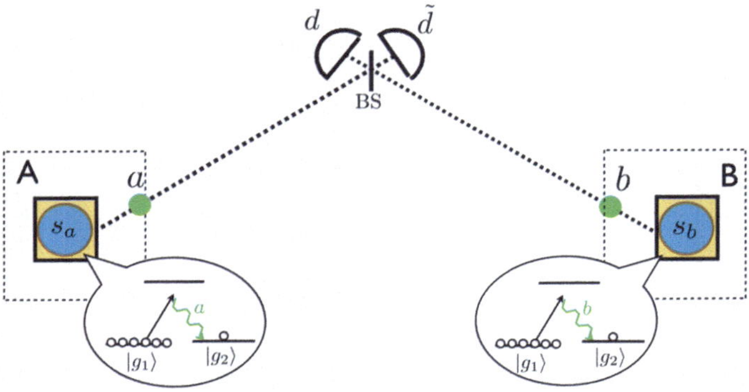

Fig. 5.4 The illustration of an entanglement swapping process between two pairs of entangled photon and atomic ensemble. (Adapted with permission from Ref. [66]. Copyright (2016) American Physical Society)

The above Sects. 5.1.1, 5.1.2, and 5.1.3 give the basic knowledge about QKD, quantum teleportation, and atom–photon entanglement. In all of these processes, high generation rate and purity of the single photons are crucial, and the two indices are in QD. Therefore, in the last two decades, several works of have been done in the fields of QKD and quantum teleportation with QD-based devices, as well as the demonstration of the entanglement between QD and photon. In the next Sects. 5.2, 5.3, and 5.4, the works in these three topics will be briefly reviewed, and the main methods or physical mechanism employed in typical works will be introduced in details.

5.2 The Progress of QKD with QD-Based Devices

As described in above section in this chapter, photons from single-photon source (SPS) have high generation rate and purity and therefore are able to lower the risk of quantum key distribution (QKD) in being attacked in the way of beam splitting [73, 74]. This advantage makes QD-based devices very crucial for QKD based on BB84 protocol. Along on this route, several QKD experiments employing optically pumped SPSs have been carried out with QD-based devices [75–80].

There are three important figures of merit that should be considered when QD-based devices are used in QKD. The first one is the efficiency. When QDs are optically pumped by light pulses, higher efficiency means higher generation rate of single photons, and this is crucial to get longer transmission distance in QKD [60]. The efficiency is determined by both of the photon generation efficiency inside the QDs and the coupling efficiency of the photon into the propagating mode outside the QDs. The second important figure of merit is the single-photon purity, which is characterized by the zero-delayed second-order correlation function $g^2(0)$, as discussed particularly in previous chapters. Low $g^2(0)$ means lower possibility of multiphoton emission and good for the security in QKD. At the last is the wavelength of the single-photon emission, which is also important because different wavelengths are preferred in QKD with different transmission routines. Specifically, quantum dots which emit photons at visible wavelengths are appropriate in free-space QKD [75, 76], while those emitting photons at telecommunication wavelengths are appropriate in fiber-based QKD [77–80]. In the following paragraph, some typical experimental progress in this field will be introduced.

In 2002, the researcher from Stanford University and NTT Basic Research Laboratories presented an experimental demonstration of QKD that uses a visible photon turnstile device based on QD [75]. The protocol used in this experiment is BB84 protocol. The experimental setup used in this work is shown in Fig. 5.5a.

As shown in Fig. 5.5a (*adapted from Fig. 1 in the supplementary material of Ref.* [75]), a single QD in a micro-post cavity forms the photon turnstile, and the fabrication processes of this structure can be found in other works [81, 82] of the group involved in this experiment. The photon turnstile is put into a helium cryostat which supplies the appropriate work temperature of 5–10 K. After excited by 3 ps laser pulses from a Ti–sapphire laser, the quantum dot emits light field at single-photon level. The radiation is then collected by a pinhole and coupled to a single-mode fiber (SMF) for the following manipulation. A combination of grating and a spectrometer slit is used for frequency filtering, and the resolution of this filter is less than 0.2 nm. Combining a polarizing beam splitter (PBS) and half-wave plate, the researchers are able to set the initial polarization of the photons. Then, an electro-optic modulator

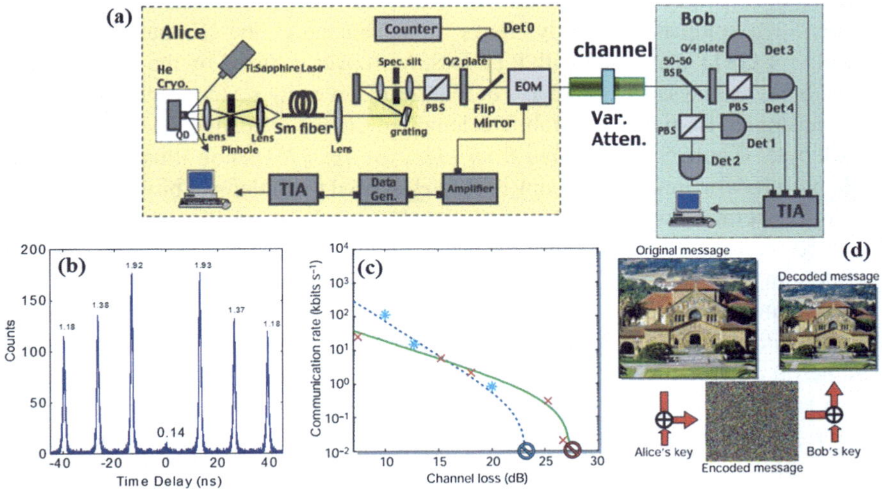

Fig. 5.5 (**a**) The experimental setup of the first experiment of QKD with photon turnstile. (**b**) The measured second-order correlation function of the photon emitted from the QD in (**a**). (**c**) The communication rate versus the channel loss for QKD with weak coherent laser and SPS of QD. (**d**) A secure exchange of an image (Stanford University's Memorial Church) with the key generated in the QKD process in (**a**). (a-d) Adapted with permission from Ref. [75]. Copyright (2002) Springer Nature

(EOM) is employed to rotate this polarization to arrive at one of the four states in BB84 protocol [19, 24]. At the terminal, Bob uses a 50/50 beam splitter to send each photon to one of two polarization analyzers. Here, the reflected photons will be measured in the H/V basis, while the transmitted photons in the R/L basis. Lastly, the photons in each polarization state are detected by silicon avalanche photodiodes (APD). The quantum efficiencies of the APDs are about 0.3, and the dark count rates are less than 100 Hz. The electrical pulses output from the detector are recorded by a time interval analyzer (TIA) and analyzed in a computer. In Alice's location, another TIA is used to record the setting of the data generator.

The measured $g^{(2)}$ of the photon turnstile is shown in Fig. 5.5b (*adapted from Fig. 2 in supplementary material of Ref.* [75]), and it can be seen that the $g^{(2)}(0)$ is about 0.14. This value means that the probability of the device emitting a multiphoton state is an order of magnitude lower than that of a laser emitting photons of the same rate, and therefore security can be improved in the presence of channel losses.

The communication rates of the QKD protocol adapting weak laser and the photon turnstile are both shown in Fig. 5.5c (*adapted from Fig. 1 in Ref.* [75]), as a function of channel loss. In the case of low channel loss, the communication rate with weak laser is higher because the turnstile device is limited by its efficiency and losses in the photon collection system. However, when channel losses get larger, the communication rate with weak laser decreases rapidly because of the high emission probability of multiphoton states. When the loss is around 16 dB, the communication rate with photon turnstile begins to get higher than that with weak laser. Especially, after loss exceeds 23 dB, weak laser no longer supports secure communication, but the single-photon source in the work can work well until the channel losses reach 28 dB. Such comparison clearly shows the advantage of the QD-based device in security for QKD in channel with high loss. Moreover, assisted by the QKD system in Fig. 5.5a, real message, the image of Stanford University's Memorial Church, was also transmitted from Alice to Bob. The length of the key exchanged in experiment is 20 kilobyte. Figure 5.5d (*adapted from Fig. 2 in Ref.* [75]) is an illustration of the exchange process. First, Alice' copy of the key is used to implement bitwise exclusive OR logic operation with each bit of the image, and the encrypted image shown in Fig. 5.5d seems like random noise for one not knowing the copy. However, after the decoding in Bob's location, the image is retrieved effectively, as shown in Fig. 5.5d.

In a quite long period, the telecommunication wavelength QKD was implemented with weak coherent pulses or heralded single photons from spontaneous parametric down conversion (SPDC) [23, 24, 83]. In 2007, a combined research group with members from Toshiba Research Europe Limited and University of Cambridge first demonstrated the QKD processes using single-photon source emitting at telecommunication wavelength [78]. Before this work, this group has demonstrated the growth of a low density of quantum dots emitting in telecommunication band by a bimodal growth technique and furtherly fabricated single-photon source [84]. The structure of the source is single InAs quantum dot embedded at the center of a spacer layer in a distributed Bragg reflector (DBR) micropillar cavity in diameter [78]. The spacer layer is made of GaAs. The DBR structure is formed by alternating layers of GaAs and $Al_{0.98}Ga_{0.02}As$ with designed thicknesses. In the cavity, there are 11 periods on top and 30 periods at the bottom.

In the QKD experiment in Ref. [78], the sample is hold in cooled in a variable temperature continuous flow liquid helium cryostat and a confocal microscope which is used to realize the effective excitation and emitted photons collection. Before the usage in the QKD experiment, the single-photon emission property of the source is first characterized by the photoluminescence and HBT experiment, and the results are shown in Fig. 5.6 (*adapted from Fig. 1 in Ref.* [78]).

In Fig. 5.6a, the photoluminescence of the source at 71 K under the excitation of laser pulses at 1064 nm is shown. Here, the excitation is below-bandgap. In this temperature, the single-photon signal from the device reached its maximum since the charged exciton was on resonance with the HE11 cavity mode. In such case, the $g^{(2)}$ was measured, and the $g^{(2)}(0)$ is 0.102 ± 0.004, as shown in Fig. 5.6b. Though this value is good for security in QKD, the emission rate is also an important quantity that must be considered. In Fig. 5.6c, the $g^{(2)}$ was measured under laser pulses stronger than those used in Fig. 5.6b, and $g^{(2)}(0) = 0.166 \pm 0.005$ was obtained.

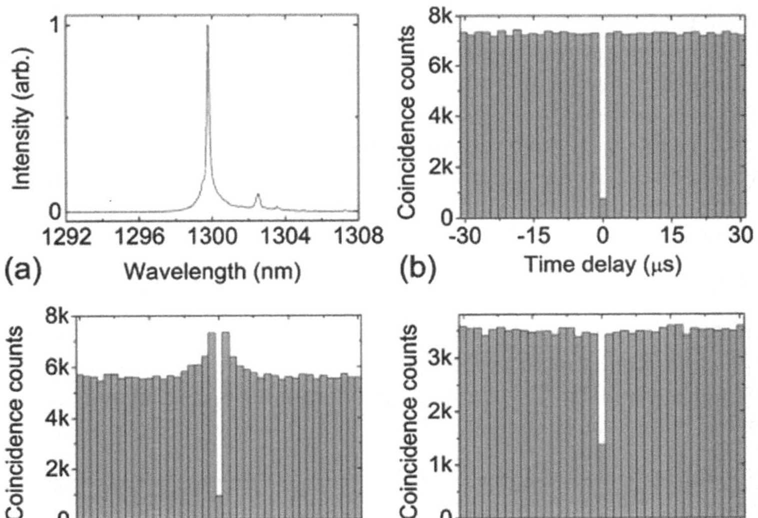

Fig. 5.6 (**a**) The normalized photoluminescence of the QD used in the QKD experiment in Ref. [78]. (**b**) The second-order correlation of the photons generated in the QD when the wavelength, repetition rate, and power in excitation pulse are 1064 nm, 80 MHz, and ~102 μW, respectively. (**c**) The second-order correlation of the photons generated in the QD when the wavelength, repetition rate, and power in excitation pulse are 1064 nm, 1 MHz, and ~2.5 μW, respectively. (**d**) The second-order correlation of the photons generated in the QD when the wavelength, repetition rate, and power in excitation pulse are 1064 nm, 1 MHz, and ~2.5 μW, respectively. The APD gating rates in (**b**), (**c**), and (**d**) are 594 kHz, 1 MHz, and 594 kHz, respectively. (a-d) Adapted with permission from Ref. [78]. Copyright (2007) AIP publishing

As well-known, the quantum dot-based single-photon source also can be excited by in above-bandgap excitation, and the $g^{(2)}$ measured under such excitation is shown in Fig. 5.6d. We can see $g^{(2)}(0)$ increases to 0.392 ± 0.011. It is obvious that the below-bandgap excitation can bring pure single-photon emission. Also, a relatively high efficiency of 5.1% was obtained, accompanying the result in Fig. 5.6c, which means the single-photon source used in this work can realize high efficiency and high single-photon purity simultaneously.

The protocol in the QKD in the work in Ref. [78] is also BB84. The coding method, however, is different from that used in Fig. 5.2, i.e., the method is phase modulation here [85]. The experimental system is shown in Fig. 5.7 (*adapted from Fig. 1 in Ref.* [78]), which is an extension of the one used to demonstrate key distribution using weak coherent pulses at 1550 nm [85]. The single-photon source is optically pumped with picosecond laser pulses at 1064 nm from a semiconductor laser diode at a frequency of 1 MHz. A laser with wavelength same to that of the single-photon source is multiplexed with the emission from the source to generate feedback signal to for minimizing any path mismatch in the two interfering arms. The bit information can be coded or decoded by the phase modulation in the two interfering arms. A 1.55 m clock laser is multiplexed with the single photon in the transmission in 35 km fiber. The interfering single photons were finally detected by InGaAs APDs, which work in gated mode at 1 MHz. The efficiencies of detectors for the photons at the emission wavelengths are around 9%.

In Fig. 5.8a (*adapted from Fig. 3 in Ref.* 78), the timely varied counts on APD1 and APD2 are shown. It can be estimated the average visibility after dark count correction was 99.7%. In the experiment, 7 kbit sifted key was distributed, and the QBER was estimated to be about 5.9%, as shown in Fig. 5.8b (*adapted from Fig. 3 in Ref.* 78). According to GLLP theory in Ref. [86], it can be concluded that the key exchange will be secure against the PNS attack given that the measured value of $g^{(2)}(0)$ is 0.166. In the analysis, Eve is assumed to be able to access all the information carried by the residual multiphoton pulses and those pulses contributing to the QBER, and the multiphoton probability is reduced from $\mu_2/2$ in the case of weak coherent pulse to $g^{(2)}(0)\mu_2/2$ when single-photon source is used. Here, μ_2 is the average number of photons per pulse.

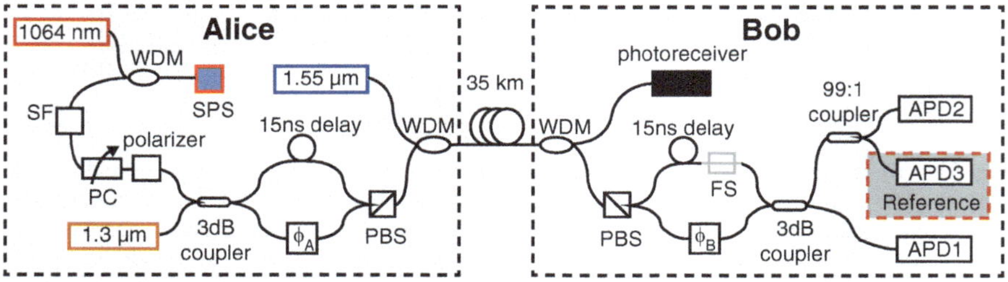

Fig. 5.7 The experimental setup used for realized phase coding BB84 protocol with a triggered QD source emitting near 1.3 μm in *Ref.* 78. *SPS* single-photon source, *SF* spectral filter, *WDM* wavelength division multiplexer, *PC* polarization controller, *PBS* polarizing beam splitter/combiner, *FS* fiber stretcher. Standard telecom fibers are used between Alice and Bob. Adapted with permission from Ref. [78]. Copyright (2007) AIP publishing

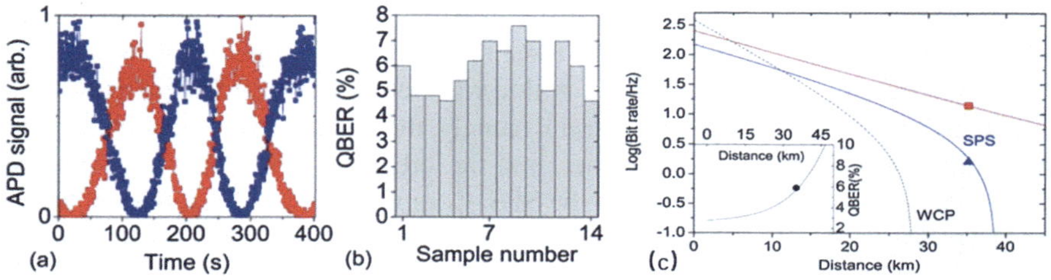

Fig. 5.8 (**a**) The counts on APD1 and APD2 versus time measured in Ref. 78. (**b**) The variation of the QBER in the process of a key transfer over 35 km of fiber. (**c**) The theoretical estimation of the PNS secure transmission distance of the SPS (solid blue curve) used in Ref. 78 and an equivalent weak coherent pulse system (dashed curve). The solid red curve is the theoretical sifted bit rate of the SPS. The data points are experimentally measured sifted bit rate (square) and secure bit rate (triangle). The inset presents the calculated value of QBER (line) with the measured value. (a-c) Adapted with permission from Ref. [78]. Copyright (2007) AIP publishing

To compare the performances of the weak coherent pulse and single-photon source in QKD. In Ref. [78], the sifted bit rate versus transmission distance was also theoretically calculated, and the results are shown in Fig. 5.8c (*adapted from Fig. 4 in Ref.* 78). In the theoretical calculations, the same detector gating conditions, dark count probability, and detection efficiency are used for the two cases. In both cases, the quantum bit errors are calculated by two parts. The first one is half the sum of the dark and stray counts, and the second one is 2.6% of the raw bit rate because of the errors in the phase modulators. In the error correction algorithm, 1.17 times the number of bits specified in the Shannon limit was used [87]. Figure 5.8c shows that QKD with single-photon source is quantitatively advantageous at longer transmission distances, when compared with QKD with uniform intensity laser pulses. Therefore, it can be concluded that a higher bit rate can be achieved with the single-photon source in Ref. [78] for transmission distances longer than 11 kms, while weak coherent pulses are favorable at short distance since the bit rates are available under high photon flux. However, this drawback of single-photon source can be overcome by improving the efficiency of the single-photon source.

An important figure of merit in QKD is the transmission distance. However, in experiment, the transmission distance in QKD with QD-based photon source is limited by the nonzero multiphoton emission probability in single-photon source and dark count in the SPD. Therefore, increasing the purity of the single-photon source and reducing the dark counts in single-photon detector can extend the transmission distance effectively. Along this way, a united researching group in Japan has made a breakthrough in 2015 [80]. They demonstrated secure QKD along fiber channel of 120 km by using an ultrahigh-purity QD-based single-photon source and a superconducting nanowire single-photon detector with ultralow dark counts.

The single-photon source used in Ref. [80] is an self-assembled InAs/InP QD with an optical horn structure [80]. To ensure that the SPS works in the region with high photon emission efficiency and low multiphoton emission probability, the QD is excited by quasi-resonant optical pulses. The excitation pulses in the experiment are generated from a DFB laser with tunable dispersion compensator. Before the QKD experiment, the emitted photon with a wavelength of 1580.5 nm was characterized by two parameters. One is n (≤ 1), namely, the average number of photons in the emitted pulse coupled to the fiber, and the other is $g^{(2)}(0)$. Both of them are dependent on the work conditions of the SPS. Compressed optical excitation pulses of ~10 ps are used in the excitation to suppress the multiphoton emissions owing to multiexciton excitations. The repetition rate of the excitation pulses is 62.5 MHz, which can effectively increase the photon emission rate when compared with pulses of low repetition rate. Under the excitation, HBT experiment is implemented. The best result obtained in the HBT measurement was $g^{(2)}(0) = 0.002$ (after background subtraction) and $n = 0.03$ when the bandwidth of the bandpass filters immediately after the SPS is 0.3 nm. However, to increase the photon emission rate in QKD, this bandwidth is increased to 0.7 nm, and slightly higher $g^{(2)}(0)$ and $n = 0.03$ are obtained. The detailed result is shown in Fig. 5.9.

In Fig. 5.9 (*adapted from Fig. 9 in Ref.* [80]), it is obvious that $g^{(2)}(0) = 0.0051$ and $n = 0.05$ are with a repetition rate of 62.5 MHz. Although the multiphoton emission cannot be suppressed optimally under this condition, the measured $g^{(2)}(0)$ is nearly one order of magnitude smaller than that in the previous experiment [79], while the averaged photon number n only gets reduced slightly ($0.06 \rightarrow 0.05$). In the

Fig. 5.9 The photon correlation measurement in Ref. [80]. The lower part is the zoom version of the central part of the upper part around $\tau \sim 0$, and the $g^{(2)}(0)$ is plotted on a logarithmic scale. Adapted with permission from Ref. [80]

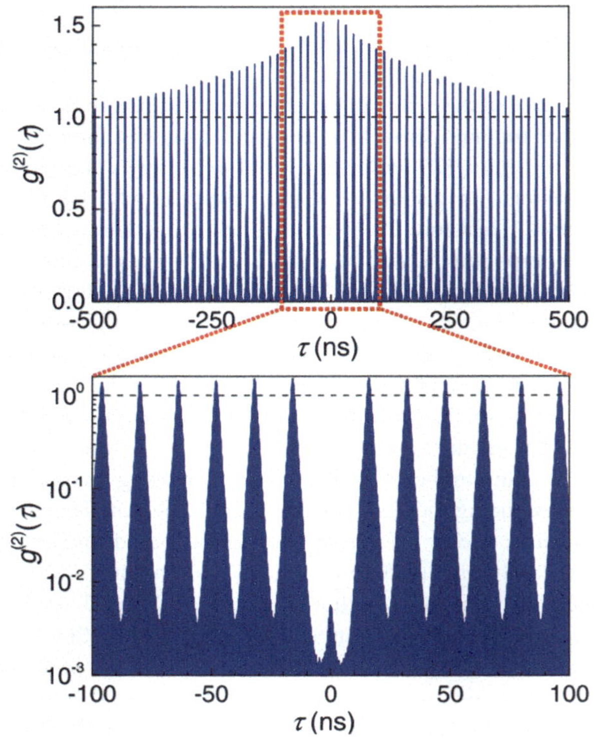

aspect of SPD, ASPD is replaced by SNSPD (SCONTEL, FCOPRS-00-15), because the latter can give higher detection efficiency and lower dark count than the former. In that experiment, the cryostat temperature and the applied bias voltage on the detector are chosen appropriately, and a dark count rate less than 20 Hz and quantum efficiency of 10% are obtained simultaneously.

A full demonstration of QKD with the SPS and SNSPD introduced above is shown in Fig. 5.10 (*adapted from Fig. 1 in Ref.* [80]). The protocol is still BB84 and the encoding bases are time-bins [88]. The unbalanced Mach-Zehnder interferometers (UMZIs) at the locations of Alice and Bob are fabricated with planar lightwave circuits. To keep a relevant common reference frame during QKD experiment [89], the two UMZIs were precisely temperature-controlled. In the following paragraph, a brief description about the implement of QKD in this work will be given.

In Fig. 5.10, first, at the location of Alice, single-photon pulses are emitted from the SPS. Each optical pulse is then converted into a pair of double pulses temporally separated by 5 ns time interval. Here, the pulse pair has fixed polarization. Subsequently, after going through a PM, the double pulses in one pair get a relative phase randomly chosen from $\theta = \{0, \pi/2, \pi, 3\pi/2\}$. The transmission fiber used in the experiment was a two-core single-mode fiber (SMF), in which one core was used for conveying single photons and the other for exchanging the synchronizing laser pulses. After the long-distance transmission, the pulse pairs arrive at the location of Bob. In this location, the polarization of photons is first randomized by using a polarization scrambler (PS). Then, the pulse pair is injected into the second UMZI. Because of the polarization mode dispersion in the waveguide, the TM and TE modes of this UMZI can be employed as an analyzer for the $X(Y)$ basis, corresponding to $\theta = \{0, \pi\}$ ($\theta = \{\pi/2, 3\pi/2\}$). Lastly, two polarization beam splitters (PBSs) are linked to the two output ports of the UMZI to distinguish the TM and TE modes. Therefore, the PS, UMZI, and PBSs constituted a qubit decoder based on a passive basis choice. The arrival port of the photon generates the chosen basis and the measurement result, and the arrived photons are detected by four SNSPDs

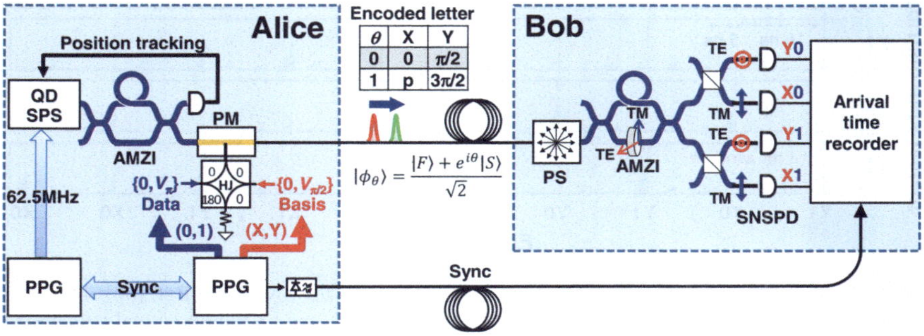

Fig. 5.10 The experimental setup used in Ref. [80] to realize long-distance QKD with ultrahigh-purity single-photon source and SNSPD. Adapted with permission from Ref. [80]

connected to the four ports. By a time interval analyzer, the arrival port and time of the photons are recorded.

In the experiment, the period is selected to be 100 bit (1.6 μs), and these bits are encoded onto a series of photon. A typical interval of arrival time distributions of the photon arrived at the four SNSPDs is shown in Fig. 5.11 (*adapted from Fig. 3 in Ref.* [80]). It is obvious that the distributions are localized around the predefined temporal positions and have finite widths. This temporal width nearly reflects the lifetime (~1 ns) of the photon emission in the QD-based SPS. Also, it is observed to be independent of the transmission distance in the experiment, which means that the dispersion–compensation for each pulse in the SPS is successful.

By post-selecting method, raw keys were generated. Specifically, useful events were selected from the distributed events shown in Fig. 5.11. In the post-selecting process, longer selection window will lead to higher raw key rate but also will bring higher error rate. Therefore, a trade-off between the two important quantities should be taken into consideration. In the experiment, a selection window with temporal width of 4 ns is used. Under such condition, the quantum efficiency η_e, dark count probability d_B per window, and S/N ratio are slightly lower than 10%, $<1 \times 10^{-7}$, and about 60.8 dB respectively. Usually, the ASPD works in the gated mode with very short temporal gating time ~1 ns. This severely limits the quantum efficiency per gate in the ASPD and results in S/N ratio <50 dB. Hence, the utilization of SNSPD, instead of ASPD in QKD experiment, can effectively increase the secure key rate.

With the high-purity QD-based SPS and the SNSPD in the experiment, the quantum bit error rate (QBER) can be reduced significantly. Figure 5.12a (*adapted from Fig. 4 in Ref.* [80]) shows the experimentally measured QBER when the lengths of the transmission fiber are 50, 100, 110, and 120 km, respectively. It can be seen that the QBER in the four cases was always below 10%, which supports the exploration of secure key distribution up to 120 km. Moreover, the utilization of high-purity QD-based SPS and SNSPD also can improve the key rates in QKD. Figure 5.12b

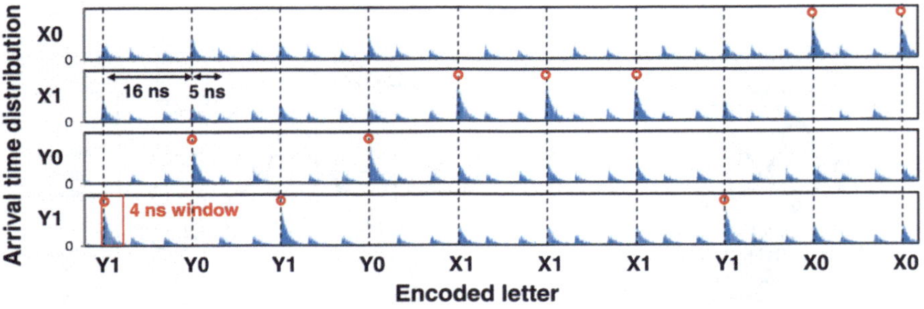

Fig. 5.11 The arrival time distribution of detected photons on the SNSPD in Ref. [80]. There are 10 bits of data shown in this figure. The encoded letters are shown on the horizontal axis by the order of time. The vertical dashed lines are the predefined temporal positions corresponding to meaningful detection events. The circles show the successfully decoded events. Adapted with permission from Ref. [80]

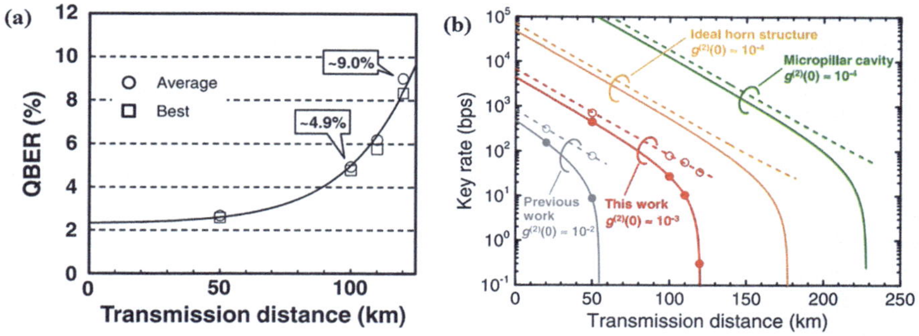

Fig. 5.12 (**a**) The QBER versus the QKD transmission distance. In experiment, the data obtained in 50, 100, 110, and 120 km transmission are measured. The rectangles are the best data of the top ten data in the experiment, while the circles represent the average of the ten data in the experiment. The solid line represents the fitted curve. (**b**) The raw key rates (red open circles and red dotted line) and the estimated secure key rates (red closed circles and red solid line) versus the QKD transmission distance. Some data from works with QDs in other structure are also shown, including the orange-dotted (solid) and green-dotted (solid) lines that represent the raw (secure) key rates for QDs in ideal horn structure and micropillar cavity, respectively. The gray open circles and dotted line represent the raw key rates and the estimated secure key rates obtained in a previous experiment with transmission distance of 50 km [79], while the closed circles and red solid line represent the secure key rates obtained there. (a-b) Adapted with permission from Ref. [80]

(*adapted from Fig. 5 in Ref.* [80]) shows the measured or estimated raw and estimated secure key rates for a series of experiments. In the experiment in Ref. [80], the measured raw (estimated secure) key rate is about ~79.9 bps (27.6 bps) and ~34.3 bps (0.307 bps) when the transmission distance is 100 km and 120 km, respectively. The theoretical analyses in Ref. [80] also suggest that the maximal range of secure QKD can be extended by ~10 km when the averaged photon number n increases from ~0.05 to ~0.1, because the main factor limiting the secure key rate in the experiment is noises in the SNSPDs.

The possible further improvement in the secure key rate by introducing QD-based SPS and SNSPD with better performance also is analyzed in Ref. [80]. In Fig. 5.12b, the orange-dotted (solid) and green-dotted (solid) lines represent the calculated raw (secure) key rate when the QD in SPS is embedded in an ideal horn structure [90] and a pillar microcavity structure [91], respectively. In the calculations about the two structures, the $g^{(2)}(0)$ is set at 10^{-4}, and this value is lower than the value of the QD SPS used in Ref. [80] by one order of magnitude. The dark count rate of the SNSPD and the system detection efficiency are set at 18 Hz and 40%, respectively. For QD in ideal horn structure, the averaged photon number is set at $n = 0.175$ when the maximal photon extraction efficiency is 35% and the coupling loss is 3 dB between the lens and the core of the single-mode fiber. When the excitation repetition rate is 62.5 MHz, the estimated secure key rate is ~40 bps with transmission distance of 150 km. Such distance rivals the longest record in phase-encoding QKD with decoy states [92]. Here, the maximal repetition rate is limited to several hundred MHz by the radiative lifetime of ~1 ns for the exciton state. Before the work in

Ref. [30], the authors have proposed a structure of hybrid pillar microcavity with which high Purcell enhancement (the enhancement factor up to 110) can be realized and output efficiency higher than 60% can be acquired. If such SPS with pillar structure is used, the temporal jitter in single-photon emission could be decreased to improve the maximal repetition rate into the regime of GHz range. In such case, the gated window in SNSPD can be narrowed, and therefore lower noise level can be realized. In Fig. 5.5, the green solid line shows the theoretically calculated secure key rate when the repetition rate is set at 1 GHz, and the temporal jitter in single-photon emission is reduced by 1/10, while $g^{(2)}(0)$ is kept at 10^{-4} and n at 0.175. This result shows that secure bit rate could reach ~100 bps at 200 km, similar to value obtained in the polarization-encoding QKD with decoy states [93]. In all the calculations, the source performance ($n = 0.175$ and $g^{(2)}(0) = 10^{-4}$) is significantly mitigated compared with ideal performance ($n = 1$ and $g^{(2)}(0) = 0$) in SPS. This indicates that $g^{(2)}(0)$ engineering for high-purity QD SPS is a very promising way to obtain unconditionally secure QKD, which also could possess high key rate when compared to the current QKD scheme with weak coherent light pulses.

5.3 The Progress of Quantum Teleportation with QD-Based Devices

QD-based devices are advantageous in quantum teleportation with photons since the multiphoton generation in preparing single photons and entangled photon pairs can be suppressed significantly. In the following paragraphs, we will show several typical results about quantum teleportation with photons generated in QD-based devices.

The full teleportation of a photon with arbitrary unknown state is important to distributed quantum computation and network. However, most explorations about this task use the light sources that produce photons with Poissonian statistics. The first experiment of quantum teleportation employing QD-based devices was implemented by the combined group with members from Toshiba Research Europe Limited and Cavendish Laboratory, University of Cambridge [94]. The experiment setup is shown in Fig. 5.13c (*adapted from Fig. 1 in Ref.* [94]).

All of the three photons states required in the teleportation experimental setup in Fig. 5.13c are electrically generated by an entangled-light-emitting diode (ELED), which emits no more than one pair of entangled photons simultaneously. In the ELED, an InAs quantum dot is embedded in a GaAs cavity, which is formed by upper *p*-doped and lower *n*-doped distributed Bragg mirrors. When current is injected into the device and pass through the GaAs *p–i–n* junction, both of the exciton (X) and biexciton (XX) stated in the quantum dot are resonantly excited. The electroluminescence spectrum of the device is shown in Fig. 5.13a (*adapted from Fig. 1 in Ref.* [94]), in which the narrow peaks correspond to radiation in the decay process of the XX state and subsequent X state.

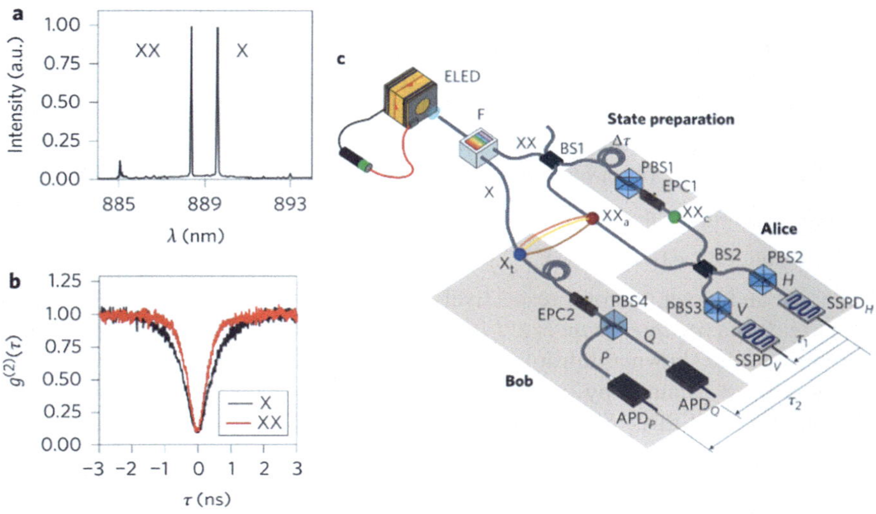

Fig. 5.13 (**a**) The electroluminescence spectrum of the ELED used in Ref. [94] when the excitation current density is 93 nA/μm^2. (**b**) The measured second-order correlation function $g^{(2)}$ (τ) for the photons generated in the X and XX transition processes, respectively. (**c**) The experimental setup used for quantum teleportation with ELED. (a-c) Adapted with permission from Ref. [94], Copyright (2013) Springer Nature

In the first step, entangled photons were distributed to the distribution of a pair of between two parties, commonly referred to as Alice and Bob. Then, a joint Bell state measurement was performed on the input (control) photons with the photon (ancilla) sent to A in the entangled photon pairs. This process would not reveal any information about the state of the control photons. In the Bell state measurement, the input control and ancilla photons are combined in a 1:1 beam splitter. The photons released from the two output ports of the of the beam splitter passed through two crossly orientated polarization beam splitters and finally detected by two single-photon detectors, respectively. This process could project the incident non-entangled two-photon state onto one of the four Bell states with probability of 1/4. If a projection is successful, the output (target) photon at the location of Bob would automatically collapse to a state, which is a trivial unitary transformation of the state of the control photon. In the configuration in Fig. 5.13c, this state will be $\alpha|H\rangle - \beta|V\rangle$ when the control photon is initially in $\alpha|H\rangle + \beta|V\rangle$.

Successful teleportation is strongly related to the purity of the single photons at the locations of Bell state measurement and Bob, indistinguishability of the control and ancilla photons, and entanglement between the two photons at Alice and Bob. To optimize the configuration in the ELED for teleportation, the two quantities were measured in the experiment. The second-order autocorrelation of the photons generated in the decay of the XX and X transitions was measured by the Hanbury Brown-Twiss interference, and the results are shown in Fig. 5.13c. In this figure, the X and XX photons show $g^{(2)}(0) = 0.105 \pm 0.008$ and 0.096 ± 0.010, respectively, and these nonzero $g^{(2)}(0)$ match well with the finite response time of the single-photon detectors. This indicated that the X and XX photons have ideal sub-Poissonian and anti-bunching properties. Since the control and ancilla photons were from a single device, the indistinguishability between them can be guaranteed. Their coherence time was 161 ± 4 ps measured by Michelson interference, and therefore two-photon interference could be observed with time-resolved post-selection. The fidelity of the entangled photons generated by the ELED was measured to be 0.777 ± 0.005 with the Bell state $\Psi^+ = (|H_a H_t\rangle + |V_a V_t\rangle)/\sqrt{2}$, in which the subscripts *a* and *t* present the ancilla and target photons, respectively.

In Fig. 5.14a (*adapted from Fig. 1 in Ref.* [94]), the results in teleporting a superposition state are shown as third-order intensity correlation functions $g^{(3)}(\tau_1, \tau_2)$. The functions were obtained by coincident photon detection measurement across the three detectors $SSPD_H$, $SSPD_V$, and APD_P (or APD_Q), shown in Fig. 5.13c. In $g^{(3)}(\tau_1, \tau_2)$, the first variable τ_1 is the delay between the photon detection events on $SSPD_H$ and $SSPD_V$, while the second variable τ_2 is the delay between the photon detection events on $SSPD_H$ and APD_P (or APD_Q). The teleported state is selected as

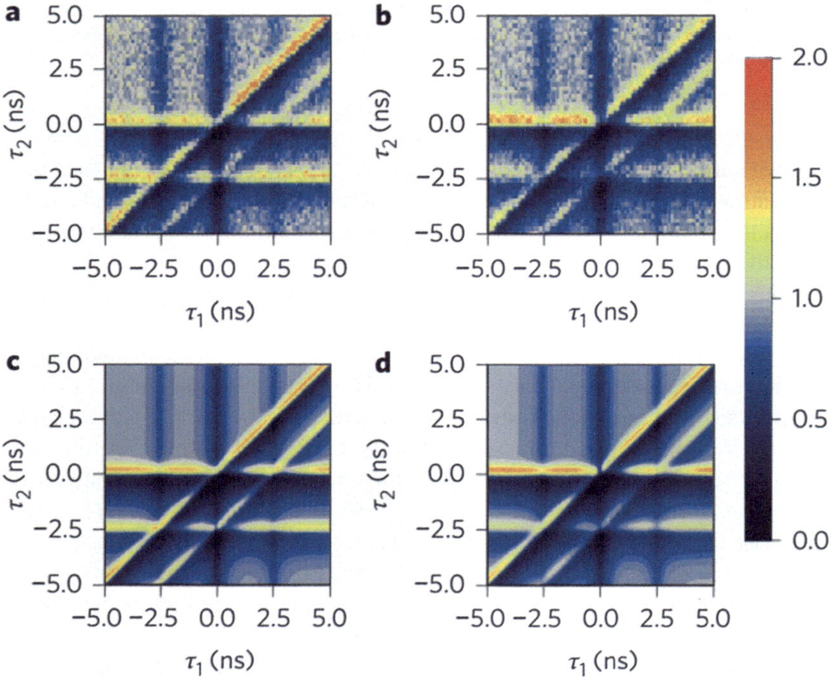

Fig. 5.14 (**a**) and (**b**) are the experimentally measured third-order correlation function $g^{(3)}$ (τ_1,τ_2) in the cases the target photon is detected at the $|+\rangle$ and $|-\rangle$, respectively. (**b**) and (**c**) are the theoretical calculations for (**a**) and (**b**), respectively. (a-d) Adapted with permission from Ref. [94], Copyright (2013) Springer Nature

$|-\rangle = (|H\rangle - |V\rangle)/\sqrt{2}$, i.e., the state of the control photon was set at $|-\rangle$. To make sure that the teleportation is successful, the photon detection at Bob is performed at the basis of $|+\rangle - |-\rangle$, where $|+\rangle = (|H\rangle + |V\rangle)/\sqrt{2}$. The most important point in Fig. 5.14a and b (*adapted from Fig. 1 in Ref.* [94]) is $\tau_1 = \tau_1 = 0$, at which three photons arrived at $SSPD_H$, $SSPD_V$, and APD_P (or APD_Q) simultaneously. In Fig. 5.14a and b, the vertical groove along $\tau_1 = 0$ is determined by the sub-Poissonian characteristic of the control and ancilla photons, while the ridges along $\tau_2 = 0$ and $\tau_1 = \tau_2$ result from the correlation between the ancilla and target photons. Comparing Fig. 5.14a and b, we can see that the groove along $\tau_1 = 0$ in (a) is filled at this point and a peak rises there, while this groove keeps stationary profile in (b). This indicates that the probability for Bob to receive photons with polarization state $|+\rangle$ is much higher than that to receive photons with polarization state $|-\rangle$ in a three-photon coincidence event. This results in the teleportation of an $|-\rangle$ polarized control photon to a $|+\rangle$ polarized target photon.

To confirm the capacity of the system in teleporting arbitrary state, the teleportations of $|H\rangle$, $|V\rangle$, $|+\rangle$, $|-\rangle$, $|L\rangle$, and $|R\rangle$ are implemented, where $|L\rangle = (|H\rangle + i|V\rangle)/\sqrt{2}$ and $|R\rangle = (|H\rangle - i|V\rangle)/\sqrt{2}$ are the left- and right-handed circular polarization state. The six states represent orthogonal pairs in three mutually unbiased bases, and the teleportation of them is sufficient to confirm the feasibility in teleporting an arbitrary control state using only real, positive, and linear combinations. To character the quality of a teleportation process, the fidelity is defined as $f_P^T = g_{P'}^{(3)} / \left(g_{P'}^{(3)} + g_{Q'}^{(3)}\right)$, where P' and Q' are the expected states of the photon detected at Bob and its orthogonal counterpart, respectively. In the experiment, the measured fidelities for the states mentioned above are $0.80 \pm 0.035, 0.833 \pm 0.027, 0.574 \pm 0.035, 0.718 \pm 0.026$, 0.646 ± 0.035, and 0.650 ± 0.032, respectively. The six values are all higher than 0.5 which corresponds to the case that no information is transferred. However, to confirm the quantum nature in the teleportation, the averaged fidelity over the set of the six initial control states should be obtained.

In Fig. 5.15a (*adapted from Fig. 3 in Ref.* [94]), this quantity is shown as a function of τ_1 and τ_2, and the cross sections along $\tau_1 = 0$ and $\tau_2 = 0$ are shown in Fig. 5.15b and c, respectively. From these plots, we can see that the averaged fidelity is $\overline{f_P^T} = 0.704 \pm 0.016$. It has been proven that the classical teleportation with best positive operator-valued measurement cannot exceed the threshold of 2/3 when the 6 states mentioned above are considered. Therefore, the $\overline{f_P^T}$ shown in Fig. 5.15 confirms that the experimental results in these teleportations have undoubted quantum nature.

After the experiment described above, several experiments have been demonstrated with ancilla photon from continuous-wave (CW) laser [95] and entangled photon pairs generated by DC electrical current-pumped [96] or CW laser-pumped QD-based devices [97]. Therefore, all of the photons in these experiments are not generated in on-demand way. This is not appropriate for real-time applications. In 2018, a group composed of scientists from Austria, Italy, and Sweden published a work in which teleportation has been implemented with all of the ancilla, control, and target photons that are generated by QD-based devices in on-demand way [98].

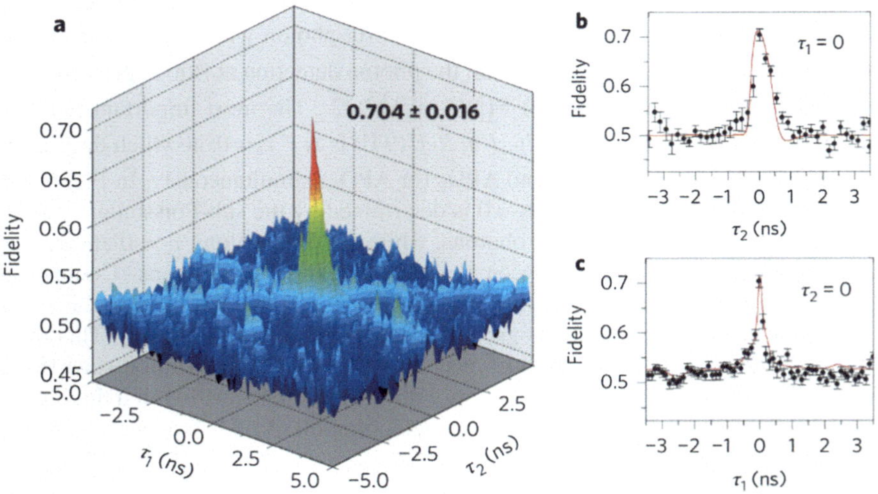

Fig. 5.15 (**a**) The experimentally measured results of the teleportation fidelity averaged over the six states of $|H\rangle$, $|V\rangle$, $|+\rangle$, $|-\rangle$, $|L\rangle$, and $|R\rangle$. (**b**) and (**c**) are the sections of the three-dimensional surface in (**a**) with $\tau_1 = 0$ and $\tau_2 = 0$, respectively. (a-c) Adapted with permission from Ref. [94], Copyright (2013) Springer Nature

Highly symmetric GaAs/AlGaAs QDs are used in the experiment, which is fabricated by droplet etching method. The QDs were embedded within a distributed Bragg reflector (DBR) cavity, which has low Q factor and therefore plays the role a solid immersion lens able to improve the single-photon collection efficiency up to ~12%. Similarly, with the works in Ref. [94], the cascaded XX and X transitions in QD shown in Fig. 5.16a (*adapted from Fig. 1 in Ref.* [98]) are employed to generate entangled photon pairs. In the pump process, a pulse with area of π can populate the XX state via two-photon excitation (TPE), as shown by the population oscillation curve in Fig. 5.16b (*adapted from Fig. 1 in Ref.* [98]). This excitation process can minimize the re-excitation processes, which will degrade the single-photon purity in the emitted photon pairs. In the experiment, the second-order autocorrelations of the XX and X photons are $0.004(0.002) \le g^{(2)}_{\mathrm{XX}}(0) \le 0.009(0.002)$ and $0.004(0.001) \le g^{(2)}_{\mathrm{XX}}(0) \le 0.349(0.005)$, respectively, which feature that the corresponding average lifetimes are $T_{1,\mathrm{XX}} \approx 140(7)$ps and $T_{1,\mathrm{X}} \approx 265(10)$ps, respectively. The two values of autocorrelations prove that the sample can generate pure single and entangled photons on demand, and the only factor limiting the on-demand emission is the extraction efficiencies far below 100%.

In the teleportation experiment, the detected Bell state is chosen to be $|\psi^-\rangle$ with expression as

$$|\psi^-\rangle = \frac{1}{\sqrt{2}}\left(|H\rangle_0 |V\rangle_1 + |V\rangle_0 |H\rangle_1\right), \tag{5.1}$$

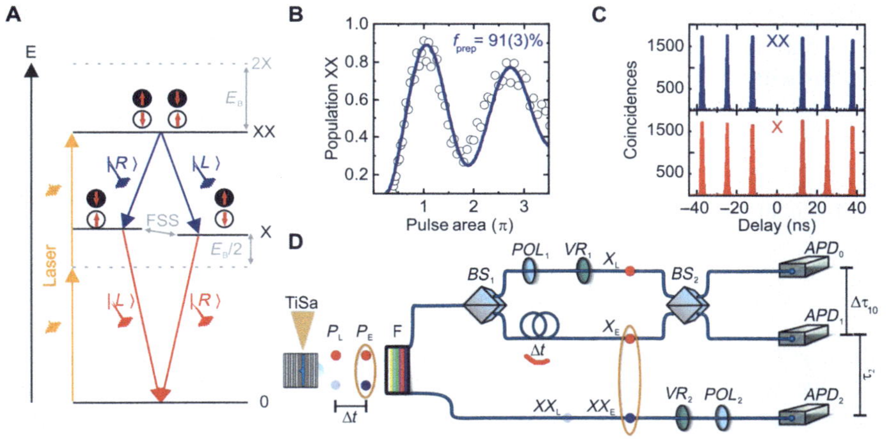

Fig. 5.16 (**a**) The level structure in QD, showing the XX and X decay process generating entangled photon pairs. FSS represents the fine structure splitting. (**b**) The population on the XX state when the pump pulse area changes. (**c**) The measured $g^{(2)}(\tau)$ for the XX and X photons. (**d**) The experimental setup for all photonic quantum teleportations using QD-based devices. (a-d) Adapted with permission from Ref. [98]

where $|H\rangle_0$ and $|V\rangle_0$ ($|H\rangle_1$ and $|V\rangle_1$) are two orthogonal linear polarization states of the photon arrived at APD_0 (APD_1). In the linear polarization basis, the polarization entangled state of photon pair generated in the XX-X transition shown in Fig. 5.16c (*adapted from Fig. 1 in Ref.* [98]) can be expressed as

$$\left|\phi^{+}(t)\right\rangle_{\mathrm{X,XX}} = \frac{1}{\sqrt{2}}\left(\left|H\right\rangle_{\mathrm{X}}\left|V\right\rangle_{\mathrm{XX}} + e^{-\mathrm{i}\Delta_{\mathrm{FFS}}t}\left|V\right\rangle_{\mathrm{X}}\left|H\right\rangle_{\mathrm{XX}}\right), \tag{5.2}$$

where Δ_{FFS} is the frequency splitting between the two bright X states formed in the fine structure splitting and the term $e^{-\mathrm{i}\Delta_{\mathrm{FFS}}t}$ presents the time evolution of the entangled state due to this splitting. When the input state of X_L is $\left|\psi\right\rangle_{\mathrm{X_L}} = \alpha\left|H\right\rangle + \beta\left|V\right\rangle$, the three-photon state of the X_E, X_L, and XX_E will be [98].

$$\begin{aligned}
\left|\psi\right\rangle_{\mathrm{X_E,X_L,XX_E}} &= \frac{1}{2}\left|\phi^{+}\right\rangle_{\mathrm{X_L,X_E}} \times\left(\alpha\left|H\right\rangle_{\mathrm{XX_E}} + e^{\delta(t)}\beta\left|V\right\rangle_{\mathrm{XX_E}}\right) \\
&+\frac{1}{2}\left|\phi^{-}\right\rangle_{\mathrm{X_L,X_E}} \times\left(\alpha\left|H\right\rangle_{\mathrm{XX_E}} - e^{\delta(t)}\beta\left|V\right\rangle_{\mathrm{XX_E}}\right) \\
&+\frac{1}{2}\left|\psi^{+}\right\rangle_{\mathrm{X_L,X_E}} \times\left(\beta\left|H\right\rangle_{\mathrm{XX_E}} + e^{\delta(t)}\alpha\left|V\right\rangle_{\mathrm{XX_E}}\right) \\
&+\frac{1}{2}\left|\psi^{-}\right\rangle_{\mathrm{X_L,X_E}} \times\left(\beta\left|H\right\rangle_{\mathrm{XX_E}} - e^{\delta(t)}\alpha\left|V\right\rangle_{\mathrm{XX_E}}\right)
\end{aligned} \tag{5.3}$$

Here, $|\psi^+\rangle$, $|\phi^+\rangle$, and ϕ^- are the three other Bell states besides $|\psi^-\rangle$. In the experiment, only the detection of $|\phi^-\rangle$ can herald the success of quantum teleportation, and this will result in a polarization state of XX_E as

$$|\psi\rangle_{XX_E}^{Tel} = \beta|H\rangle_{XX_E} - e^{i\Delta_{FFS}t}\alpha|V\rangle_{XX_E} \tag{5.4}$$

and the teleportation fidelity can be defined as

$$f_T = \mathrm{Tr}\left(\rho_{XX_E}|\psi\rangle_{XX_E}^{Tel}\,{}_{XX_E}^{Tel}\langle\psi|\right), \tag{5.5}$$

where ρ_{XX_E} is the density matrix of the reconstructed state of the XX_E photon.

In the experiment, the three-photon correlations are recorded when the input polarization state of the XX_E is set at several different states. To construct a coincidence histogram showing the experimental results, a parameter $\Delta\tau_{10}$ is introduced to present the time difference of the photon arriving times on the detector APD_0 and APD_1. This means that we can expect a successful detection of $|\psi^-\rangle_{X_L,X_E}$ only around $\Delta\tau_{10} = 0$. Also, the relative arrival time τ_2 of the teleported XX_E photon is set at $\tau_2 = 0$. Since QD is excited by consecutive double pulses, there will be 12 possible time correlation humps in the histogram, and the spacing between every adjacent two humps is 12.5 ns. In Fig. 5.17 (*adapted from Fig. 1 in Ref.* [98]), the experimental results are shown when the input state is set at $|\psi\rangle_{X_L} = |D\rangle_{X_L} = \left(|H\rangle_{X_L} + |V\rangle_{X_L}\right)/\sqrt{2}$ and the corresponding XX detection bases are co-polarized and cross-polarized with the input states, respectively.

In Fig. 5.17a and b, only the hump at $\Delta\tau_{10} = \tau_2 = 0$ contain the information about quantum teleportation, and other side peaks are not important because they result from the correlations between photons not interfering at BS2 or stem from XX_L photons and therefore are not related to the quantum teleportation. The comparison

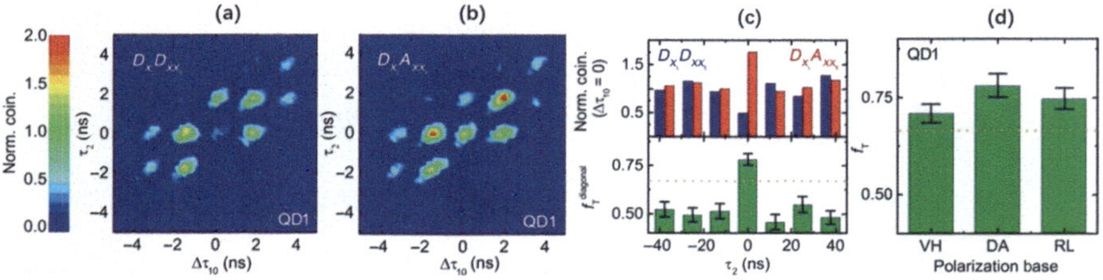

Fig. 5.17 The third-order correlation measured in teleportation experiment with a *D*-polarized X input photon, when the XX photons are detected at co-polarized (**a**) and cross-polarized (**b**) basis. (**c**) The integrated coincidences for both the D and A detection bases at different excitation cycles (top) and the corresponding calculated teleportation fidelity (bottom). (**d**) The measured teleportation fidelities for three pairs of orthogonal input states VH, DA, and RL. The dashed orange line represents classical limit in teleportation experiment. (a-d) Adapted with permission from Ref. [98]. In This figure, the Figs. (a) and (b) are adapted from the Fig. 3 (a) in Ref. [98], while the Figs. (c) and (d) are adapted from the Figs. 3 (b) and (c) in Ref. [98], respectively

between Fig. 5.17a and b indicates that the XX_E photons have high probability to be at the state $|\psi\rangle_{XX_E} = |A\rangle_{XX_E} = \left(|H\rangle_{XX_E} - |V\rangle_{XX_E}\right)/\sqrt{2}$, while the probability to be at the state $|D\rangle_{XX_E}$ is negligible. This is consistent with the theoretical prediction in Eq. (5.4). Furthermore, when $|\psi^-\rangle_{X_L,X_E}$ is detected in the BSM, the fidelity of teleportation can be extracted from the normalized third-order correlations with XX_E photon being at the states of $|A\rangle_{XX_E}$ and $|D\rangle_{XX_E}$ by the following expression [98].

$$f_T^{D-A} = \frac{g_{|A\rangle}^{(3)}(0)}{g_{|D\rangle}^{(3)}(0) + g_{|A\rangle}^{(3)}(0)}. \tag{5.6}$$

This expression gives a fidelity as high as 78%, as shown in Fig. 5.15c. To confirm the quantum nature in the teleportation experiment, similar measurements should be implemented under several different input state polarizations. In this experiment, three different input states equally spaced on the Bloch sphere are selected, and the calculated teleportation fidelities are shown in Fig. 5.17d. According to the relevant theory, the averaged fidelities for these three states cannot exceed 2/3 if only classical resources are employed in the teleportation experiment. However, it is clear in Fig. 5.17d that the averaged fidelities calculated according to the experimental data violate the classical bound, and this result can exclude any bias stemming from the experimental setup. Therefore, the experimental result undoubtedly indicates the quantum nature of the teleportation experiment with photons from single QDs.

The above results verify that QD photons can be used to teleport arbitrary input states, and the relevant conclusion can be extended to different QDs. The main reasons for such extension include two aspects. First, this extension can confirm that the conclusions obtained above are general and not QD-specific. Second, this extension is necessary in determining what degree of entanglement and photon indistinguishability is needed to successfully overcome the classical limit in teleportation. In many different QDs, the entanglement fidelity and two-photon interference visibility are measured, and the result is shown in the following Table 5.1.

Table 5.1 For QD labeled No. 1-5

QD [#]	V_{Hom} [%]	FSS [μeV]	f_E [%]	f_T^{exp}[%]	f_T^{theory}[%]
1	65 (2)	1.15 (0.16)	93 (0.3)	74.5 (1.6)	74.0
2	62 (3)	1.24 (0.12)	91 (0.3)	71.6 (0.8)	72.5
3	57 (4)	1.76 (0.20)	86 (0.3)	67.6 (0.7)	68.3
4	57 (4)	2.52 (0.30)	82 (0.3)	63.8 (0.8)	64.6
5	58 (4)	2.98 (0.14)	78 (0.3)	61.3 (1.3)	63.2

Note: This table shows the measured two-photon interference visibility. V_{Hom}, fine structure splitting; FSS, corresponding entanglement fidelity; f_E, average teleportation fidelity; f_T^{exp}, and the corresponding theoretically calculated average teleportation fidelity; f_T^{theory}. Reprinted with permission from [98]

To explain the data in the above table, a theoretical model is developed, which can reproduce the dependence of the experimental teleportation fidelities on the two-photon interference visibility. Though the teleportation fidelities can be calculated by Eq. (5.5), the density matrix $\rho_{\mathrm{XX_E}}$ should be obtained first. In Ref. [98], the author developed an expression

$$\rho_{\mathrm{XX_E}} = \int_{t_{\min}}^{t_{\max}} P(t) \sum_B p_{|B\rangle} |\psi\rangle_{\mathrm{XX_E}}^{B} {}_{\mathrm{XX_E}}^{B}\langle\psi|(t)\,dt, \tag{5.7}$$

to calculate $\rho_{\mathrm{XX_E}}$. In this equation, $P(t)=(1/T_{1\mathrm{X}})e^{-t/T_{1\mathrm{X}}}$ is the probability density of entangled photon pair generated at time t within the X lifetime $T_{1\mathrm{X}}$, while $p_{|B\rangle}$ represents the success of Bell state B detection and is related to the HOM visibility. Since the measurement time in the experiment is much longer than the X lifetime and temporal filtering is not used, the $t_{\min}$ and $t_{\max}$ can be set as 0 and infinite, respectively. Combined with the evolution of the entangled state, the probability density $P(t)$, as introduced for the output photon in Eq. 5.4, scales with the entanglement fidelity of the two-photon state $|\phi^+\rangle_{\mathrm{X_E},\mathrm{XX_E}}(t)$. Moreover, the possible polarization rotations induced by the experimental setup must be taken into consideration to correctly compare the measured teleportation fidelities with the prediction in Eq. 5.5. For this reason, a full tomography of XX_E on a second QD is implemented to obtain the single-photon density matrices.

Figure 5.18a (*adapted from Fig. 4 in Ref.* [98]) shows the real and imaginary parts of the density matrix obtained in the tomography when the input state of XX_E is a diagonal state. The teleportation fidelity extracted from this density matrix turns out to be 75.5 ± 1.5%. For all the other input states, similar measurements are also implemented, and an average fidelity of 71.8 ± 0.9% is obtained. This value agrees well with the theoretically calculated value 71.6 ± 0.8% obtained according to Eqs. (5.5), (5.6), and (5.7). The imaginary part of the density matrices shown in Fig. 5.18a reveals that there exist relative phase shifts between $|\psi\rangle_{\mathrm{X_L}}$ and $|\psi\rangle_{\mathrm{XX_E}}$. These phase shifts can be attributed to the polarization-dependent imperfections of the optical elements used in the experiment. The scattered laser photons could also lower the $g^{(2)}(0)$, but such effect is very weak here. In the simulation, such imperfections can be taken into consideration by extracting the Stokes parameters S_X and S_Y from the off-diagonal matrix elements, and then the angular offset of the Stokes vector with respect to S_Z can be calculated. Such angular offset is then applied as a rotation matrix on the corresponding input state to account for the mentioned phase shift.

With the theory, the dependence of the teleportation fidelity on the HOM visibility V_{HOM} and FSS can be calculated, in both of the ideal case and real case in which the imperfections of the experimental setup are taken into consideration. The relevant calculated results are shown in Fig. 5.18b (*adapted from Fig. 4 in Ref.* [98]). In Fig. 5.18, it is obvious that the HOM visibility should be higher than 50% for the violation of classical limit. In contrast, the teleportation fidelity is more tolerant to the FSS under V_{HOM} = 1, and the classical limit is violated even under the FSS = 5 eV. This can be attributed to the fact that the lifetime of the X state in the QDs used in the experiment is rather short, about $T_{1\mathrm{X}}$ = 265 ps, which can lead to

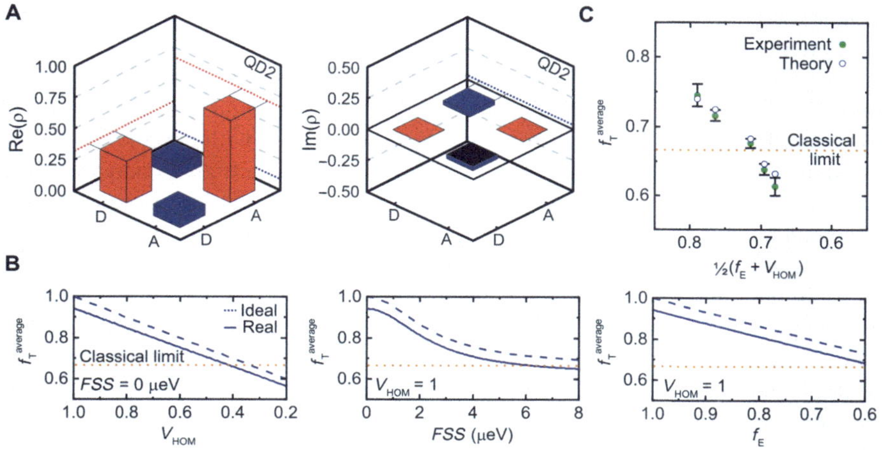

Fig. 5.18 The left and right parts in (**a**) show the measured real and imaginary parts, respectively, of the density matrix in the quantum teleportation experiment with QD2. In both (**a**) and (**b**), the reference frame is chosen in a respective from the diagonal input state. (**b**) The calculated average teleportation fidelity according to Eqs. (5.5), (5.6), and (5.7). From left to right, the three subfigures show the teleportation fidelity versus V_{HOM} with vanishing FSS, teleportation fidelity versus FSS with $V_{HOM} = 1$, and teleportation fidelity versus f_E with $V_{HOM} = 1$. In each subfigure, the expected ideal behavior is shown as dashed line, while the solid line represents results with real experimental conditions taken into consideration. (**c**) The experimental (green dot) and theoretical (blue circle) average teleportation fidelity obtained for all the five measured QDs. (a-c) Reprinted with permission from Ref. [98]

probability density function decaying rapidly when compared with the time evolution of the entangled state. To verify the validity of the theoretical mode in Eq. (5.7), several QDs are measured, and the average teleportation fidelities with them are shown as a function of their photon indistinguishability V_{HOM} and entanglement fidelity f_E in Fig. 5.18c. For the sake of simplicity, data are shown as a function of $1/2(f_E + V_{HOM})$, although it has been predicted in Fig. 5.18b that V_{HOM} brings stronger impact on the teleportation fidelity when compared with f_E. It is obvious that there exists an excellent quantitative agreement between experiments and theory, which means that the theory developed above could cover the physics determining the measured values of teleportation fidelities on semiconductor QDs. The small discrepancy between the theoretical and experimental results in Fig. 5.18b can be mainly attributed to the uncertainty in the FSS that can be as large as 0.3 meV.

5.4 The Progress of Atom–Photon Entanglement with QD-Based Devices

In 2012, two research groups, one composed of members from the United States, the United Kingdom, Japan, and Germany and the other one from Switzerland, independently demonstrated the entanglement between photon and quantum dot for the first time [51, 52, 99]. In more specific words, they have realized the quantum entanglement between the photon generated in the quantum dot and the spin state in the quantum dot.

To generate the photon, both of the groups employed the optical excitation method. First, they use the optical pumping technique to transfer the spin state in the QD from the downstate to the upstate. Then, a laser pulse with appropriate wavelength and duration was used to drive the quantum dot to a higher-energy-excited state, which can then decay to either one of the two lower-energy spin states. Accompanying this relaxation process, a photon is spontaneously emitted. This spontaneous emission brings the photon-QD system into a superposition of two states. In the first (second) state, the electron in QD has up (down)-orientated spin, and the photon is blue (red) and vertically (horizontally) polarized. Therefore, the state of the emitted photon gets quantum correlated (entangled) with the spin state of the electron in QD. However, one problem with this superposition state is that the electron spin is correlated with both of the color and polarization of the photon. In practical applications involving entanglement, the electron spin should be correlated in only one of the two degrees of freedom. To overcome this problem, the two groups mentioned above use two different methods.

In the work of K. De Greve and colleagues (combined group with four countries), the photon generated in the spontaneous emission in QD is down-converted by a second-order nonlinear process, namely, the difference frequency generation (DFG) [51]. By this process, the quantum correlation between the electron spin and photon color could be erased since the new photons generated in the DFG are in a broadened frequency range, instead of two sharp ones not overlapping with each other. Then, the electron spin in QD gets quantum entangled only with the photon polarization. Moreover, the new frequency formed in DFG is in telecommunication band, which is appropriate for long-distance photon propagation in optical fiber. This experiment shows that QD photon entanglement is promising in achieving practical quantum networks.

In another work, Gao and colleagues (from Switzerland) select frequency to be the only one degree of the freedom entangled with the electron spin in QD [52]. This is achieved by using a frequency filter in the collection of the photons generated in QD and only letting the anticlockwise circularly polarized photon pass the photon collection system. This eliminates the correlation with polarization, leaving an entangled state of photon color and electron spin. In the following paragraphs, this work will be taken as an instance to introduce the principle of QD photon entanglement.

In this work, a single-electron-charged, self-assembled InGaAs quantum dot was utilized. The whole structure of the device is shown in Fig. 5.19a.

As shown in Fig. 5.19a (*adapted from Fig. S1 in the supplementary material of Ref.* [52]), the QD is embedded in a Schottky diode structure. It is separated by a 40 nm tunneling barrier from a layer of n++-doped GaAs, which plays the role of back contact. A structured metallic gate is deposited on the surface, and its thickness is about 6 nm to guarantee that it is transparent for photon emitted from the QD. To enhance the photon collection efficiency, the Schottky diode structure is grown on the top of a 28-layer distributed Bragg reflector (DBR) mirror. The DBR mirror and GaAs layer above the QD can form a bad cavity with $Q < 10$. Since the QD emission

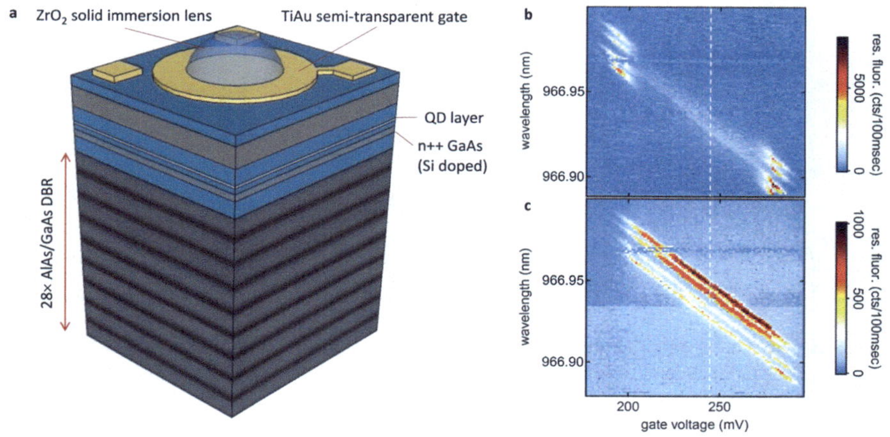

Fig. 5.19 (**a**) The structure of the QD-based device used in Ref. [52]. (**b**) The measured resonance fluorescence generated by continuous-wave pump light versus the gate voltage and the pump wavelength. (**c**) The measured resonance fluorescences under a pulse sequence shown in the following Fig. 5.20 (**b**) versus the gate voltage and the wavelength of the continuous-wave pump light. (a-c) Adapted with permission from Ref. [52]. Copyright (2012) Springer Nature

can be the redirected into the vertical mode of the cavity, an approximately fourfold enhancement in the QD resonance fluorescence (RF) emission is observed in the characterization of the QD. Besides, a ZrO_2 solid immersion lens (SIL) is fabricated on the top of the Schottky diode structure to further enhance the collection efficiency. In the experiment, two million counts are obtained per second when the QD is resonantly excited by laser power above the QD saturation power. The photons originated from the decay of a trion state, and the lifetime of this state is about 0.65 ns. This means that the photon generation rate in the structure in Fig. 5.19a is about 7.7×10^8 Hz above saturation. An overall collection efficiency of 0.26% can therefore be derived. After the APD efficiency (25%), fiber coupling efficiency (40%), and the ratios on the beam splitter (50%) and polarizer (50%) are corrected, an extraction efficiency $\geq$10% can be derived.

The QD charge state is controlled by the gate voltage between the surface gate and the back contact. All the experiments are implemented at the center of the gate voltage region where the QD contains a single electron in the ground state. Such region is called single-electron charging plateau. Figure 5.19b (*adapted from Fig. S1 in the supplementary material of Ref.* [52]) shows the measured resonance fluorescence versus the gate voltage and laser wavelength when continuous-wave laser is used to resonantly excite the QD. At the edges of the one-electron charge plateau, there exist four peaks in the spectrum. However, at the center of the charge plateau, these peaks disappear due to spin pumping.

The relevant energy-level diagram is shown in Fig. 5.20a (*adapted from Fig. 1 in Ref.* [52]), as well as the allowed optical transitions in Voigt configuration in which an external magnetic field (B_x) is applied perpendicular to the QD's growth

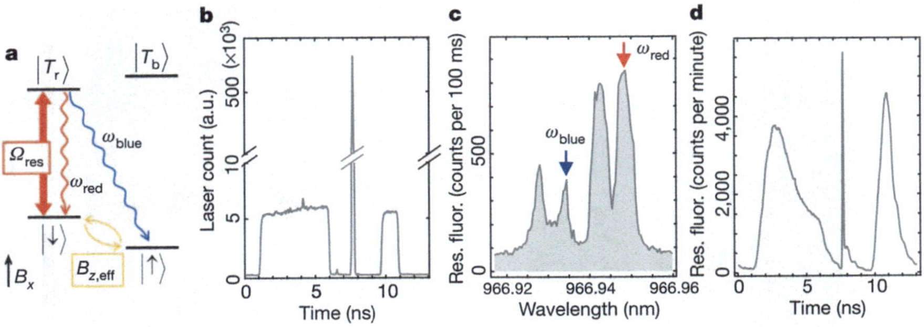

Fig. 5.20 (**a**) The level structure and the excitation/emission processes in the QD used in Ref. [52]. (**b**) The pulse sequence used to characterize the quantum dot transitions. A 5-ns resonant laser excitation is first used, and then a 4-ps p-pulse is applied. Lastly, a 1.2-ns resonant laser excitation comes. The repetition period of 13 ns. (**c**) The measured spectrum of the resonance fluorescence when the pulse sequence in b is used and the laser frequency ω_L is scanned at Bx~0.7 T. (**d**) The measured time-varying resonance fluorescence when $\omega_L = \omega_{red}$, indicated by the red arrow in c. (a-d) Adapted with permission from Ref. [52]. Copyright (2012) Springer Nature

direction. The oscillator strengths of all the four transitions between the two ground states and the two excited states in Fig. 5.16a are nearly identical, and the polarizations of the photon absorbed or emitted in the transitions are linear, parallel (H), or perpendicular (V) to the external magnetic field. The two ground states are labeled as $|\uparrow\rangle$ and $|\downarrow\rangle$, respectively, denoting the spin state parallel and antiparallel to the external magnetic field, respectively. By spontaneously emitting an H (V) polarized photon at frequency ω_{red} (ω_{blue}), the QD can relax from trion state $|T_r\rangle$ to $|\downarrow\rangle$ ($|\uparrow\rangle$). In addition to a laser which can resonantly drive QD from $|\downarrow\rangle$ to $|T_r\rangle$, non-resonant, right-hand circularly (σ^+) polarized, 4-ps-long mode-locked laser pulses are also used. This laser is red-detuned from the trion transitions by 210 GHz and therefore can coherently rotate the electron spin while minimal trion excitation is induced in this process. In Fig. 5.20a, these rotation pulses are labeled as $B_{z,eff}$, namely, time-dependent effective magnetic field along the z-direction.

The principle behind the deterministic generation of a spin–photon entangled state can be summarized simply. After the QD is exited to trion state, the spontaneous emission happens in the decay of this state, and it can project the hybrid QD/photon system into an entangled state [52].

$$|\psi\rangle = \frac{1}{\sqrt{2}}\left(|\downarrow\rangle|\omega_{red};H\rangle + i|\uparrow\rangle|\omega_{blue};V\rangle\right), \tag{5.8}$$

where $|\omega_{red}; H\rangle$ and $|\omega_{blue}; V\rangle$ represent the state of the photon generated in the spontaneous emission. The electronic Zeeman energy satisfies $\omega_Z = \omega_{blue} - \omega_{red}$.

The resonance fluorescence is employed to resolve the optical transitions of the quantum dot. The non-resonant spin rotation pulses are necessary in observing the resonance fluorescence [100]. To characterize the quantum dot, a pulse sequence as shown in Fig. 5.20b (*adapted from Fig. 1 in Ref.* [52]) is used, in which a 4-ps spin

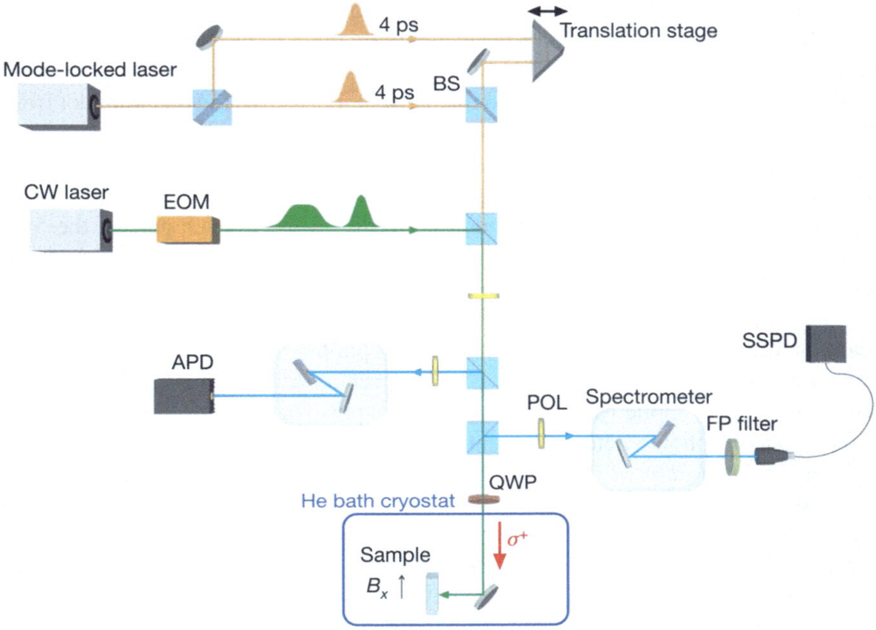

Fig. 5.21 The experimental setup used for the generation and observation of the entanglement between photon and spin in quantum dot. Adapted with permission from Ref. [52]. Copyright (2012) Springer Nature

rotation pulse is sandwiched within two resonant laser pulses with temporal length of 1.2 ns. The resonance fluorescence spectrum measured in the experiment is shown in Fig. 5.20c (*adapted from Fig. 1 in Ref.* [52]) when the resonant laser is scanned across the four allowed quantum dot transitions shown in Fig. 5.20a. The single-photon detection system with ultrafast response can allow the measurement of the time-dependent resonance fluorescence with high resolution. The result is shown in Fig. 5.20d (*adapted from Fig. 1 in Ref.* [52]), in which the photons emitted during the excitation of the two resonant lasers can be easily distinguished from those emitted due to the unexpected trion excitation in the duration of spin rotation pulse and the background photons from this pulse.

The experimental setup is shown in Fig. 5.21 (*adapted from Fig. 1 (e) in Ref.* [52]). In this setup, it is realizable to selectively detect the resonance fluorescence generated in different time windows or at different frequencies. To suppress the σ^+-polarized background laser light, a polarizer is used, which can project the polarization of the transmitted resonance fluorescence photons to σ^-(left-hand circular). By this way, the correlation between the polarization state of the photon and the spin state in the QD is erased, and the residual entanglement is between the spin state of the QD and the center frequency of the single-photon pulse. In this entangled state, the photonic states can be written as $|\omega_{\text{red}}\rangle$ and $|\omega_{\text{blue}}\rangle$.

To acquire spin–photon entanglement, the quantum dot is prepared into the state $|\uparrow\rangle$ with a probability higher than 87% by applying the 5-ns laser field resonant with the transition between $|\downarrow\rangle$ and $|T_r\rangle$. This pulse is called measurement/preparation pulse. Then, a 4-ps π-pulse is added to flip the spin, i.e., drives the quantum dot from $|\uparrow\rangle$ to $|\downarrow\rangle$. After the spin flip, a 1.2-ns resonant laser pulse is used to excite the QD to $|T_r\rangle$, and an entangled spin–photon pair is then generated in the decay process. The pulse in the last step is called entanglement pulse. The entire pulse sequence is depicted in Fig. 5.21a, and its repetition period is 13 ns. In each period, the 5-ns resonant laser also can implement the spin measurement for the preceding pulse sequence. The average number of photons emitted during the spin measurement/pumping time window is about 1, and a relatively modest combined collection and detection efficiency is about 0.1%. Therefore, a photon detection event at the APD can herald that the QD is in spin state $|\downarrow\rangle$. However, an additional π-pulse with duration of 3 ns can be applied after the entanglement pulse in one period is enabled, and this will make a photon detection event at the APD herald the spin state $|\uparrow\rangle$ in QD.

Figure 5.22a (*adapted from Fig. 2 in Ref.* [52]) shows the results of the time-resolved photon counts of resonance fluorescence at frequencies ω_{blue} and ω_{red}. The onset of entanglement pulse is set as $t = 0$, and the measured resonance fluorescence starts at this time. As shown in Fig. 5.21, a Fabry–Pérot interferometer is used to select out the photons at ω_{blue} or ω_{red} before the superconducting single-photon detector (SSPD). Without any conditional signal, the photon counts of the resonance fluorescence at the two frequencies have nearly identical intensity. This indicates that the oscillator strengths of the transitions from $|\uparrow\rangle$ and $|\downarrow\rangle$ to the excited state $|T_r\rangle$ are nearly equal. The shape of pulses formed by resonance fluorescence photons is determined by the time evolution of the trion population, which is related to the time evolution excitation laser intensity and the lifetime (~650 ps) of $|T_r\rangle$.

The normalized photon autocorrelation function is shown in Fig. 5.22c (*adapted from Fig. 2 in Ref.* [52]), for photons emitted within the time interval from 1.2 ns to 1.64 ns shown by the dashed green box in Fig. 5.22b (*adapted from Fig. 2 in Ref.* [52]). It can be seen that the $g^{(2)}(0)$ is about 0.02±0.01. The nonzero $g^{(2)}(0)$ results from the two-photon emission events that can lead to a small degradation of entanglement. When the dashed green box in Fig. 5.22b extends to the whole pulse of entanglement pulses, the measured $g^{(2)}(0)$ turns to be 0.16±0.01, shown by the black bars in Fig. 5.22c. This nonzero value is limited by the residual laser photons and the nonzero multiple excitation possibility of the QD during the entanglement pulse. Thus, the time interval from 1.2 ns to 1.64 ns is more appropriate for observing the spin–photon entanglement, and the deterioration of the entanglement fidelity due to the fluctuating nuclear spin environment can also be avoided in such interval. It is necessary to measure the classical correlation between the spin direction and the photon frequency before the demonstration of the spin–photon correlations in the computational non-rotated basis.

To perform this measurement, the coincidence counts are recorded between the resonance fluorescence photons (at ω_{blue} or ω_{red}) generated during the measurement/preparation pulse and those generated immediately after the entanglement pulse. Conditional on the detection of the spin state $|\downarrow\rangle$, it is clear that the probability in

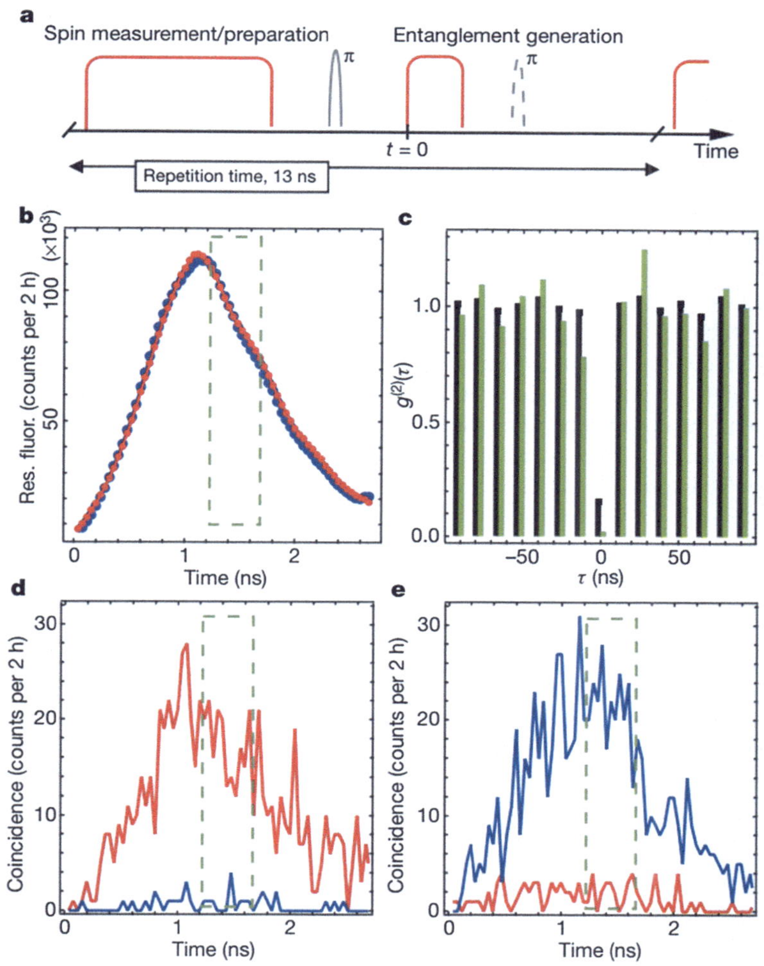

Fig. 5.22 (**a**) The pulse sequence used for observing the correlations in between the photon and spin in QD. Here the correlation is the non-rotated basis. (**b**) The measured time-resolved resonance fluorescence at frequencies of ω_{red} (red) and ω_{blue} (blue) using the pulse sequence shown in (**a**). (**c**) The measured self-correlation of resonance fluorescence photons generated in a 3-ns time window starting at the onset of the entanglement pulse ($t = 0$). (**d**) The measured time-dependent resonance fluorescence at ω_{red} (red) and ω_{blue} (blue), when a photon is detected during the subsequent measurement/preparation pulse. The physical meaning of the curves in (**e**) is similar with those in (**c**), except that a π-pulse is applied before the measurement/preparation pulse. (a-e) Adapted with permission from Ref. [52]. Copyright (2012) Springer Nature

detecting a ω_{blue} photon is significantly lower than that in detecting a ω_{red} photon, as shown in Fig. 5.22e. Specifically, in the time interval from 1.2 ns to 1.64 ns, the ratio of the red to blue photons is around 21:3. A 4-ps π-pulse at $t = 3$ ns is introduced to make it possible to herald the detection of $|\uparrow\rangle$ by single-photon color measurements.

As a result, the detection events of photons with ω_{red} are strongly suppressed, as shown in Fig. 5.22f. However, in this case, the ratio of the blue to red photons is only 10:3 because of the rotation spin pulse-induced unexpected trion excitation that lowers the fidelity in the subsequent spin measurement. By the two measurements, a fidelity of the classical correlations is determined to be F_1~0:87±0:09.

In order to verify the entanglement in the generated spin–photon, it is necessary to prove there exist correlations between the spin orientation in QD and photons color in a rotated basis. To construct the rotated basis, it is crucial to have a relative phase between the two components of the entangled state in Eq. (5.1). After generated at time t_g, the blue and red frequency components of the photonic mode get different phases in their propagation. By this way, the two components in Eq. (5.8) get a time-dependent relative phase. In the experiments, the generation time can be estimated by the photon detection time t_d on the SSPD by the expression $t_g \sim t_d - L/c$, where L is the distance between the QD and the detector and c is the speed of light. Hence, when time-dependent photon is detected at t_d, the spin–photon entangled state can be expressed as [52].

$$|\psi(t)\rangle = \frac{1}{\sqrt{2}}\left(|\downarrow\rangle|\omega_{\mathrm{red}};H\rangle e^{-i\omega_Z(t-t_g)} + i|\uparrow\rangle|\omega_{\mathrm{blue}};V\rangle\right). \tag{5.9}$$

After a $\pi/2$ pulse or a $3\pi/2$ pulse at $t = t_1 = 1.7$ ns is used, the electron spin in QD can be projected into the state $\left(|\downarrow\rangle - i|\uparrow\rangle\right)/\sqrt{2}$ or $\left(|\downarrow\rangle + i|\uparrow\rangle\right)/\sqrt{2}$, respectively, by a photon detection event at the APD within the duration of the following measurement/preparation pulse. Correspondingly, the photonic state consistent with the spin measurement after the $\pi/2$ pulse can be expressed as

$$\frac{1}{\sqrt{2}}\left(|\omega_{\mathrm{red}};H\rangle e^{-i\omega_Z(t-t_g)} - |\omega_{\mathrm{blue}};V\rangle\right). \tag{5.10}$$

The polarizer in Fig. 5.21 can map all the photons to the σ^--polarized mode. Thus, after passing the polarizer, the photonic state changes to

$$|\varphi\rangle = \frac{1}{\sqrt{2}}\left(|\omega_{\mathrm{red}}\rangle e^{-i\omega_Z(t-t_g)} - i|\omega_{\mathrm{blue}}\rangle\right). \tag{5.11}$$

In the next step, the Fabry–Pérot interferometer is removed, and the single-photon pulses generated by spontaneous emission enter into a fast SSPD with a jitter around 60 ps. When the jitter of the SSPD is smaller than $1/\omega_Z$, the photon emission time t_g is well-defined, and the time dependence in Eq. (5.9) can be resolved. This means that the photonic state can be projected unambiguously onto a superposition state of two frequency components. In such case, the probability of coincidence event is $P \approx \{1 + \sin[\omega_Z(t_1 - t_g)]\}/2$. The time-dependent relative phase makes it possible to infer the overlap of two single-photon state generated at different times with two orthogonal states. The projective measurements into

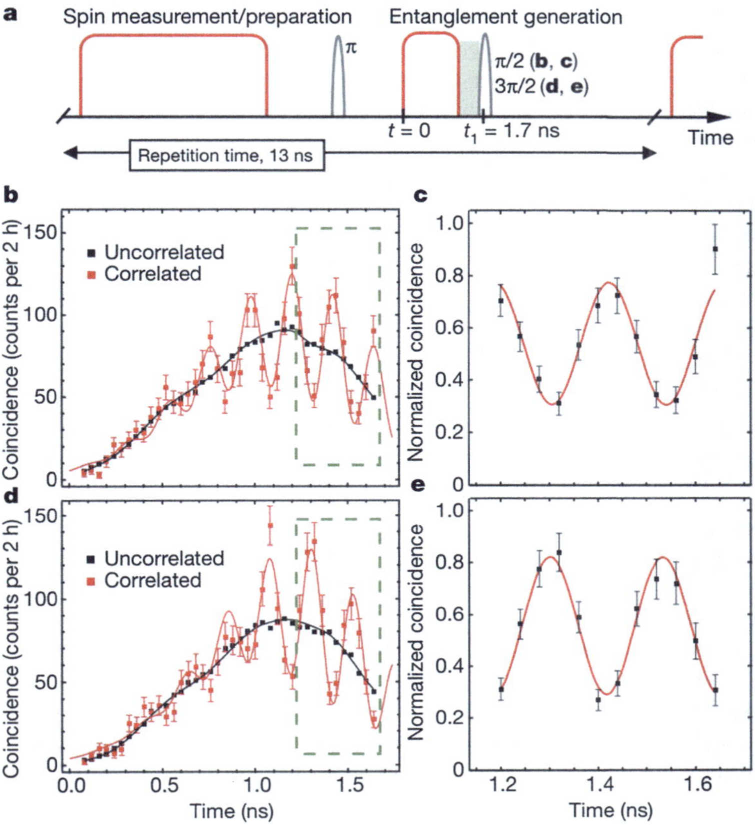

Fig. 5.23 (**a**) The pulse sequence used for observing the correlations in between the photon and spin in QD. Here the correlation is in the rotated basis. (**b**) The time-resolved coincidence counts between single-photon detection event after an entanglement pulse and single-photon detection event during the first measurement/preparation pulse after a $\pi/2$ pulse is applied (red squares). An average of coincidence events taking place in 25 different excitation/preparation cycles is also shown for comparison (black squares). The red curve is the fitted curve for the experimental data, while the black curve represents the rescaled single-photon pulse shown in (**b**). (**c**) The normalized coincidence within the temporal window (dashed box in b). Here, normalization is implemented by dividing the number of counts due to spin–photon correlation by twice the number of accidental counts in the corresponding time-bin. The physical meanings in (**d**) and (**e**) are similar with those in (**b**) and (**c**), respectively, except that a $3\pi/2$ pulse is applied. (a-e) Adapted with permission from Ref. [52]. Copyright (2012) Springer Nature

orthogonal states, such as $|\omega_{\mathrm{red}}\rangle \pm |\omega_{\mathrm{blue}}\rangle$, can be performed effectively based on random photon detection times and the fact that the single-photon pulse width $t > 1/\omega_z$.

The coincidence counts shown in Fig. 5.23b (*adapted from Fig. 3 in Ref.* [52]) are between the single photons induced by the entanglement pulse and the photons generated during the measurement/preparation pulse, following a $\pi/2$-pulse. Compared with the unconditioned single-photon detection events shown in

Fig. 5.22b, the conditioned single-photon detection events in Fig. 5.23b show remarkable oscillations with a period of $2\pi/\omega_Z = 238$ ps. Such phenomenon results from the quantum interference in the entangled spin–photon system. The averaged values of the coincidence count between the spin and photon detection events distributing in different excitation/preparation periods are also shown in Fig. 5.23c (*adapted from Fig. 3 in Ref.* [52]). Since there is no quantum interference related to this data, they show no oscillation behavior. However, these coincidences can be used in the calculation of the degree of quantum correlation. The visibility V_- for the observed oscillations in Fig. 5.23c is determined to be 45%±6%, after the coincidence counts of quantum correlation in spin–photon pairs are normalized and the time interval from 1.2 to 1.64 ns is selected out, as shown in Fig. 5.23b. Figure 5.23d (*adapted from Fig. 3 in Ref.* [52]) shows the time-dependent coincidence counts in another case, in which a $3\pi/2$ pulse is applied before the measurement/preparation pulse. The normalized values of $V_+ = 47\%\pm4\%$ in the interval with time from 1.2 to 1.64 ns are shown in Fig. 5.23e, and the oscillation is for spin detection along $\left(|\downarrow\rangle + i|\uparrow\rangle\right)/\sqrt{2}$. With the two visibility in Fig. 5.23c and e, the fidelity of the quantum correlations in the rotated basis can be calculated as $F_2 = (V_+ + V_-)/2 = 0.46\pm0.04$. Then, the overall entanglement fidelity can be calculated as $F \geq (F_1 + F_2)/2 = 0.67 \pm 0.05$. Here, the jitter in SSPD is an important factoring limiting the entanglement fidelity.

The work described in detail is one of the milestones in realizing the light–matter quantum interface by quantum dots for the application in future network. Based on the techniques developed in the work introduced above, the entanglement between the spin states in two quantum dots can be further realized by using the entanglement swapping protocol in Sect. 5.1.3. To realize this protocol, the Bell state projection of the photons generated in two QDs is a key step, behind which the main physical mechanism is the interference between single photons generated from two independent QDs. Such interference has been demonstrated in several works [101, 102]. In 2015, the group having demonstrated the spin–photon entanglement described above first extends that work to entangled hole spins in distant QDs [103]. Increasing the QD–photon entanglement distance is also important in developing practical quantum network, and down-converting the visible photon in spin/photon entanglement to telecommunication band via second-order nonlinear process is also an effective method for this goal [51, 104].

5.5 Summary

In this chapter, we summarize some typical applications of QD-based devices in quantum information and the recent experimental progresses in this topic. In Sect. 5.1, a brief introduction of the quantum communication is in three directions, including QKD (Sect 5.1.1), quantum teleportation (Sect. 5.1.2), and atom–photon entanglement (Sect. 5.1.3). In Sects. 5.2 and 5.3, the experimental progresses in

QKD and quantum teleportation with photons generated in QD-based devices are reviewed. In Sect. 5.4, the experimental demonstrations of quantum entanglement between photons and spin in QD are introduced. In all of the Sects. 5.2, 5.3, and 5.4, one or two typical references are introduced in detail to show the physical principle and technologies involved in relevant experiments.

References

1. Antonio, A., Bloch, I., Buhrman, H., et al. (2018). The quantum technologies roadmap: A European community view. *New Journal of Physics, 20*, 080201.
2. Guo, G. C., & Ying, M. S. (2019). Preface to special topic on quantum computing. *National Science Review, 6*, 20.
3. Raymer, M., & Monroe, C. (2019). The US national quantum initiative. *Quantum Science Technology, 4*, 020504.
4. Yamamoto, Y., Sasaki, M., & Takesue, H. (2019). Quantum information science and technology in Japan. *Quantum Science Technology, 4*, 020502.
5. Dowling, J., & Milburn, G. (2003). Quantum technology: The second quantum revolution. *Philosophical Transactions of the Royal Society of London. Series A: Mathematical, Physical and Engineering Sciences, 361*, 1655.
6. Ladd, T., Jelezko, F., Laflamme, R., et al. (2010). Quantum computers. *Nature, 464*, 45.
7. Shor, P. (1997). Polynomial-time algorithms for prime factorization and discrete logarithms on a quantum computer. *SIAM Journal on Computing*, 1484.
8. Deutsch, D., & Jozsa, R. (1992). Rapid solution of problems by quantum computation. *Philosophical Transactions of the Royal Society of London. Series A: Mathematical, Physical and Engineering Sciences, 439*, 553–558.
9. Grover, L. K. (1996). A fast quantum mechanical algorithm for database search. *Physical Review Letters, 78*, 212–219.
10. Krenn, M., Malik, M., Scheidl, T., Ursin, R., & Zeilinger, A. (2016). Quantum communication with photons. In M. Al-Amri, M. El-Gomati, & M. Zubairy (Eds.), *Optics in our time*. Cham: Springer.
11. Gisin, N., & Thew, R. (2007). Quantum communication. *Nature Photonics, 1*, 165–171.
12. Yuan, Z. S., Bao, X. H., Lu, C. Y., Zhang, J., Peng, C. Z., & Pan, J. W. (2010). Entangled photons and quantum communication. *Physics Reports, 497*(1).
13. Khan, I., Elser, D., Dirmeier, T., Marquardt, C., & Leuchs, G. (2017). Quantum communication with coherent states of light. *Philosophical Transactions of the Royal Society of London. Series A: Mathematical, Physical and Engineering Sciences, 375*, 20160235.
14. Kimble, H. J. (2008). The quantum internet. *Nature, 453*, 1023.
15. Wehner, S., Elkouss, D., & Hanson, R. (2018). Quantum internet: A vision for the road ahead. *Science, 362*, 303.
16. Pant, M., Krovi, H., Towsley, D., et al. (2019). Routing entanglement in the quantum internet. *npj Quantum Information, 5*, 25.
17. Dür, W., Lamprecht, R., & Heusler, S. (2017). Towards a quantum internet. *European Journal of Physics, 38*, 043001.
18. Gisin, N., Ribordy, G., Tittel, W., & Zbinden, H. (2002). Quantum cryptography. *Reviews of Modern Physics, 74*, 145.
19. Bennett, C. H., & Brassard, G. (1984). Quantum cryptography: Public key distribution and coin tossing. *Proceedings of IEEE International Conference on Computer Systems and Signal Process, 175*, 179.
20. Bennett, C. H., Brassard, G., & Ekert, A. (1992). Quantum cryptography. *Scientific American, 267*, 50.

21. Bennett, C. H. (1992). Quantum cryptography using any two nonorthogonal states. *Physical Review Letters, 68*, 3121.
22. Biham, E., & Mor, T. (1997). Security of quantum cryptography against collective attacks. *Physical Review Letters, 78*, 2256.
23. Townsend, P. (1997). Quantum cryptography on multiuser optical fiber networks. *Nature (London), 385*, 47.
24. Brėguet, J., Muller, A., & Gisin, N. (1994). Quantum cryptography with polarized photons in optical fibers: Experimental and practical limits. *Journal of Modern Optics, 41*, 2405.
25. Buttler, W. T., Hughes, R. J., Kwiat, P. G., et al. (1998). Practical free-space quantum key distribution over 1 km. *Physical Review Letters, 81*, 3283.
26. Buttler, W. T., Hughes, R. J., Lamoreaux, S. K., et al. (2000). Daylight quantum key distribution over 1.6 km. *Physical Review Letters, 84*, 5652.
27. Ekert, A. K. (1991). Quantum cryptography based on Bell's theorem. *Physical Review Letters, 67*, 661.
28. Jennewein, T., Simon, C., Weihs, G., Weinfurter, H., & Zeilinger, A. (2000). Quantum cryptography with entangled photons. *Physical Review Letters, 84*, 4729–4732.
29. Boaron, A., Boso, G., Rusca, D., et al. (2018). Secure quantum key distribution over 421 km of optical fiber. *Physical Review Letters, 121*, 190502.
30. Bennett, C. H., Brassard, G., Crėpeau, C., et al. (1993). Teleporting an unknown quantum state via dual classical and Einstein-Podolsky-Rosen channels. *Physical Review Letters, 70*, 1895.
31. Bouwmeester, D., Pan, J. W., Mattle, K., et al. (1997). Experimental quantum teleportation. *Nature, 390*, 575.
32. Ursin, R., Jennewein, T., Aspelmeyer, M., et al. (2004). Quantum teleportation across the Danube. *Nature, 430*, 849.
33. Boschi, D., Branca, S., Martini, D., et al. (1998). Experimental realisation of teleporting an unknown pure quantum state via dual classical and Einstein–Podolski–Rosen channels. *Physical Review Letters, 80*, 1121–1125.
34. Jin, X.-M., Ren, J. G., Yang, B., et al. (2010). Experimental freespace quantum teleportation. *Nature Photonics, 4*, 376.
35. Kim, Y. H., Kulik, S. P., & Shih, Y. (2001). Quantum teleportation of a polarisation state with complete bell state measurement. *Physical Review Letters, 86*, 1370.
36. Yin, J., Ren, J. G., & Lu, H. (2012). Quantum teleportation and entanglement distribution over 100kilometre freespace channels. *Nature, 488*, 185–188.
37. Ma, X. S., Herbst, T., Scheidl, T., et al. (2012). Quantum teleportation over 143 kilometres using active feedforward. *Nature, 489*, 269.
38. Marcikic, I., de Riedmatten, H., Tittel, W., et al. (2003). Long distance teleportation of qubits at telecommunication wavelengths. *Nature, 421*, 509.
39. Wang, X. L., Cai, X. D., Su, Z. E., et al. (2015). Quantum teleportation of multiple degrees of freedom of a single photon. *Nature, 518*, 516.
40. Furusawa, A., Sørensen, J. L., Braunstein, S. L., et al. (1998). Unconditional quantum teleportation. *Science, 282*, 706.
41. Takei, N., Yonezawa, H., Aoki, T., et al. (2005). High fedelity teleportation beyond the nocloning limit and entanglement swapping for continuous variables. *Physical Review Letters, 94*, 220502.
42. Sherson, J. F., Krauter, H., Olsson, R. K., et al. (2006). Quantum teleportation between light and matter. *Nature, 443*, 557–560.
43. Valivarthi, R., Grimau Puigibert, M., Zhou, Q., et al. (2016). Quantum teleportation across a metropolitan-area fibre network. *Nature Photonics, 10*, 676.
44. Sun, Q. C., Mao, Y. L., Chen, S. J., et al. (2016). Quantum teleportation with independent sources and prior entanglement distribution over a network. *Nature Photonics, 10*, 671.
45. Moehring, D. L., Maunz, P., & Olmschenk, S. (2007). Entanglement of single-atom quantum bits at a distance. *Nature, 449*, 68.

46. Stute, A., Casabone, B., Schindler, P., et al. (2012). Tunable ion–photon entanglement in an optical cavity. *Nature, 485*, 482.
47. Volz, J., et al. (2006). Observation of entanglement of a single photon with a trapped atom. *Physical Review Letters, 96*, 030404.
48. Volz, J., Weber, M., Schlenk, D., et al. (2006). Observation of entanglement of a single photon with a trapped atom. *Physical Review Letters, 96*, 030404.
49. Ritter, S., Nölleke, C., Hahn, C., et al. (2012). An elementary quantum network of single atoms in optical cavities. *Nature, 484*, 195–200.
50. Togan, E., Chu, Y., Trifonov, A. S., et al. (2010). Quantum entanglement between an optical photon and a solid-state spin qubit. *Nature, 466*.
51. De Greve, L. Y., McMahon, P. L., et al. (2012). Quantum–dot spin–photon entanglement via frequency down-conversion to telecom wavelength. *Nature, 491*, 421.
52. Gao, W. B., Fallahi, P., Togan, E., et al. (2012). Observation of entanglement between a quantum dot spin and a single photon. *Nature, 491*, 426.
53. Bock, M., Eich, P., Kucera, S., et al. (2018). High-fidelity entanglement between a trapped ion and a telecom photon via quantum frequency conversion. *Nature Communications, 9*, 1998.
54. Matsukevich, D. N., & Kuzmich, A. (2004). Quantum state transfer between matter and light. *Science, 306*, 663.
55. Matsukevich, D. N., Chaneliere, T., Bhattacharya, M., et al. (2005). Entanglement of a photon and a collective atomic excitation. *Physical Review Letters, 95*, 040405.
56. Farrera, P., Heinze, G., de Riedmatten, H., et al. (2018). Entanglement between a photonic time-bin qubit and a collective atomic spin excitation. *Physical Review Letters, 120*, 100501.
57. Li, L., Dudin, Y. O., & Kuzmich, A. (2013). Entanglement between light and an optical atomic excitation. *Nature, 498*, 466–469.
58. de Riedmatten, H., Laurat, J., Chou, C. W., et al. (2006). Direct measurement of decoherence for entanglement between a photon and stored atomic excitation. *Physical Review Letters, 97*, 113603.
59. Chen, S., Chen, Y. A., Zhao, B., et al. (2007). Demonstration of a stable atom-photon entanglement source for quantum repeaters. *Physical Review Letters, 99*, 180505.
60. Lodahl, P. (2018). Quantum-dot based photonic quantum networks. *Quantum Science Technology, 3*, 013001.
61. R. L. Rivest, , A. Shamir, and L. M. Adleman, "A method of obtaining digital signatures and public-key cryptosystems," Communications of the ACM 21, 120 (1978).
62. Carrasco-Casado, A., Fernandez Marmol, V., & Denisenko, N. (2016). Free-space quantum key distribution. In M. Uysal, C. Capsoni, Z. Ghassemlooy, et al. (Eds.), *Optical wireless communications–an emerging technology*. Springer International Publishing.
63. Deutsch, D., Ekert, A., Jozsa, R., et al. (1996). Quantum privacy amplification and the security of quantum cryptography over noisy channels. *Physical Review Letters, 77*, 2818.
64. Ursin, R., Tiefenbacher, F., Schmitt-Manderbach, T., et al. (2007). Entanglement-based quantum communication over 144 km. *Nature Physics, 3*, 481.
65. Pirandola, S., Eisert, J., Weedbrook, C., Furusawa, A., & Braunstein, S. L. (2015). Advances in quantum teleportation. *Nature Photonics, 9*, 641.
66. Sangouard, N., Simon, C., de Riedmatten, H., & Gisin, N. (2011). Quantum repeaters based on atomic ensembles and linear optics. *Reviews of Modern Physics, 83*, 33.
67. Duan, L.-M., Lukin, M. D., Cirac, J. I., & Zoller, P. (2001). Long-distance quantum communication with atomic ensembles and linear optics. *Nature (London), 414*, 413.
68. Briegel, H.-J., Dür, W., Cirac, J. I., & Zoller, P. (1998). Quantum repeaters: The role of imperfect local operations in quantum communication. *Physical Review Letters, 81*, 5932.
69. W. Tittel, , M. Afzelius, T. Chaneliére, R. Cone, S. Kröll, S. Moiseev, and M. Sellars, "Photon-echo quantum memory in solid state systems", Laser & Photonics Reviews 4, 244 (2009).
70. Afzelius, M., Simon, C., de Riedmatten, H., & Gisin, N. (2009). Multimode quantum memory based on atomic frequency combs. *Physical Review A, 79*, 052329.

71. Clausen, C., Usmani, I., Bussières, F., et al. (2011). Quantum storage of photonic entanglement in a crystal. *Nature, 469*, 508.
72. Afzelius, M., Gisin, N., & de Riedmatten, H. (2015). Quantum memory for photons. *Physics Today, 68*, 42.
73. Scarani, V., Behmann-Pasquinucci, H., Cerf, N. J., et al. (2009). The security of practical quantum key distribution. *Reviews of Modern Physics, 81*, 1301.
74. Wang, X. B. (2005). Beating the photon-number-splitting attack in practical quantum cryptography. *Physical Review Letters, 94*, 230503.
75. Waks, E., Inoue, K., Santori, C., Fattal, D., Vučković, J., Solomon, G. S., & Yamamoto, Y. (2002). Quantum cryptography with a photon turnstile. *Nature, 420*, 762.
76. Heindel, T., Kessler, C., & Rau, M. (2012). Quantum key distribution using quantum dot single-photon emitting diodes in the red and near infrared spectral range. *New Journal of Physics, 14*, 083001.
77. Intallura, P. M., Ward, M. B., Karimov, O. Z., et al. (2009). Quantum communication using single photons from a semiconductor quantum dot emitting at a telecommunication wavelength. *Journal of Optics A: Pure and Applied Optics, 11*, 054005.
78. Intallura, P. M., Ward, M. B., Karimov, O. Z., et al. (2007). Quantum key distribution using a triggered quantum dot source emitting near 1.3 μm. *Applied Physics Letters, 91*, 161103.
79. Takemoto, K., Nambu, Y., Miyazawa, T., et al. (2010). Transmission experiment of quantum keys over 50 km using high-performance quantum-dot single-photon source at 1.5 μm wavelength. *Applied Physics Express, 3*, 092802.
80. Takemoto, K., Nambu, Y., Miyazawa, T., et al. (2015). Quantum key distribution over 120 km using ultrahigh purity single-photon source and superconducting single-photon detectors. *Scientific Reports, 5*, 14383.
81. Santori, C., Pelton, M., Solomon, G., Dale, Y., & Yamamoto, Y. (2001). Triggered single photons from a quantum dot. *Physical Review Letters, 86*, 1502.
82. Santori, C., Fattal, D., & Vučković, J. (2002). Indistinguishable photons from a single-photon device. *Nature, 419*, 594.
83. Soujaeff, A., Nishioka, T., Hasegawa, T., et al. (2007). Quantum key distribution at 1550 nm using a pulse heralded single photon source. *Optics Express, 15*, 726.
84. Ward, M. B., Farrow, T., See, P., et al. (2007). Electrically driven telecommunication wavelength single-photon source. *Applied Physics Letters, 90*, 063512.
85. Gobby, C., Yuan, Z. L., & Shields, A. J. (2004). Quantum key distribution over 122 km of standard telecom fiber. *Applied Physics Letters, 84*, 3762.
86. Gottesman, D., Lo, H.-K., Lütkenhaus, N., & Preskill, J. (2004). Security of quantum key distribution with imperfect devices. *Quantum Information and Computation, 4*, 325.
87. Brassard, G., & Savail, L. (1994). Secret-key reconciliation by public discussion. *Lecture Notes in Computer Science, 765*, 410.
88. Yoshino, K., Fujiwara, M., Tanaka, A., et al. (2012). High-speed wavelength-division multiplexing quantum key distribution system. *Optics Letters, 37*, 223–225.
89. Tang, Y. L., Yin, H. L., Ma, X. F., et al. (2013). Source attack of decoy-state quantum key distribution using phase information. *Physical Review A, 88*, 022308.
90. Takemoto, K., Takatsu, M., Hirose, S., & Yokoyama, N. (2007). An optical horn structure for single-photon source using quantum dots at telecommunication wavelength. *Journal of Applied Physics, 101*, 081720.
91. Song, H. Z., Takemoto, K., Miyazawa, T., et al. (2013). Design of Si/SiO2 micropillar cavities for Purcell-enhanced single photon emission at 1.55 μm from InAs/InP quantum dots. *Optics Letters, 38*, 3241.
92. Rosenberg, D., Peterson, C., Harrington, J., et al. (2009). Practical long-distance quantum key distribution system using decoy levels. *New Journal of Physics, 11*, 045009.
93. Liu, Y., Chen, T. Y., Wang, J., et al. (2010). Decoy-state quantum key distribution with polarized photons over 200 km. *Optics Express, 18*(51).

94. Nilsson, J., Stevenson, R. M., Chan, K. H. A., et al. (2013). Quantum teleportation using a light-emitting diode. *Nature Photonics, 7*, 311.
95. Stevenson, R., Nilsson, J., Bennett, A., et al. (2013). Quantum teleportation of laser-generated photons with an entangled-light-emitting diode. *Nature Communications, 4*, 2859.
96. Varnava, C., Stevenson, R., Nilsson, J., et al. (2016). An entangled-LED-driven quantum relay over 1km. *npj Quantum Information, 2*, 16006.
97. Huwer, J., Stevenson, R., Skiba-Szymanska, J., et al. (2017). Quantum dot-based telecommunication-wavelength quantum relay. *Physical Review Applied, 8*, 024007.
98. Reindl, M., Huber, D., Schimpf, C., et al. (2018). All-photonic quantum teleportation using on-demand solid-state quantum emitters. *Science Advances, 4*, 1255.
99. Economou, S. (2012). Putting a spin on photon entanglement. *Nature, 491*, 343.
100. Xu, X. D., Wu, Y. W., Sun, B., et al. (2007). Fast spin state initialization in a singly charged InAs-GaAs quantum dot by optical cooling. *Physical Review Letters, 99*, 097401.
101. Flagg, E., Muller, A., Polyakov, S., et al. (2010). Interference of single photons from two separate semiconductor quantum dots. *Physical Review Letters, 104*, 137401.
102. Patel, R., Bennett, A., Farrer, I., et al. (2010). Two-photon interference of the emission from electrically tunable remote quantum dots. *Nature Photonics, 4*, 632.
103. Delteil, A., Sun, Z., Gao, W. B., et al. (2016). Generation of heralded entanglement between distant hole spins. *Nature Physics, 12*, 218.
104. De Greve, K., McMahon, P., Yu, L., et al. (2013). Complete tomography of a high-fidelity solid-state entangled spin–photon qubit pair. *Nature Communications, 4*, 2228.

Chapter 6
Cesium Lead Halide Perovskite Quantum Dots in the Limelight: Dynamics and Applications

Xinping Zhai, Yifan Huang, Zhanzu Feng, Xiaodong Zhang, and Qiang Wang

Abstract All-inorganic metal halide perovskites are a new emerging star following organometallic halide perovskites, thanks to their outstanding optoelectronic properties and relatively higher stability at the same time. Our group has contributed in this field through developing novel fabrication methods for perovskite micro/nano crystals, cesium lead halide perovskite quantum dots ($CsPbX_3$ QDs) in particular, and investigating their photoinduced excited-state dynamics via various time-resolved spectroscopic techniques. Therefore in this chapter, we focus on the review of (nonlinear) optical properties and dynamics of $CsPbX_3$ QDs. The chapter is organized as the followings: immediately after Sect. 6.1 of *Introduction*, Sect. 6.2 summarizes current preparation methods and focuses on mechanosynthesis (ball milling technique strategy), a methodology we are among the earliest groups to advocate. Section 6.3 details the photoinduced excited-state dynamics of perovskite QDs as fast as femtosecond (fs) scale, through different time-resolved spectroscopic tools including fluorescence lifetime measurement, fs transient absorption spectroscopy, streak camera, etc. Section 6.4 reviews the applications of $CsPbX_3$ QDs stemming from their unique optical/electronic properties: optoelectronic devices, photocatalysis, nanolasers, and particularly nonlinear optics. The final Sect. 6.5 is *Perspectives*, which gives the thoughts for the future trend of the field from the authors. To sum up, all-inorganic perovskite QDs have thrived in the optoelectronic field. Numerous new

X. Zhai · Q. Wang (✉)
State Key Laboratory of Applied Organic Chemistry (SKLAOC), Key Laboratory of Nonferrous Metal Chemistry and Resources Utilization of Gansu Province, College of Chemistry and Chemical Engineering, Key Laboratory of Special Function Materials and Structure Design, Ministry of Education, Lanzhou University, Lanzhou, China
e-mail: qiangwang@lzu.edu.cn

Y. Huang (✉)
School of Electronic and Information Engineering, Lanzhou City University, Lanzhou, China
e-mail: yifanhuang@lzcu.edu.cn

Z. Feng · X. Zhang
National Key Laboratory of Materials Behavior and Evaluation Technology in Space Environment, Harbin, China

P. Yu, Z. M. Wang (eds.), *Quantum Dot Optoelectronic Devices*, Lecture Notes in Nanoscale Science and Technology 27,
https://doi.org/10.1007/978-3-030-35813-6_6

experimental findings and applications of them are springing up every day. As an exciting new territory, there is much to be explored through joint efforts of researchers from varied background of physics, chemistry, optics, electronics, engineering, etc., through employing powerful methodologies like ultrafast spectroscopy.

Keywords Perovskite quantum dots · Cesium lead halide · Excited-state dynamics · Ultrafast spectroscopy · Optoelectronic devices · Nonlinear optics

6.1 Introduction

Perovskite originally refers to a mineral of calcium titanium oxide with the chemical formula of $CaTiO_3$, which was first discovered by Gustav Rose in the Ural Mountains of Russia in 1839 and named after the Russian scientist Lev Perovski. Today the word "perovskite" has been generally used to describe the type of materials having the same crystal structure as that of $CaTiO_3$. Perovskite compounds are widespread on the earth, among which the reserves of magnesium silicate perovskite ($MgSiO_3$) are the most abundant [1, 2].

The general chemical formula of perovskite is ABX_3, where A and B represent two types of cations with different ionic radii and X represents an anion associated with them. In the oxide perovskite (ABO_3), A is a divalent metal cation (e.g., Mg^{2+}, Ca^{2+}, Sr^{2+}, Ba^{2+}, Pb^{2+}). B is a tetravalent cation (e.g., Ti^{4+}, Si^{4+}, Fe^{4+}), and X is O^{2-} ion with two negative charges. For an ideal cubic crystal perovskite, B cation is combined with 6 anions to form an octahedral anion, and A cation is a 12-coordinate structure (see Fig. 6.1a). To maintain the stability of the cubic structure, strict requirements in terms of the radii of all the ions constituting the perovskite must be satisfied. Even a slight distortion or deformation will reduce the coordination number of A or B cation or both of them and result in a decrease in the symmetry of the perovskite. Interestingly, many of the physical properties of perovskites, especially their electronic, magnetic, and dielectric properties, are determined by these slight distortions, which thus provide a useful tool to accurately modulate the physicochemical properties of perovskites by replacing the cations to generate distortions. However, these replacement cations must meet certain conditions to form a stable perovskite crystal form, which is called the tolerance factor t, as shown in Eq. (6.1) [3, 4]

$$t = \left(R_B + R_X\right) / \sqrt{2}\left(R_A + R_X\right) \tag{6.1}$$

Among them, R_A, R_B, and R_X represent the ionic radii of A, B, and X, respectively. It is possible to form a stable perovskite crystal when t is between 0.8 and 1. If the tolerance factor condition is not satisfied, the structure extension of the octa-

(a)

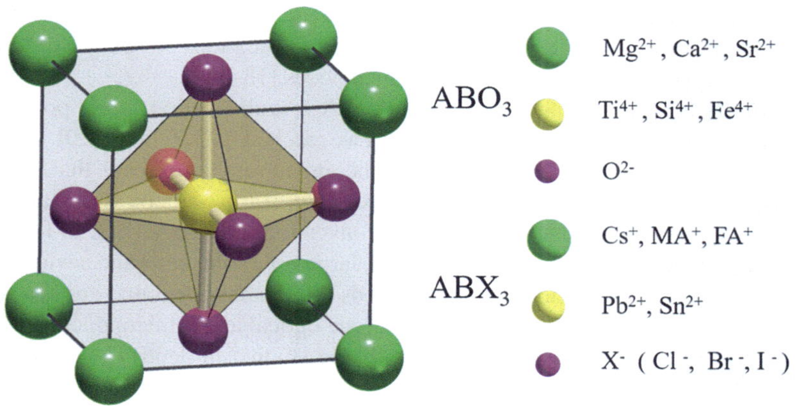

(b)

Perovskite Crystalline System (ABX_3)			
Inorganic Oxide Perovskite (ABO_3)		Halide Perovskite (ABX_3)	
Intrinsic Perovskite (ABO_3)	Doped Perovskite ($A_xB_{1-x}M_yN_{1-y}O_3$)	Organic Halide Perovskite (ABX_3)	Inorganic Halide Perovskite (ABX_3)

Fig. 6.1 (**a**) Crystal structure of ABX_3 cubic perovskites. (**b**) Classification of perovskites

hedron frame with common vertex may collapse. For instance, an NH_4CdCl_3 structure with co-edge connection is formed if $t < 0.8$; and when $t > 1$, a $CsNiBr_3$ structure with coplanar connection is formed [5–7]. In order to obtain the $[BX_6]^{4-}$octahedral structure, the octahedral factor of μ as shown in Eq. (6.2) requires an effective overlap between X and B ions to coordinate.

$$\mu = R_B / R_X \tag{6.2}$$

When $\mu < 0.41$, the radius of B cation is too small to form the octahedral structure [4].

$BaTiO_3$ was the first oxide perovskite exhibiting photocurrent response, and the history dated back to 1956 [1]. Similar phenomena were also observed subsequently in $LiNbO_3$ perovskites as an intrinsic property of ferroelectricity, arising from the space charge of the crystal [8]. At present, the application of perovskites in the field of optoelectronics is in full swing [9–12], a large part of which is contributed from the halide perovskites.

Halide perovskites are a huge branch of the perovskite family and can be divided into organometallic halide perovskites and all-inorganic metal halide perovskites (Fig. 6.1b). Compared with inorganic oxide perovskites, organometallic halide perovskites (OHPs) appearing in the past 20 years are indeed younger [13], and much remains to be discovered. In OHPs, A cation could be an aliphatic amine salt or an aromatic amine salt, and the divalent B metallic cation may be Cu^{2+}, Ni^{2+},

Co^{2+}, Fe^{2+}, Mn^{2+}, Pd^{2+}, Cd^{2+}, Ge^{2+}, Sn^{2+}, Pb^{2+}, Eu^{2+}, etc. Among these combinations, there are currently three major categories (when B cations are Ge^{2+}, Sn^{2+}, Pb^{2+}) that exhibit great potential for optoelectronic devices [13–15].

For typical organometallic perovskites (A is amine cation), their power conversion efficiency (PCE) in the solar cell has reached ~23.9% [16–20] from 3.81% within a few years [16]. The incredible increment rate far exceeds that of other traditional semiconductor materials. For example, it took silicon materials for more than 20 years to achieve a 20% power conversion efficiency [21]. In recent years, organometallic perovskite-based flexible, large-area solar cells and a wide variety of devices have been fabricated to meet needs in scalable integrated circuits and other applications [22, 23]. In addition, the unique optical and semiconducting properties of organometallic perovskites are also being studied in depth [24] and applied to various fields, such as photodetectors (PDs), light-emitting diodes (LEDs) [25–27], field-effect transistors (FETs), and lasers [28–30].

However, organometallic perovskites still have many unresolved problems, especially in terms of stability. Because of the presence of organic functional groups, they are easily decomposed in exposure to photoirradiation or heat like many other organic materials [31, 32]. Organometallic perovskites are also extremely sensitive to oxygen and moisture in the air, and hence devices based on them are subject to severe environmental constraints. Since 2016, numerous research groups have made relentless efforts to improve the performance of perovskite devices, including adding other species with suitable tolerance factors (such as Cs^+, formamidinium (FA)) to enhance stability [32], surface modification to increase water resistance [33], polymer encapsulation [34], and the replacement of the organic electron transport layer with inorganic oxides [35] to improve device performance. Despite advances in these areas, the presence of organic groups itself had seriously hampered the practical application of such devices.

One effective method to improve perovskite stability on the premise of preserving its excellent optical and electronic properties was to replace the organic functional groups with inorganic components. The idea was firstly proposed by Protesescu et al. in 2015 [36, 37], and an all-inorganic metal halide perovskite ($CsPbX_3$) quantum dots (QDs) was synthesized. This all-inorganic material had a higher melting point (>500 °C) [38, 39] and better light stability, which made them more suitable in the optoelectronic field. These QDs usually possessed near-unity photoluminescence quantum yield, e.g., α-$CsPbI_3$ colloidal QDs [40]. More interestingly, their emission spectra were adjustable within a broad range through tuning the halide composition and size of the QDs [36, 41, 42]. The full width at half maximum (FWHM) was narrow (12–40 nm), and the color gamut was wide (~140%). In this sense, inorganic perovskite QDs have shown even better performance than conventional cadmium sulfide (CdS) QDs, proving its promise for use in next-generation LEDs [38, 43].

The research on inorganic halide perovskites has been more than 50 years. However, the majority of the previous work focused on the study of their crystal structure, such as phase transition and ionic conductivity [44–47]. There were also plenty of reports on the fluorescent properties of bulk inorganic perovskites, but not of QDs. In fact, many research groups believed that the fluorescence center existed in CsX crystals [48–50]. Overall, the optical properties of the bulk inorganic perovskite published were not a focus at that time. The old semiconductor materials experience renaissance until the invention of inorganic perovskite QDs, which have thrived in a wide range of optoelectronic applications shortly after its birth: Zeng's group produced LEDs with inorganic perovskite QDs [37]; the observation of amplified stimulated radiation (ASE) at low threshold with wavelength-adjustable QDs was reported by Kovalenko's group [51, 52], just to name a few. Apart from the QDs, perovskites with a variety of morphology like nanowires, nanosheets, and even larger particles have been reported to show excellent photo physical/chemical properties [53–55]. Undoubtedly, all-inorganic metal halide perovskites are a new emerging star following organic metal halide perovskites, thanks to their outstanding optoelectronic properties and relatively higher stability at the same time. Accordingly, there have been excellent reviews on the synthesis, properties, and applications on all-inorganic perovskites. The readers are referred to previous literatures for specific purpose [21, 38, 39, 42, 43, 56–62]. Our group has contributed in this field through developing novel fabrication methods for perovskite micro/nano crystals, $CsPbX_3$ QDs in particular, and investigating their photoinduced excited-state dynamics via various time-resolved spectroscopic techniques. Therefore in this chapter, we focus on the review of (nonlinear) optical properties and dynamics of $CsPbX_3$ QDs.

6.2 Fabrication Methods of Cesium Lead Halide Perovskite QDs

6.2.1 Summary of Traditional Synthesis Methods

As one type of "soft" crystal [63, 64], $CsPbX_3$ QDs have rich conformational structures. Orthogonal, tetragonal, and cubic phases are the most common for classical $CsPbX_3$ crystals, which may change over the ambient temperature and total energy of the system [39, 49], and different crystal structures can lead to large differences in properties [65, 66]. Herein we use $CsPbBr_3$ as a representative to describe their crystal structures. As the ideal cubic system (Pm3m), $CsPbBr_3$ also adopts the crystal structure of the general perovskite as shown in Fig. 6.1a. The cesium ions occupy the apex of the cube, and the octahedral structure $[PbBr_6]^{4-}$ that combines lead with bromide ions resides at the center of the cube.

The first synthesis of $CsPbX_3$ QDs was via the hot-injection method [36]. Since then, a large number of synthesis methods simpler and more controllable have been developed [67–69]. For the time being, there are two main types of preparation methods: thermal injection and room temperature synthesis. The room temperature synthesis includes anti-solvent precipitation, ultrasonic synthesis [70], mechanochemical synthesis, and alike. Microwave-assisted synthesis is also seen in literatures [71].

6.2.1.1 Hot-Injection Synthesis

At present, hot-injection is one popular approach to create perovskite crystals. This method is the basis to dope perovskite with other metal cations, such as Bi^{3+}, Sn^{4+}, Cd^{2+}, alkaline-earth metal, and lanthanide [72–76]. In this method, cationic precursors (PbX_2 powder is dissolved in a solvent (usually 1-octadecene) containing a surfactant (generally oleic acid or oleylamine)) are prepared firstly. Then, the preheated Cs precursor (generally Cs-oleate) is injected into cationic precursors at a specific temperature and maintained under inert atmosphere (e.g., N_2) for a certain period of time to form the products. Like other nanoparticles, the size of perovskite QDs synthesized by the hot-injection method can be adjusted by changing the temperature at the time of heat injection, the reaction time, the surfactants, the concentration of the precursor, etc. For example, Deng's group added surfactants significantly influenced the size and morphology of the produced QDs [68].

6.2.1.2 Anti-Solvent Precipitation

Anti-solvent precipitation for synthesizing perovskite QDs at room temperature is one convenient method. Firstly, CsX and PbX_2 (X = Cl, Br, I or a mixture thereof) are dissolved in good solvents such as dimethyl sulfoxide (DMSO) or N,N-dimethylformamide (DMF). Ligands such as oleic acid (OA) or oleylamine (OAm) are subsequently added. Then, a small quantity of the above solution is added dropwise to poor solvents such as toluene or n-hexane, where those ions have very low solubility. As DMF/DMSO is miscible with toluene/n-hexane, the mixing causes a high degree of supersaturation immediately and leads to a burst of nucleation, until the seeds proceed to form nanoparticles. Surfactants help regulate the size of perovskite QDs and make them dispersed in nonpolar solvents easier. All operations are carried out at room temperature without inert gas protection. The optical properties of the obtained perovskite QDs are comparable to those synthesized by thermal injection [77].

6.2.1.3 Microwave-Assisted Synthesis

Microwave-assisted synthesis is one fast, low-cost, and environment-friendly strategy to prepare nanoparticles, which features attractive merits of homogeneous heating, high heating rate, selective reaction, and scalable production. Previously, our group had successfully fabricated carbon QDs and metal organic framework materials using this method [78, 79]. At present, works based on this method to prepare perovskite QDs are in the ascendant [71, 80, 81]. Generally, Cs-oleate solution and PbX_2 precursor are mixed in a container and then transferred to microwave oven, irradiated at a certain power for some minutes with microwave. Crude $CsPbX_3$ QDs are then collected after cooling by ice-water bath.

6.2.1.4 Ultrasonic Synthesis

Ultrasonic synthesis of perovskite QDs was also reported [70]. The Cs precursor ($CsCO_3$) and PbX_2 were dissolved in a solvent (mineral oil or 1-octadecene) with surfactants (OA, OAm). Then the mixture reacted for a period of time to generate the QDs under certain power of ultrasound. In this way, $CsPbX_3$ QDs were synthesized in a large quantity. The fluorescence QY reached 92% and exhibited good stability. By adjusting the ratio of Cs and Pb of raw materials, perovskite QDs of different thicknesses were obtained [70].

6.2.2 *Novel Mechanosynthesis Method*

The abovementioned methods all require the use of organic solvent, and hot injection even demands of excessive Pb salts, which consumes substantial reagents and is environment-unfriendly. To address this issue, our group took the lead in applying the non-solvent mechanochemical synthesis method to the fabrication of perovskite QDs [82].

Mechanochemistry refers to induce the chemical reaction between solids and solids by simple grinding/milling [83]. Historically, it has been one of the least noticeable synthetic methods in chemical synthesis. However, it has received increasing attention in recent decades for making various materials, e.g., metal-organic frameworks (MOFs) [84], organohalide perovskites [85], and black phosphorus (BP) [86], as the method allows reaction of a large quantity of raw reagent and hence potentially for scalable production. More importantly, mechanochemical synthesis without solvents can perfectly avoid the environmental pollution and safety problems caused by purification and recovery of solvents [87].

As shown in Fig. 6.2, through directly grinding CsX and PbX_2 salts based on stoichiometric ratio in the absence of solvent, where X refers to the halogen element of Cl, Br, I, or their mixture, a series of $CsPbX_3$ QDs with tunable composition and corresponding optical properties can be facilely fabricated similar to assembling Lego building blocks.

Many experimental parameters influence the morphology of the final products during the grinding process, such as rotational speed, ball milling time, ball ratio, and so on. Change in one parameter frequently requires corresponding adjustment of other relevant parameters. For instance, with the other experimental conditions being the same, only different ball milling speeds lead to QDs with different qualities (Fig. 6.3a, b): those made at 500 r/min were more uniform in shape and size than that at 150 r/min, and $CsPbBr_3$ QDs were not obtained when the rotation speed was lower than 100 r/min.

Despite of the difficulty, Buzaglo et al. proposed an empirical formula (6.3) that can roughly describe the connection of the relevant parameters [88]:

$$E_t = K \cdot w_P^3 \cdot t \tag{6.3}$$

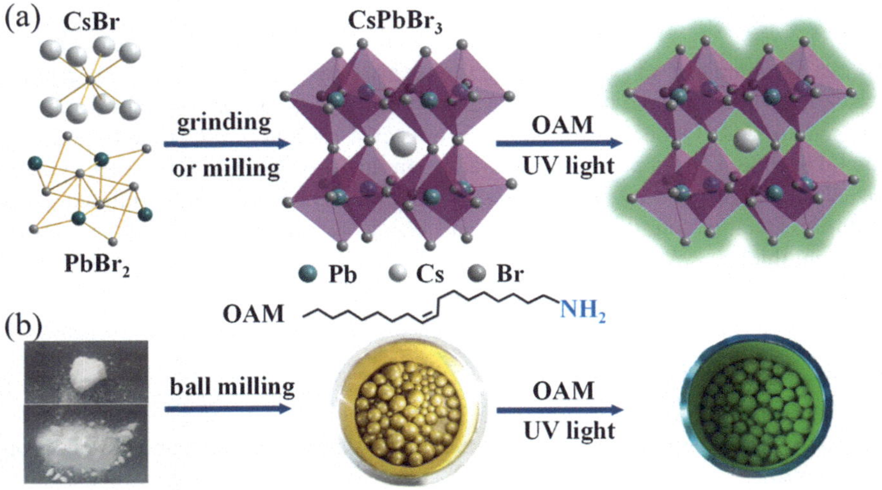

Fig. 6.2 (**a**) Schematic diagram of synthesis procedure for $CsPbX_3$ QDs and (**b**) corresponding photographs. (Reproduced with permission [82]. Copyright © 2017, American Chemical Society)

where E_t is the energy produced during ball milling, w_P is the ball milling speed, t is the ball milling time, and K is a constant associated with the ball ratio.

Other conditions, including milling time, the ratio of the ball to the materials, the number and the type of ligands, all affect the morphology and properties of the resulting perovskite QDs. Through a series of permutations and combinations, it is like stacking Lego bricks to obtain perovskite QDs with adjustable wavelength and composition as shown in Fig. 6.3c.

The quality of the perovskite QDs obtained by this method emulates that of the conventional method, while at the same time has the advantages of rapidness and environment-friendliness. The as-prepared QDs showed comparable stability and PLQY to those of quantum dots prepared by solution processing.

6.3 Photoinduced Excited-State Dynamics of Perovskite QDs

The origin of the unique optoelectronic properties of perovskite QDs, which direct their broad applications in a wide range of fields such as display, photovoltaics, and nonlinear optics, is closely related to their excited-state dynamics upon photoexcitation. Therefore, the rich ammunition arsenal of time-resolved spectroscopic techniques have been employed to disentangle the excited-state dynamics of perovskite QDs, which oftentimes involves multiple time scales from ultrafast fs to longer ms or even second scale. Moreover, the quality of the nanocrystals (NCs) as aroused by surface defects and various ligands used also adds difficulty to the investigation.

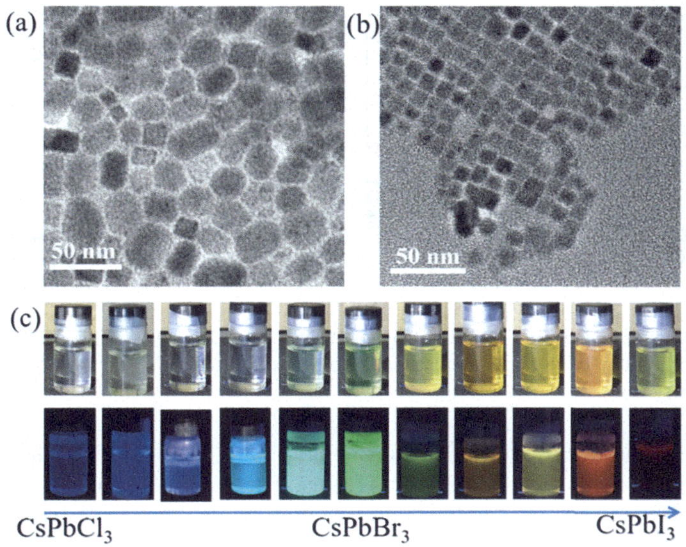

Fig. 6.3 TEM images of $CsPbBr_3$ QDs obtained under ball milling at different rotation speeds: 150 r/min (**a**) and 500 r/min (**b**). (**c**) Photographs of the $CsPbX_3$ QDs with tunable halide ion composition, without (upper) and with (down) UV light irradiation (dispersed in toluene). (Adapted with permission [82]. Copyright © 2017, American Chemical Society)

The following parts reviewed several popular techniques for extracting photoinduced fast dynamics of QDs.

6.3.1 Fluorescence Lifetime Measurement

Fluorescence lifetime is closely related to the structure of the fluorescent substance and the environment in which it is located. It is powerful in elucidating the changes of the samples themselves or interaction with other substances. Hence the fluorescence lifetime measurement is one effective way to disclose the emission properties of the usually highly luminescent perovskite QDs. In combination with information from steady-state fluorescence measurements (QY), radiative and nonradiative rates can be extracted. And using Stern-Volmer plots, fluorescence quenching mechanism can be clarified [89]. At present, fluorescence lifetime measurements have been widely used in the study of perovskite QDs for gaining a deep understanding of the carrier recombination dynamics [51, 77, 90, 91]. Time-correlated single-photon counting (TCSPC) is the most popular technique applied for the measurement, with the advantages of simplicity and accuracy [89].

We list our previous work as an example to illustrate how the fluorescence lifetime measurement can be used for the dynamic analysis of perovskite QDs [82].

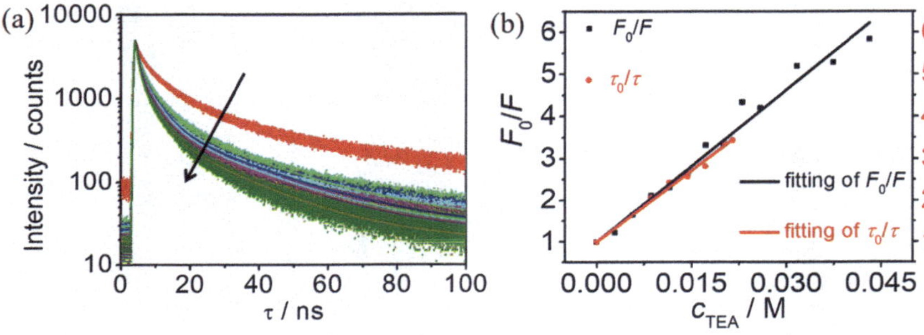

Fig. 6.4 (**a**) Fluorescence lifetime measurement of $CsPbBr_3$ QDs with addition of TEA (shown by the arrows); (**b**) Stern-Volmer plots. (Adapted with permission [82]. Copyright © 2017, American Chemical Society)

Based on the decay profile shown in Fig. 6.4a, $CsPbBr_3$ QDs exhibited three lifetime components of 1.7, 8.3, and 41.0 ns. The short (1.7 and 8.3 ns) lifetime components were attributed to the bound excitons while the long 41.0 ns one to free excitons [77, 92]. The averaged lifetime was 10.1 ns. When $CsPbBr_3$ QDs were mixed with triethylamine (TEA), a frequently used reducing agent ("sacrificial reagent") for photocatalysis [93], its averaged lifetime was significantly shortened. The results suggested the occurrence of energy and/or charge transfer between the two species, while the former can be excluded as there was no overlap between absorption spectra of TEA and emission spectra of $CsPbBr_3$ QDs [82].

Stern-Volmer plots were then drawn to derive the charge transfer rates (Fig. 6.4b). The averaged lifetime component versus increased TEA concentration (τ_0/τ vs. c_{TEA}) exhibited linear fittings, overlapping with that from steady-state fluorescence measurement (F_0/F vs. c_{TEA}, F is emission intensity). The observation confirmed a dynamic quenching of perovskite QDs by TEA. The charge transfer rate constant was estimated to be as large as ~1.1–1.2 × 10^{10} M^{-1} s^{-1}, within the diffusion-controlled limit [94, 95]. The observed efficient charge transfer between perovskite QDs and electron donor/acceptors indicated the potential applications of the QDs in photocatalysis and optoelectronic devices [96].

6.3.1.1 Streak Camera Measurement

Streak camera is another time-resolved spectral detection device, which can simultaneously obtain spatial and time information of a sample, but the current instrumental cost is beyond many research groups. Compared with common fluorescence lifetime instruments based on TCSPC, its outstanding features are wide time window of ps-ms and ultrahigh (fs ~ ps order) temporal resolution. It has been used to identify ultrafast stimulated radiation process, providing valuable information for developing low-threshold short-pulse lasers based on perovskite QDs. Tang et al. investigated the excited-state dynamics of $CsPbX_3$ microsphere by a streak camera

[97]. They found a narrower emission band (~0.09 nm) for the material, and a very fast decay time (30.5 ± 0.9 ps) dominated the dynamics. The results indicated a transform from spontaneous emission to stimulated emission, and the threshold was extracted to be as low as ~0.42 μJ·cm^{-2}. Similar phenomena were also observed on perovskite nanorods [98]. Streak cameras can also be used to investigate the carrier dynamics crucial for photovoltaic devices. For instance, based on the streak camera measurement, Han et al. derived the carrier mobility over 100 cm^2 V^{-1} S^{-1} for $CsPbBr_3$ microcrystals through the one-dimensional diffusion equation [99].

6.3.2 *Transient Absorption Spectroscopy*

TCSPC-based fluorescence lifetime measurement and streak cameras apply to luminescence materials. In this meaning, transient absorption spectroscopy is a more general technique as it applies to both luminescent and non-emissive samples. Based on the pump-probe scheme, it can give a time-resolution up to the chemical reaction limit of fs time scale. Apart from measuring emission dynamics, transient absorption method gives rich information on the absorption of samples in the excited state. Transient absorption spectroscopy proves to be a powerful tool to disentangle the series of complicated dynamic behaviors of the perovskite QDs after irradiation, carrier diffusion, relaxation, Auger recombination, interfacial charge transfer, etc. (Fig. 6.5).

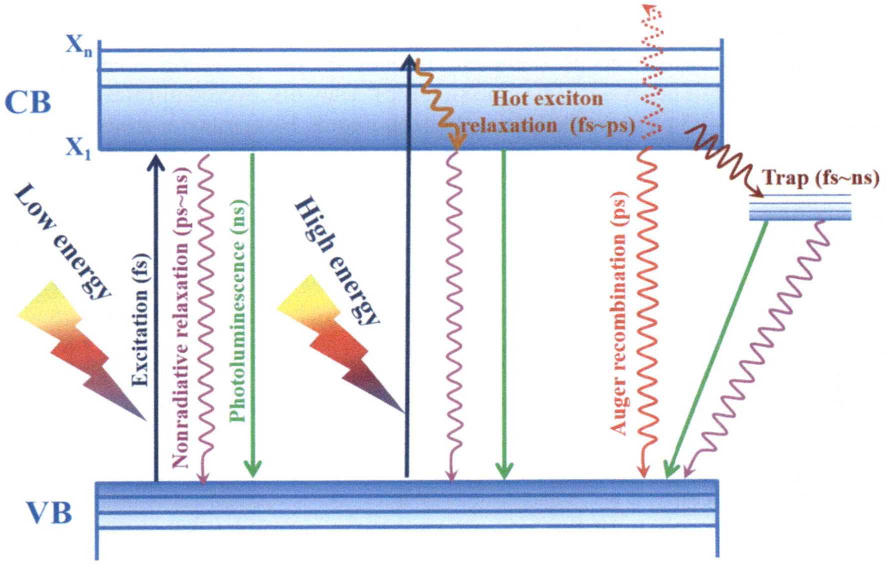

Fig. 6.5 Schematic illustration of the various dynamic processes within perovskite QDs upon photoexcitation. *CB* conduction band, *VB* valence band

Transient absorption spectroscopy can be used to identify the various dynamic steps of perovskite QDs after photoexcitation. Liu et al. [100] studied the ultrafast dynamic processes of $CsPbI_3$ perovskite NCs under low and high pump photon energy. In the transient absorption spectra of $CsPbI_3$, the broadband of 600–700 nm was attributed to the overlap of ground state bleaching (GSB) and simulated emission (SE). And the bands around 575 and 725 nm were assigned to the absorption of excited state (EA). Upon high energy excitation, the fast rise of ~5 ps in the early time revealed that the hot exciton relaxation was involved. From the global fitting results, they also confirmed the participation of Auger recombination process [100]. Likewise, Samanta's group studied the short-lived photoinduced transients of $CsPb(Br_xI_{1-x})_3$ [101]. They attributed the ~700 fs rise to the state filling of the exciton dissociation. And the lifetime components at 35–65 ps and ~ns level were attributed to the charge carrier trapping and recombination, respectively.

The defects act as trap states and introduce a competition process against recombination. The trapping and recombination usually relax on a time scale of ps and ns, respectively. Samanta et al. studied the transient absorption of $CsPbX_3$ and extracted 45 ps and 215 ps electron trapping rates for $CsPbBr_3$ and $CsPbI_3$ NCs, respectively [101]. They assigned the different trap rates to the distinct locations of the trap states, which were close to the higher excitonic states in $CsPbI_3$ while near the band edge in $CsPbBr_3$. Consequently, the trapping rate was significantly slower in $CsPbI_3$, and the PLQY is much higher than that of $CsPbBr_3$. The common method to decrease the defects or trap states is passivation, which reduces the nonradiative recombination and improves PLQY of PQDs [102].

Transient absorption spectroscopy is advantageous in extracting ultrafast energy/charge transfer rates between perovskite QDs and electron acceptors/donors. Lian and colleagues found, as a result of electrostatic interaction, the electron acceptors/donors (benzoquinone and phenothiazine) bound on the QD surface and caused static quenching of perovskite QD emission [96]. They discovered ultrafast electron transfer rate of ~65 ps from the QDs to benzoquinone. Hole transfer from the QDs to phenothiazine also occurred at a fast rate of ~49 ps. Similarly, Begum et al. disclosed Bi^{3+}-doped $CsPbBr_3$ QDs underwent fast electron transfer on the picosecond time scale between tetracyanoethylene (TCNE) and $PC_{61}BM$ [103].

In addition to the direct use of transient absorption spectroscopic techniques to study the excited-state dynamics of perovskite QDs, its combination with other in situ methods may provide multidimensional information. For instance, Kamat's group developed an in situ electrochemical transient absorption measurement to study the influence of electrochemical bias on the charge carrier recombination of $CsPbBr_3$ films [104]. The spectroelectrochemical measurement showed the charge carrier lifetimes of the films decreased with the added bias from −0.6 to +0.6 V (vs. Ag/AgCl). The method opened one avenue to investigate the charging effects of perovskite solar cells.

At present, the study of the excited-state dynamics of perovskite QDs is still quite incomplete. More experimental data at improved accuracy, in combination with computational results, may facilitate the identification of the complicated dynamic behaviors of the perovskite QDs, which involve various photophysical/

photochemical processes and span a large time window from ultrafast femtosecond to longer time scale beyond second.

6.4 Applications

$CsPbX_3$ QDs possess unique optical and electronic properties, such as high fluorescence QY, single-photon emission, quantum confinement effect, third-order nonlinear optical response, high carrier mobility, etc., which enable their wide range of applications in optoelectronic field.

High Quantum Yield The fluorescence QY of $CsPbX_3$ QDs can easily reach 90% even without surface treatment. The halogen atoms in $CsPbX_3$ can be adjusted without a phase change. Therefore it is possible to tune the emission spectra from ultraviolet to near infrared by varying the type and proportion of the halogen elements. In addition, $CsPbX_3$ QDs have narrower FWHM and are advantageous in wide color gamut display. Moreover, the fluorescence peaks of the $CsPbX_3$ QDs almost remain the same in a broad temperature range of 25–100 °C, which further corroborates its excellent optical properties [43].

Defect Tolerance For traditional semiconductors, surface defects and trap state are disadvantageous for their optoelectronic performance even at low concentrations [105]. Although the low formation energy and high specific surface make the generation of defects easy, the photoelectric performance of the perovskite QDs still performs well, such as near-unity PLQY and narrow FWHM [106, 107]. Perovskite QDs exhibited large defect tolerance because the defect was present in the conduction band or valence band of PQDs instead of forming mid-gap trap states [42, 106, 108]. Defect tolerance is critical for photovoltaic applications, e.g., open-circuit voltage in photovoltaics based on α-$CsPbI_3$ colloidal QDs was approaching the Shockley-Queisser limit (1.2 V) [109, 110].

Single-Photon Emission $CsPbX_3$ QDs are promising single-photon emission fluorescent materials for potential applications in quantum technologies such as quantum cryptography and quantum communication [111–113]. In 2015, Xiao's group observed $CsPbX_3$ with single-photon emission, which exhibited much larger absorption cross sections and faster output of single photons (as indicated by the dramatically shortened radiative lifetime) than metal-chalcogenide nanomaterials [114]. Gabriele et al. discovered single-photon emission of $CsPbX_3$ NCs with ultrafast decay time of 250 ps [115]. In a latest report, Bawendi's group found coherent single-photon emission from $CsPbBr_3$ QDs with long optical coherence times of 80 picoseconds [113]. However to pave the way for quantum-light applications, it is essential to overcome the obstacles of photo-blinking and photoinstability and further understand the transition mechanism (such as the Auger process) and realize room-temperature single-photon emission of $CsPbX_3$ NCs [42, 116].

Third-Order Nonlinear Optical Property Perovskite QDs also exhibit third-order nonlinear optical properties. For instance, third-order nonlinear absorption of perovskite QDs in the forms of two/multiphoton absorption and reversible absorption have been reported [117–120]. These effects can be harnessed for biological imaging, nanolasing, optical limiting, Q-switch frequency modulation, etc.

Sun et al. studied the nonlinear optical properties of $CsPbBr_3$ QDs, including their multiphoton absorption, related fluorescence spectroscopy, and stimulated emission [121]. The QDs showed bright green under the excitation of 800 nm femtosecond laser beam, and the relationship between the emission intensity and the input light energy proved the occurrence of two-photon excited fluorescence. Their PL spectra induced by one-, two-, and three-photon absorption were nearly identical, consistent with our observation for $CsPbBr_3$ QDs [82]. Increased pump energy led to stimulated emission, the threshold of which was 2.5 mJ cm^{-2}, several times lower than that of CdSe QDs, and demonstrated great potential for nanolasers.

High Carrier Mobility Generally speaking, it requires both n-type and p-type light absorbing materials to construct a p-n junction to effectively separate electrons and holes for a photo-generated voltage, while organometallic perovskites like $CH_3NH_3PbI_3$ can be simultaneously used as a light trapping and charge transporting material without the use a p-n junction due to their balanced electron and hole transport length [122, 123]. Likewise, charge mobility and lifetime of all-inorganic perovskite QDs are comparable to those of organometallic perovskites and are thus suitable for similar applications [124]. Gurivi et al. measured remarkably high carrier mobility based on $CsPbBr_3$ NCs using time-resolved terahertz (THz) spectroscopy [125]. The mobility up to ~4500 $cm^2\ V^{-1}\ s^{-1}$ is one order of magnitude higher than that of organometallic perovskites and several times higher than that of bulk $CsPbBr_3$ single crystal (1000 $cm^2\ V^{-1}\ s^{-1}$) [39]. Sanehira et al. reported a surface-modified method to enhance the mobility of $CsPbI_3$ QD film [110]. They treated the film with A-site cation halide salts, and the mobility of the $CsPbI_3$ QD film improved from 0.23 to 0.50 $cm^2\ V^{-1}\ s^{-1}$, which was significantly advantageous over PbS and PbSe QD films. Han et al. on the other hand prepared $CsPbBr_3$ microcrystals with carrier mobility over 100 $cm^2\ V^{-1}\ s^{-1}$ [99].

Endowed with the abovementioned unique features, $CsPbX_3$ QDs are promising materials for optoelectronic applications. For instance, high QY enables their use for light-emitting diodes (LEDs); high carrier mobilities and strong photo absorption covering a broad spectrum are beneficial for photovoltaic devices and photodetectors; and stimulated emission feature is potential for fabrication of nanolasers. The following parts summarize the recent advancement.

6.4.1 Optoelectronic Devices

LED and Display The basic working principle of LEDs is to inject electrons and holes into the cathode and anode of the device, respectively, by direct-current voltage, and then the carriers are recombined in the active layer, accompanied by the release of excessive energy in the form of light, a process also called electroluminescence. In the field of LED and display, $CsPbX_3$ QDs have many advantages: near 100% fluorescence QY, narrow FWHM, easy acquisition of red-green-blue tricolor, saturated emission colors, and free of or reduced photo-blinking. Besides, like organic LEDs (OLED), the QDs-LEDs (QLED) can self-illuminate without backlight. Together with the outstanding solution processability of the QDs, QLED are thus strong candidates for future flexible high-definition display. In 2015, Zeng's group took the lead in constructing tricolor QLED devices by $CsPbX_3$ QDs through tailoring the halide composite and the size of the QDs, achieving an impressive ~25 nm FWHM for the device, but the luminance and external quantum efficiency (EQE) values is far from satisfaction [37]. In 2016, they increased the EQE of green light QLED to 6.3% by 50 times through a surface purification method to optimize the surface ligands density of $CsPbBr_3$ QDs, which was the highest value at that time [91]. In 2017, Kido's group combined surface control with the optimization of the interfacial energy band and further improved the efficiency of $CsPbBr_3$ QD-based QLED devices to 8.7%. Meanwhile, a low turn-on voltage of 2.60 V and a maximum power efficiency of 31.7 lm W^{-1} were achieved [126]. They continued to update their records to 21.3% in 2018 and obtained high color purity with red emission by $CsPb(Br/I)_3$ QDs, which were prepared by anion exchange from pristine $CsPbBr_3$ QDs using alkyl ammonium iodine salt [127].

Solar Cells Solar cells generate electron-hole pairs when a certain frequency of light irradiates the p-n junction. The generated holes are concentrated toward the p-type semiconductor. Meanwhile, the electrons move to the n-type semiconductor, thereby generating a photo-generated voltage. When there is an external circuit, current is generated. This conversion is also called photovoltaic effect.

Hodes and the coworkers compared the power conversion efficiency (PCE) of solar cells based on inorganic perovskites and organometallic perovskites in 2015 [124]. Interestingly, they found the all-inorganic perovskites performed even better with excellent temperature stability and higher PCE of 5.95%, which spurred the research of using inorganic perovskites as solar materials. To further improve the PCE, the inorganic perovskite-based solar cells have to be optimized in many aspects: the band gap remains to be optimized, and lattice distortion, surface defects, and poor interface contact should also be taken into consideration, instead of purely dependent on improving the stability of the materials [21]. For example, Snaith et al. applied $CsPbBrI_2$ QDs to solar cells with a PCE of up to 9.8%. The Luther group found that $CsPbI_3$ QDs treated with ethyl acetate formed a stable cubic phase solid. Even after 2 months, no evolution of orthogonal phase was observed from XRD patterns. The PCE of the solar cell based on this material reached as high as 10.77% [128]. In 2017, the same research team reported $CsPbI_3$ QDs film treated with A-site cation halide salt, which greatly improved the electrical conductivity and fluorescence lifetime of the material, and the PCE of the device reached the highest value of 13.43% among the NREL-certified QDs solar cells [110]. Hitherto although the inorganic perovskite solar cells are not as good as the organometallic perovskite ones (>20%) in terms of PCE, their relative stability makes them worthy of further research and exploration.

Photodetectors The working principle of the photodetector is similar to that of the solar cells. Lee's group first reported $CsPbI_3$ QDs for a photodetector in 2016, which exhibited a large on/off photocurrent ratio of 10^5 and rapid rise and decay time (24 and 29 ms, respectively) [129]. The long radiative lifetime of the $CsPbI_3$ QDs was beneficial for producing large photocurrents [129]. Zeng's group improved the quality of perovskite QDs films and thus the performance of the as-fabricated photodetectors through annealing [130]. The on/off ratio could also be enhanced by synergetic effect of introduced additives such as TiO_2 and plasmonic Au NPs [131, 132]. A recent development was the expansion of $CsPbCl_3$ NCs/GFET heterojunction photodetectors to the UV light detection [133], but more efforts are still needed to further expand the detection range of perovskite photodetectors.

Field-Effect Transistors (FET) Different from LEDs or solar cells which have two-terminal electrodes, field-effect transistors usually have three electrodes of source, drain, and gate. The injection of carriers from the source, the transport of carriers at the semiconductor material and the dielectric interface, and the collection of carriers at the drain are controlled by regulating the voltages of the three electrodes.

Without considering ion shielding effect, perovskites are ideal materials for FETs. Firstly, perovskites have different dimensions and morphologies and can easily meet processing requirement. Moreover, in terms of electronic properties, perovskites have very high intrinsic carrier mobility and bipolar characteristics; in addition, perovskites have high light absorption coefficients and PCE, which is very suitable for the development of highly photosensitive FETs. Unfortunately, perovskite materials have the characteristics of ion semiconductors. Ion shielding effect is formed at the interface between the perovskite material and the dielectric when a certain voltage is applied to the gate, which disables the gate voltage and fails to regulate the transport of carriers in the transistor channel. The review by Zeng's group systematically described the progress of perovskite FETs in recent years and summarized methods of optimizing or eliminating the ion shielding effect by improvement of the material structure and composition. Perovskite FETs working stably at room temperature would be the ultimate goal [134].

Inorganic perovskite QDs have demonstrated excellent performance on traditional photoelectric conversion devices. Recently their applications have expanded to some new fields that are worthy of notice in particular. The following parts give brief introduction on the new trends of perovskite QDs, such as being used in nonlinear optics, photocatalysis, and scintillator fields.

6.4.2 Nanolasers

The absorption and radiation of photons by matter manifest in three fundamental processes, stimulated absorption (i.e., absorption), spontaneous emission, and stimulated radiation, and the latter is an essential requirement for generating laser output.

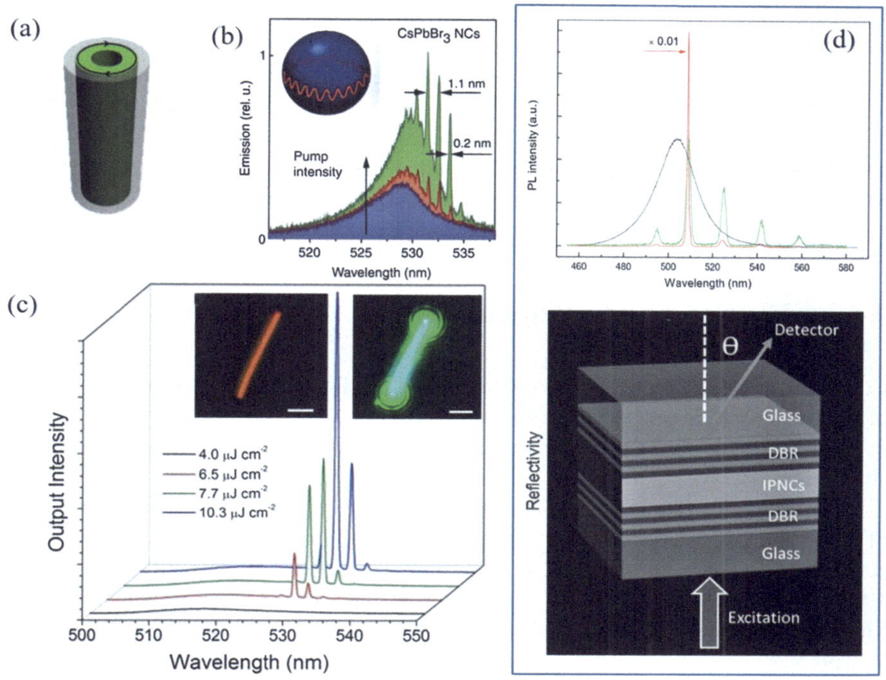

Fig. 6.6 (**a**) Lasing of $CsPbBr_3$ QDs in WGM mode in capillary (Adapted with permission [51]. Copyright © 2015, John Wiley and Sons). (**b**) $CsPbBr_3$ QDs packaged on silicon microspheres for WGM mode lasing (Adapted with permission [52]. Copyright © 2015, Nature Publication Group). (**c**) Lasing cavity between the intrinsic two end-facets of perovskite nanowires (Adapted with permission [138]. Copyright © 2016, National Academy of Sciences). (**d**) $CsPbBr_3$ crystals lasing in VCSE mode (Adapted with permission [139]. Copyright © 2017, John Wiley and Sons)

Nakagawa's group pioneered in the study of stimulated radiation of $CsPbBr_3$ crystals in 2007. But unfortunately, their results were not satisfactory because the fluorescence QY of then $CsPbBr_3$ microcrystals was not high enough. Later on, Kovalenko's group reported low-threshold stimulated emission of $CsPbBr_3$ QDs [52]. $CsPbBr_3$ QDs were also found to exhibit better photo stability. Under high-intensity laser excitation, the emission intensity of the QDs still maintained up to 90% of the initial value after 1.6×10^7 laser shots, which was an order of magnitude higher than that of conventional QDs and the most recent organometallic perovskite QDs [51].

There are several configuration selections to test the stimulation of perovskite NCs [135, 136]. In Fig. 6.6a, the $CsPbBr_3$ QDs was infiltrated into a hollow capillary to form a whispering gallery mode (WGM) laser [51]. The QDs can also be attached to the silicon microspheres to build a high-quality WGM nanolaser, as shown in Fig. 6.6b [47]. WGM is one of the simplest laser configurations, where the output beam direction is not fixed [137]. Another mode is employment of $CsPbX_3$ nanowires [138] or nanorods [98], using their intrinsic cavities formed between two end-facets (Fig. 6.6c). Additionally, Sun's group successfully obtained vertical cavity surface emitting (VCSE) laser of single/multi-longitudinal mode by sandwiching solution-treated $CsPbBr_3$ crystals between two reflecting surfaces of highly reflective media [139]. It was characterized by a narrow spectrum, a nonlinear enhancement of fluorescence intensity, a shorter fluorescence lifetime, and a significant reduction in divergent beam energy (Fig. 6.6d). The success of VCSE lasers based on inorganic perovskite NCs proves the feasibility of independent adjustment of the cavity and NCs, pushing forward the practical application of perovskite crystals in laser field.

6.4.3 Nonlinear Optical Applications

Compared with research of perovskite QDs in the linear optical region, the studies of nonlinear optical properties of the materials are still in the early stage, especially in terms of their third-order nonlinear responses such as reversible and saturable absorption.

Currently one effective method to derive third-order nonlinear optical parameters of materials, such as nonlinear absorption coefficient and two-photon absorption coefficient, is the single-beam Z-scan technique proposed by Sheik-Bahae et al. in 1989 [140]. The method has the advantages of easy operation and high sensitivity. Moreover, for the samples, the real and imaginary parts of the nonlinear polarizability as well as their corresponding positive and negative signs can be readily obtained. Figure 6.7a shows the schematic diagram of the Z-scan setup. The incident pulsed laser is split into two beams by a beam splitter (BS): one is received by the detector D1 and used as a reference light to monitor the fluctuation of the laser energy; the other beam passes through the focusing lens and the sample. While the sample moves left and right around the focus in the direction of the Z-axis (laser propagation direction), the transmitted light is received by the detector D2. Whenever the sample moves to a different Z position, the beam intensity on D1, D2 will be recorded to obtain two experimental curves: curve of $I_{D2/D1}$ reflects the nonlinear absorption intensity at different positions of the sample, which we call the "open"-aperture Z-scan curve; when an aperture is added in front of D2, another curve of $I_{D2*/D1}$ is obtained, which is related to both nonlinear absorption and nonlinear refraction intensity of the sample, and "closed" aperture Z-scan curve is named.

Hence, the Z-scan technique can measure the normalized transmittance of samples as a function of position (Z). With the open-aperture Z-scan mode, two types of curves were typically provided: those samples with nonlinear reversible saturable absorption show normalized transmittance below 1.0 around the focus point ($z \rightarrow 0$). One valley appears at the focus, like the curve in Fig. 6.7b (left), which indicates the sample transmission decreases with increased input pulse intensity. This phenomenon is named optical limiting and is greatly desired for smart laser protection. On the contrary, for the samples demonstrating saturable absorption, the Z-scan curve will manifest itself as a peak at the focus (Fig. 6.7b right), as the sample transmission increases with increased input pulse intensity. Materials with such valuable

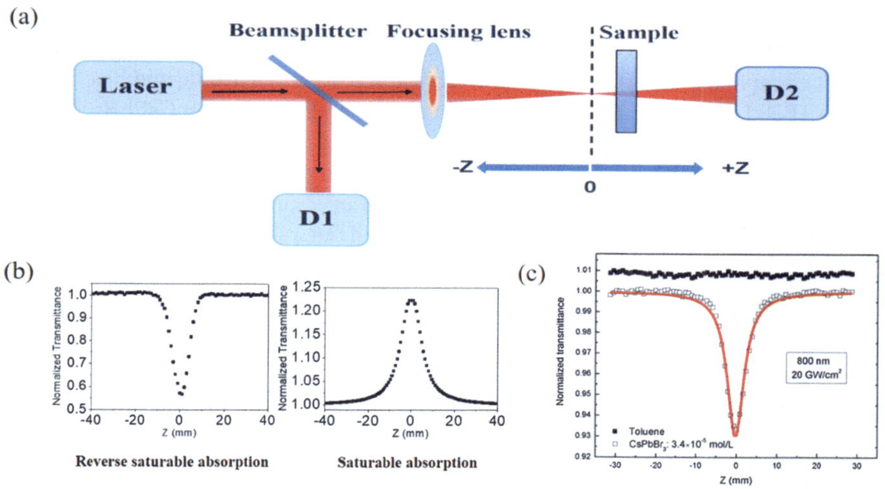

Fig. 6.7 (**a**) Schematic diagram of the Z-scan setup. (**b**) Typical Z-scan data of reverse saturable absorption (left) and saturable absorption (right) under open aperture. (**c**) Optical limiting effects of $CsPbBr_3$ QDs (black line is from the solvent reference). (Reproduced with permission [121]. Copyright © 2016, American Chemical Society)

properties can be made into saturable absorbers to shorten laser output pulse in the way of *Q*-switch or mode-locking.

6.4.3.1 Optical Limiting

In several previous studies, perovskite QDs showed upconversion fluorescence induced by two-photon absorption (TPA), which was first discovered in organic-inorganic perovskites and later introduced to all-inorganic perovskite QDs [141, 142]. TPA process may also lead to nonlinear optical limiting effect, in addition to other mechanism like nonlinear scattering, nonlinear refraction, and nonlinear reflection. For example, Fig. 6.7c shows the optical limiting effect of $CsPbBr_3$ QDs stemming from TPA; the cross-sectional absorption area was calculated to be 1.2×10^5 GM [121]. Optical limiting based on third-order nonlinear responses of materials is one of the most active areas in laser protection, and $CsPbX_3$ QDs may play an important role in this field. Currently there are rare studies, and the research on the mechanism is still in its infancy.

6.4.3.2 Saturable Absorbers

Saturable absorbers open up an effective way to generate ultrashort optical pulses through mode-locking. The absorption of the saturable absorber increases with the increase of incident light intensity. When the saturation value is reached, the saturable absorber is transparent to the input laser beam. Many different types of materials have been developed for saturable absorbers, including carbon nanotubes (CNTs), graphene, transitional metallic dichalcogenides (TMDCs), and black phosphorus (BP) [143–145]. Tang's group developed a fiber laser using $CsPbBr_3$ NCs as a saturable absorber, and the modulation depth and saturation intensity reached 13.1% and 10.7 MW cm^{-2}, respectively. The maximum average output power of single pulses with duration of ~ 216 ps was 10.5 mW, and the laser spectrum was centered at ~1076 nm [146]. Zhang's group carried out a similar work with $CsPbBr_3$ QDs. A laser output at 522 nm with maximum average output of 9.11 mW was achieved, and the corresponding shortest pulse width was 653 ns [147].

6.4.4 Photocatalysis

In our research as discussed in Sect. 6.3.1, both lifetime measurement and steady-state fluorescence measurement demonstrated TEA effectively quenched the emission of $CsPbBr_3$ QDs. Combined with the valence and conduction band information, the photoinduced dynamic study revealed the hole transfer process of QDs toward electron donors [82]. As TEA is oftentimes applied as reducing agent during photocatalysis, the findings implied the potential of perovskite QDs in photocatalytic reactions.

Kovalenko et al. have also suggested the possibility of perovskite QDs in photocatalysis in nonaqueous media, which was an unexplored area at that time [108]. Now studies have begun to show up and develop rapidly. Kuang's group reported for the first time the photocatalytic reduction of CO_2 by $CsPbBr_3$ QDs. Further combining QDs and graphene oxide, the photocatalytic performance was improved by 25.5%. Steady-state fluorescence measurement and transient fluorescence decay curves revealed that the introduction of graphene oxide accelerated charge transfer from $CsPbBr_3$ QDs to graphene oxide and benefitted the CO_2 reduction step [148]. Sun's group got similar results [149]. Recently, new breakthroughs have been made in catalyzing organic reactions using perovskite NCs. With $CsPbBr_3$ and $MAPbBr_3$ NCs, Yan's group demonstrated α-alkylation of aldehydes with a turnover number (TON) of over 52,000 under visible light illumination [150].

6.4.5 Scintillator

Scintillators give off light upon the excitation of ionizing radiation. Therefore they can be used for radiation detection and safety protection and are usually processed into crystals in applications. They play irreplaceable roles in nuclear medicine imaging equipment such as X-ray and CT, security inspection, nondestructive testing, oil logging, radioactivity detection, and environmental monitoring. Conventional scintillators are generally synthesized by crystallization at a high temperature, and their radioluminescence is difficult to tune across the visible spectrum. Moreover, the conversion efficiency of X-ray photon energy is limited.

Bulk organic-inorganic hybrid perovskites have been successfully applied to convert X-ray photons into charge carriers [151] due to the presence of heavy Pb atom [152] and large electron-hole diffusion lengths of the perovskites [24]. Based on these studies, Liu and co-workers discovered that $CsPbX_3$ perovskite NCs can be used as new scintillators with strong X-ray absorption, high-efficiency triplet luminescence, tunable electron energy level, and faster radiant luminescence rate [153]. What is even more attractive is that these perovskite NCs are prepared by a solution method at a relatively low synthesis temperature, and their emission is tunable in the visible region by simple modification of the anions in the precursors. The detection limit can be as low as 13 nGy s^{-1}, which is equivalent to a few hundredth of the ordinary medical imaging radiation dose. In addition, the discovery of this type of perovskite nanocrystalline scintillator also provides the possibility to prepare large-area flexible scintillator membranes, which can greatly improve X-ray detection and imaging sensitivity, and reduce X-ray dose in medical diagnosis and safety inspection, which makes X-ray-based applications safer.

6.5 Perspectives

Researchers have relentlessly explored in the fantastic perovskite QD world. New phenomena and physical concepts of the QDs have kept emerging. For instance, the observations of their triplet states [154], the synthesis and application of 2D perovskites are all latest developments [155]. Besides the new trends, the following issues plaguing and hampering the even broader applications of perovskite QDs still need more researchers to join the efforts. The breakthroughs of the shackles in these fields may eventually embrace a "phoenix nirvana" type of rebirth for perovskite QDs.

Stability Issues Stability remains the foremost challenge for perovskite QDs. The poor structural, optical, environmental, and thermal stability still seriously restrict their applications in the field of optoelectronics. The paradox for the QDs is that easy as it is to make them, it is just as easy to break them. Strong ionic bond properties and low lattice energy are disadvantageous to its storage in polar solvents. At present, the general measure is to stabilize them with surface adsorbents, such as oleic acid or oleylamine. However, their weak adsorption ability is vulnerable to dielectric environment. Alternatively, insoluble salts or metal oxides, which are insensitive to dielectric environment, may be used as protective layers [42]. Encapsulation with polymer materials or other transparent inorganic materials like SiO_2 is also possible solutions [42, 156].

Toxicity Issues The harmfulness of lead to human body and environment is also a critical factor to be considered. Seeking low toxic and highly stable non-lead perovskite materials is an important direction to conquer these problems. Hence, many efforts have been dedicated to develop $CsPbX_3$ QDs with the replacement of Pb by Bi [157, 158], Sb [159], Sn [160], etc. For example, $CsSnX_3$ has shown promising potential as complementary for lead-containing perovskites in optoelectronics. Orthogonal $CsSnI_3$ QDs have shown high conductivity due to inherent defects with Sn vacancies [160], and their narrowband gap of 1.3 eV is of great value in the fields of solar cells, near-infrared LEDs, photodetectors, biomedical imaging, telecommunications, and sensors [161]. However, currently the QY of $CsSnX_3$ still falls behind their lead-bearing perovskite counterpart. Optoelectronic devices assembled with tin-based perovskites still have much room to be improved [162, 163].

Practical Applications Nowadays the two major problems of stability and toxicity seriously limit the large-scale industrial application of the materials. In addition to these issues, the costs of production, packaging, installation, maintenance, and so on are all among those to be considered. Compared with the relatively mature II–VI family of semiconductor QDs, and particularly the already commercialized CdSe quantum dots in QLED display [164, 165], can perovskite QDs compete and prevail with the above-discussed unique merits?

Other issues, including the tunable lattice structure of perovskite QDs originating from intrinsic "soft crystal" nature [63, 64], controversial luminescent mechanism of some new types of perovskites like zero-dimensional Cs_4PbBr_6 [166–168], extended potential of perovskites for spintronics [169], and the newly discovered chiral perovskites [170], are all intriguing fundamental questions to be explored and will probably bring new physical insights into this field. To sum up, $CsPbX_3$ QDs are a new rising star and have thrived in the optoelectronic field with their unique properties. Numerous new experimental findings and applications of them are springing up every day. As an exciting new territory, there is much to be explored through joint efforts of researchers from various backgrounds of physics, chemistry, optics, electronics, engineering, etc., through employing powerful methodologies like ultrafast spectroscopy.

Acknowledgments This work is supported by the National Natural Science Foundation of China (NSFC. 21673106) and Open Fund of National Key Laboratory of Materials Behavior and Evaluation Technology in Space Environment (ZZYY-KFJJ-LXTZ03).

References

1. Chynoweth, A. G. (1956). Surface space-charge layers in barium titanate. *Physics Review, 102*, 705–714.
2. Mitzi, D. B. (2001). Templating and structural engineering in organic–inorganic perovskites. *Journal of the Chemical Society, Dalton Transactions, 1*, 1–12.
3. Goldschmidt, V. M. J. N. (1926). Die Gesetze der Krystallochemie. *Naturwissenschaften, 14*, 477–485.
4. Protesescu, L., Yakunin, S., Kumar, S., Bar, J., Bertolotti, F., Masciocchi, N., Guagliardi, A., Grotevent, M., Shorubalko, I., Bodnarchuk, M. I., Shih, C. J., & Kovalenko, M. V. (2017).

Dismantling the "Red Wall" of colloidal perovskites: Highly luminescent formamidinium and formamidinium-cesium lead iodide nanocrystals. *ACS Nano, 11*, 3119–3134.
5. Glazer, A. M. (1975). Simple ways of determining perovskite structures. *Acta Crystallographica Section A, 31*, 756–762.
6. Glazer, A. M. (1972). The classification of tilted octahedra in perovskites. *Acta Crystallographica Section B, 28*, 3384–3392.
7. Goodenough, J. B. (2004). Electronic and ionic transport properties and other physical aspects of perovskites. *Reports on Progress in Physics, 67*, 1915–1993.
8. Chen, F. S. (1969). Optically induced change of refractive indices in $LiNbO_3$ and $LiTaO_3$. *Journal of Applied Physics, 40*, 3389–3396.
9. Choi, T., Lee, S., Choi, Y. J., Kiryukhin, V., & Cheong, S. W. (2009). Switchable ferroelectric diode and photovoltaic effect in $BiFeO_3$. *Science, 324*, 63–66.
10. Alexe, M., & Hesse, D. (2011). Tip-enhanced photovoltaic effects in bismuth ferrite. *Nature Communications, 2*, 256.
11. Kreisel, J., Alexe, M., & Thomas, P. A. (2012). A photoferroelectric material is more than the sum of its parts. *Nature Materials, 11*, 260.
12. Yang, S. Y., Seidel, J., Byrnes, S. J., Shafer, P., Yang, C. H., Rossell, M. D., Yu, P., Chu, Y. H., Scott, J. F., Ager, J. W., 3rd, Martin, L. W., & Ramesh, R. (2010). Above-bandgap voltages from ferroelectric photovoltaic devices. *Nature Nanotechnology, 5*, 143–147.
13. Papavassiliou, G. C. (2006). Synthetic three-and lower-dimensional semiconductors based on inorganic units. *Molecular Crystals and Liquid Crystals Science and Technology, Section A, 286*, 231–238.
14. Ishihara, T., Takahashi, J., & Goto, T. (1989). Exciton-state in two-dimensional perovskite semiconductor $(C_{10}H_{21}NH_3)_2PbI_4$. *Solid State Communications, 69*, 933–936.
15. Mitzi, D. B. (1996). Synthesis, crystal structure, and optical and thermal properties of $(C_4H_9NH_3)_2MI_4$ (M = Ge, Sn, Pb). *Chemistry of Materials, 8*, 791–800.
16. Kojima, A., Teshima, K., Shirai, Y., & Miyasaka, T. (2009). Organometal halide perovskites as visible-light sensitizers for photovoltaic cells. *Journal of the American Chemical Society, 131*, 6050–6051.
17. Lee, Y., Kwon, J., Hwang, E., Ra, C.-H., Yoo, W. J., Ahn, J.-H., Park, J. H., & Cho, J. H. (2015). High-performance perovskite-graphene hybrid photodetector. *Advanced Materials, 27*, 41–46.
18. Lim, K.-G., Ahn, S., Kim, Y.-H., Qi, Y., & Lee, T.-W. (2016). Universal energy level tailoring of self-organized hole extraction layers in organic solar cells and organic–inorganic hybrid perovskite solar cells. *Energy & Environmental Science, 9*, 932–939.
19. Luo, D., Yang, W., Wang, Z., Sadhanala, A., Hu, Q., Su, R., Shivanna, R., Trindade, G. F., Watts, J. F., Xu, Z., Liu, T., Chen, K., Ye, F., et al. (2018). Enhanced photovoltage for inverted planar heterojunction perovskite solar cells. *Science, 360*, 1442–1446.
20. Shen, H., Duong, T., Peng, J., Jacobs, D., Wu, N., Gong, J., Wu, Y., Karuturi, S. K., Fu, X., Weber, K., Xiao, X., White, T. P., & Catchpole, K. (2018). Mechanically-stacked perovskite/CIGS tandem solar cells with efficiency of 23.9% and reduced oxygen sensitivity. *Energy and Environmental Science, 11*, 394–406.
21. Li, X., Cao, F., Yu, D., Chen, J., Sun, Z., Shen, Y., Zhu, Y., Wang, L., Wei, Y., Wu, Y., & Zeng, H. (2017). All inorganic halide perovskites nanosystem: Synthesis, structural features, optical properties and optoelectronic applications. *Small, 13*, 1603996.
22. Docampo, P., Ball, J. M., Darwich, M., Eperon, G. E., & Snaith, H. J. (2013). Efficient organometal trihalide perovskite planar-heterojunction solar cells on flexible polymer substrates. *Nature Communications, 4*, 2761.
23. Mei, A., Li, X., Liu, L., Ku, Z., Liu, T., Rong, Y., Xu, M., Hu, M., Chen, J., Yang, Y., Gratzel, M., & Han, H. (2014). A hole-conductor-free, fully printable mesoscopic perovskite solar cell with high stability. *Science, 345*, 295–298.
24. Dong, Q., Fang, Y., Shao, Y., Mulligan, P., Qiu, J., Cao, L., & Huang, J. (2015). Solar cells. Electron-hole diffusion lengths > 175 mum in solution-grown $CH_3NH_3PbI_3$ single crystals. *Science, 347*, 967–970.

25. Cho, H., Jeong, S. H., Park, M. H., Kim, Y. H., Wolf, C., Lee, C. L., Heo, J. H., Sadhanala, A., Myoung, N., Yoo, S., Im, S. H., Friend, R. H., & Lee, T. W. (2015). Overcoming the electroluminescence efficiency limitations of perovskite light-emitting diodes. *Science, 350*, 1222–1225.
26. Byun, J., Cho, H., Wolf, C., Jang, M., Sadhanala, A., Friend, R. H., Yang, H., & Lee, T. W. (2016). Efficient visible quasi-2D perovskite light-emitting diodes. *Advanced Materials, 28*, 7515–7520.
27. Peng, M., Wen, W., Chen, S., Chen, B., Yan, K., Hu, H., Dong, B., Gao, X., Yu, X., Jiang, X., & Zou, D. (2016). Organolead trihalide perovskite materials for efficient light emitting diodes. *Science China: Chemistry, 59*, 653–658.
28. Dou, L., Yang, Y. M., You, J., Hong, Z., Chang, W. H., Li, G., & Yang, Y. (2014). Solution-processed hybrid perovskite photodetectors with high detectivity. *Nature Communications, 5*, 5404.
29. Tan, Z. K., Moghaddam, R. S., Lai, M. L., Docampo, P., Higler, R., Deschler, F., Price, M., Sadhanala, A., Pazos, L. M., Credgington, D., Hanusch, F., Bein, T., Snaith, H. J., et al. (2014). Bright light-emitting diodes based on organometal halide perovskite. *Nature Nanotechnology, 9*, 687–692.
30. Zhang, Q., Ha, S. T., Liu, X., Sum, T. C., & Xiong, Q. (2014). Room-temperature near-infrared high-Q perovskite whispering-gallery planar nanolasers. *Nano Letters, 14*, 5995–6001.
31. Nie, W., Blancon, J. C., Neukirch, A. J., Appavoo, K., Tsai, H., Chhowalla, M., Alam, M. A., Sfeir, M. Y., Katan, C., Even, J., Tretiak, S., Crochet, J. J., Gupta, G., et al. (2016). Light-activated photocurrent degradation and self-healing in perovskite solar cells. *Nature Communications, 7*, 11574.
32. Yang, S., Wang, Y., Liu, P. R., Cheng, Y. B., Zhao, H. J., & Yang, H. G. (2016). Functionalization of perovskite thin films with moisture-tolerant molecules. *Nature Energy, 1*, 15016.
33. Kim, Y. H., Cho, H., Heo, J. H., Kim, T. S., Myoung, N., Lee, C. L., Im, S. H., & Lee, T. W. (2015). Multicolored organic/inorganic hybrid perovskite light-emitting diodes. *Advanced Materials, 27*, 1248–1254.
34. Zhao, Y., Wei, J., Li, H., Yan, Y., Zhou, W., Yu, D., & Zhao, Q. (2016). A polymer scaffold for self-healing perovskite solar cells. *Nature Communications, 7*, 10228.
35. You, J., Meng, L., Song, T. B., Guo, T. F., Yang, Y. M., Chang, W. H., Hong, Z., Chen, H., Zhou, H., Chen, Q., Liu, Y., De Marco, N., & Yang, Y. (2016). Improved air stability of perovskite solar cells via solution-processed metal oxide transport layers. *Nature Nanotechnology, 11*, 75–81.
36. Protesescu, L., Yakunin, S., Bodnarchuk, M. I., Krieg, F., Caputo, R., Hendon, C. H., Yang, R. X., Walsh, A., & Kovalenko, M. V. (2015). Nanocrystals of cesium lead halide perovskites ($CsPbX_3$, X = Cl, Br, and I): Novel optoelectronic materials showing bright emission with wide color gamut. *Nano Letters, 15*, 3692–3696.
37. Song, J., Li, J., Li, X., Xu, L., Dong, Y., & Zeng, H. (2015). Quantum dot light-emitting diodes based on inorganic perovskite cesium lead halides ($CsPbX_3$). *Advanced Materials, 27*, 7162–7167.
38. Veldhuis, S. A., Boix, P. P., Yantara, N., Li, M., Sum, T. C., Mathews, N., & Mhaisalkar, S. G. (2016). Perovskite materials for light-emitting diodes and lasers. *Advanced Materials, 28*, 6804–6834.
39. Stoumpos, C. C., Malliakas, C. D., Peters, J. A., Liu, Z., Sebastian, M., Im, J., Chasapis, T. C., Wibowo, A. C., Chung, D. Y., Freeman, A. J., Wessels, B. W., & Kanatzidis, M. G. (2013). Crystal growth of the perovskite semiconductor $CsPbBr_3$: A new material for high-energy radiation detection. *Crystal Growth & Design, 13*, 2722–2727.
40. Liu, F., Zhang, Y., Ding, C., Kobayashi, S., Izuishi, T., Nakazawa, N., Toyoda, T., Ohta, T., Hayase, S., Minemoto, T., Yoshino, K., Dai, S., & Shen, Q. (2017). Highly luminescent phase-stable CsPbI3 perovskite quantum dots achieving near 100% absolute photoluminescence quantum yield. *ACS Nano, 11*, 10373–10383.
41. Nedelcu, G., Protesescu, L., Yakunin, S., Bodnarchuk, M. I., Grotevent, M. J., & Kovalenko, M. V. (2015). Fast anion-exchange in highly luminescent nanocrystals of cesium lead halide perovskites ($CsPbX_3$, X = Cl, Br, I). *Nano Letters, 15*, 5635–5640.

42. Akkerman, Q. A., Raino, G., Kovalenko, M. V., & Manna, L. (2018). Genesis, challenges and opportunities for colloidal lead halide perovskite nanocrystals. *Nature Materials, 17*, 394–405.
43. Swarnkar, A., Chulliyil, R., Ravi, V. K., Irfanullah, M., Chowdhury, A., & Nag, A. (2015). Colloidal $CsPbBr_3$ perovskite nanocrystals: Luminescence beyond traditional quantum dots. *Angewandte Chemie International Edition, 54*, 15424–15428.
44. Møller, C. K. (1958). Crystal structure and photoconductivity of cæsium plumbohalides. *Nature, 182*, 1436–1436.
45. Narayan, R. L., Sarma, M. V. S., & Suryanarayana, S. V. (1987). Ionic-conductivity of $CsPbCl_3$ and $CsPbBr_3$. *Journal of Materials Science Letters, 6*, 93–94.
46. Hirotsu, S., & Suzuki, T. (1978). Critical thermodynamic properties around the successive phase transitions of $CsPbBr_3$ and $CsPbCl_3$. *Ferroelectrics, 20*, 179–180.
47. Zeng, Q., Baker, I., & Yan, Z.-C. (2006). Nanostructured Mn–Al permanent magnets produced by mechanical milling. *Journal of Applied Physics, 99*, 08E902.
48. Babin, V., Fabeni, P., Nikl, M., Pazzi, G. P., Sildos, I., Zazubovich, N., & Zazubovich, S. (1999). Polarized luminescence of $CsPbBr_3$ nanocrystals (quantum dots) in CsBr: Pb single crystal. *Chemical Physics Letters, 314*, 31–36.
49. Kondo, S., Suzuki, K., Saito, T., Asada, H., & Nakagawa, H. (2004). Photoluminescence and stimulated emission from microcrystalline $CsPbCl_3$ films prepared by amorphous-to-crystalline transformation. *Physical Review B, 70*, 205322.
50. Nikl, M., Nitsch, K., Mihóková, E., Polák, K., Fabeni, P., Pazzi, G. P., Gurioli, M., Santucci, S., Phani, R., Scacco, A., & Somma, F. (1999). Luminescence of $CsPbBr_3$-like quantum dots in CsBr single crystals. *Physica E (Amsterdam, Netherlands), **4***, 323.
51. Wang, Y., Li, X., Song, J., Xiao, L., Zeng, H., & Sun, H. (2015). All-inorganic colloidal perovskite quantum dots: A new class of lasing materials with favorable characteristics. *Advanced Materials, 27*, 7101–7108.
52. Yakunin, S., Protesescu, L., Krieg, F., Bodnarchuk, M. I., Nedelcu, G., Humer, M., De Luca, G., Fiebig, M., Heiss, W., & Kovalenko, M. V. (2015). Low-threshold amplified spontaneous emission and lasing from colloidal nanocrystals of caesium lead halide perovskites. *Nature Communications, 6*, 8056.
53. Akkerman, Q. A., Motti, S. G., Srimath Kandada, A. R., Mosconi, E., D'Innocenzo, V., Bertoni, G., Marras, S., Kamino, B. A., Miranda, L., De Angelis, F., Petrozza, A., Prato, M., & Manna, L. (2016). Solution synthesis approach to colloidal cesium lead halide perovskite nanoplatelets with monolayer-level thickness control. *Journal of the American Chemical Society, 138*, 1010–1016.
54. Liu, H., Huang, Z., Wei, S., Zheng, L., Xiao, L., & Gong, Q. (2016). Nano-structured electron transporting materials for perovskite solar cells. *Nanoscale, 8*, 6209–6221.
55. Zhang, D., Yang, Y., Bekenstein, Y., Yu, Y., Gibson, N. A., Wong, A. B., Eaton, S. W., Kornienko, N., Kong, Q., Lai, M., Alivisatos, A. P., Leone, S. R., & Yang, P. (2016). Synthesis of composition tunable and highly luminescent cesium lead halide nanowires through anion-exchange reactions. *Journal of the American Chemical Society, 138*, 7236–7239.
56. Zhang, D., Eaton, S. W., Yu, Y., Dou, L., & Yang, P. (2015). Solution-phase synthesis of cesium lead halide perovskite nanowires. *Journal of the American Chemical Society, 137*, 9230–9233.
57. Zhou, R., Yang, Z., Xu, J., & Cao, G. (2018). Synergistic combination of semiconductor quantum dots and organic-inorganic halide perovskites for hybrid solar cells. *Coordination Chemistry Reviews, 374*, 279–313.
58. He, X., Qiu, Y., & Yang, S. (2017). Fully-inorganic trihalide perovskite nanocrystals: A new research frontier of optoelectronic materials. *Advanced Materials, 29*, 1700775.
59. Sutherland, B. R., & Sargent, E. H. (2016). Perovskite photonic sources. *Nature Photonics, 10*, 295–302.
60. Chang, S., Bai, Z., & Zhong, H. (2018). In situ fabricated perovskite nanocrystals: A revolution in optical materials. *Advanced Optical Materials, 6*, 1800380.

61. Weidman, M. C., Goodman, A. J., & Tisdale, W. A. (2017). Colloidal halide perovskite nanoplatelets: An exciting new class of semiconductor nanomaterials. *Chemistry of Materials, 29*, 5019–5030.
62. Fu, P., Shan, Q., Shang, Y., Song, J., Zeng, H., Ning, Z., & Gong, J. (2017). Perovskite nanocrystals: Synthesis, properties and applications. *Scientific Bulletin, 62*, 369–380.
63. Kong, Q., Lee, W., Lai, M., Bischak, C. G., Gao, G., Wong, A. B., Lei, T., Yu, Y., Wang, L.-W., Ginsberg, N. S., & Yang, P. (2018). Phase-transition–induced p-n junction in single halide perovskite nanowire. *PNAS, 115*, 8889–8894.
64. Lai, M., Obliger, A., Lu, D., Kley, C. S., Bischak, C. G., Kong, Q., Lei, T., Dou, L., Ginsberg, N. S., Limmer, D. T., & Yang, P. (2018). Intrinsic anion diffusivity in lead halide perovskites is facilitated by a soft lattice. *PNAS, 115*, 11929–11934.
65. Jellicoe, T. C., Richter, J. M., Glass, H. F., Tabachnyk, M., Brady, R., Dutton, S. E., Rao, A., Friend, R. H., Credgington, D., Greenham, N. C., & Bohm, M. L. (2016). Synthesis and optical properties of lead-free cesium tin halide perovskite nanocrystals. *Journal of the American Chemical Society, 138*, 2941–2944.
66. Maitra, U., Gupta, U., De, M., Datta, R., Govindaraj, A., & Rao, C. N. (2013). Highly effective visible-light-induced H_2 generation by single-layer 1T-MoS_2 and a nanocomposite of few-layer 2H-MoS_2 with heavily nitrogenated graphene. *Angewandte Chemie International Edition, 52*, 13057–13061.
67. Chen, X., Peng, L., Huang, K., Shi, Z., Xie, R., & Yang, W. (2016). Non-injection gram-scale synthesis of cesium lead halide perovskite quantum dots with controllable size and composition. *Nano Research, 9*, 1994–2006.
68. Sun, S., Yuan, D., Xu, Y., Wang, A., & Deng, Z. (2016). Ligand-mediated synthesis of shape-controlled cesium lead halide perovskite nanocrystals via reprecipitation process at room temperature. *ACS Nano, 10*, 3648–3657.
69. Wei, S., Yang, Y., Kang, X., Wang, L., Huang, L., & Pan, D. (2016). Room-temperature and gram-scale synthesis of $CsPbX_3$ (X = Cl, Br, I) perovskite nanocrystals with 50-85% photoluminescence quantum yields. *Chemical Communications, 52*, 7265–7268.
70. Tong, Y., Bladt, E., Aygueler, M. F., Manzi, A., Milowska, K. Z., Hintermayr, V. A., Docampo, P., Bals, S., Urban, A. S., Polavarapu, L., & Feldmann, J. (2016). Highly luminescent cesium lead halide perovskite nanocrystals with tunable composition and thickness by ultrasonication. *Angewandte Chemie International Edition, 55*, 13887–13892.
71. Wangyang, P., Gong, C., Rao, G., Hu, K., Wang, X., Yan, C., Dai, L., Wu, C., & Xiong, J. (2018). Recent advances in halide perovskite photodetectors based on different dimensional materials. *Advanced Optical Materials, 6*, 1701302.
72. Parobek, D., Dong, Y., Qiao, T., & Son, D. H. (2018). Direct hot-injection synthesis of Mn-doped CsPbBr3 nanocrystals. *Chemistry of Materials, 30*, 2939–2944.
73. Milstein, T. J., Kroupa, D. M., & Gamelin, D. R. (2018). Picosecond quantum cutting generates photoluminescence quantum yields over 100% in ytterbium-doped $CsPbCl_3$ nanocrystals. *Nano Letters, 18*, 3792–3799.
74. Cai, T., Yang, H., Hills-Kimball, K., Song, J. P., Zhu, H., Hofman, E., Zheng, W., Rubenstein, B., & Chen, O. (2018). Synthesis of all-inorganic Cd-doped $CsPbCl_3$ perovskite nanocrystals with dual-wavelength emission. *The Journal of Physical Chemistry Letters, 9*(24), 7079–7984.
75. Shao, H., Bai, X., Cui, H., Pan, G., Jing, P., Qu, S., Zhu, J., Zhai, Y., Dong, B., & Song, H. (2018). White light emission in Bi^{3+}/Mn^{2+} ion co-doped CsPbCl3 perovskite nanocrystals. *Nanoscale, 10*, 1023–1029.
76. Wang, A., Yan, X., Zhang, M., Sun, S., Yang, M., Shen, W., Pan, X., Wang, P., & Deng, Z. (2016). Controlled synthesis of lead-free and stable perovskite derivative Cs_2SnI_6 nanocrystals via a facile hot-injection process. *Chemistry of Materials, 28*, 8132–8140.
77. Li, X., Wu, Y., Zhang, S., Cai, B., Gu, Y., Song, J., & Zeng, H. (2016). $CsPbX_3$ quantum dots for lighting and displays: Room-temperature synthesis, photoluminescence superiorities, underlying origins and white light-emitting diodes. *Advanced Functional Materials, 26*, 2435–2445.

78. Huang, Y.-F., Zhou, X., Zhou, R., Zhang, H., Kang, K.-B., Zhao, M., Peng, Y., Wang, Q., Zhang, H.-L., & Qiu, W.-Y. (2014). One-pot synthesis of highly luminescent carbon quantum dots and their nontoxic ingestion by zebrafish for in vivo imaging. *Chemistry – A European Journal, 20*, 5640–5648.
79. Xu, J.-Y., Zhai, X.-P., Gao, L.-F., Chen, P., Zhao, M., Yang, H.-B., Cao, D.-F., Wang, Q., & Zhang, H.-L. (2016). In situ preparation of a MOF-derived magnetic carbonaceous catalyst for visible-light-driven hydrogen evolution. *RSC Advances, 6*, 2011–2018.
80. Liu, W., Zheng, J., Cao, S., Wang, L., Gao, F., Chou, K.-C., Hou, X., & Yang, W. (2018). General strategy for rapid production of low-dimensional all-inorganic $CsPbBr_3$ perovskite nanocrystals with controlled dimensionalities and sizes. *Inorganic Chemistry, 57*, 1598–1603.
81. Pan, Q., Hu, H., Zou, Y., Chen, M., Wu, L., Yang, D., Yuan, X., Fan, J., Sun, B., & Zhang, Q. (2017). Microwave-assisted synthesis of high-quality "all-inorganic" $CsPbX_3$ (X = Cl, Br, I) perovskite nanocrystals and their application in light emitting diodes. *Journal of Materials Chemistry C, 5*, 10947–10954.
82. Zhu, Z.-Y., Yang, Q.-Q., Gao, L.-F., Zhang, L., Shi, A.-Y., Sun, C.-L., Wang, Q., & Zhang, H.-L. (2017). Solvent-free mechanosynthesis of composition-tunable cesium lead halide perovskite quantum dots. *Journal of Physical Chemistry Letters, 8*, 1610–1614.
83. James, S. L., Adams, C. J., Bolm, C., Braga, D., Collier, P., Friscic, T., Grepioni, F., Harris, K. D., Hyett, G., Jones, W., Krebs, A., Mack, J., Maini, L., et al. (2012). Mechanochemistry: Opportunities for new and cleaner synthesis. *Chemical Society Reviews, 41*, 413–447.
84. Friscic, T., Reid, D. G., Halasz, I., Stein, R. S., Dinnebier, R. E., & Duer, M. J. (2010). Ion- and liquid-assisted grinding: Improved mechanochemical synthesis of metal-organic frameworks reveals salt inclusion and anion templating. *Angewandte Chemie International Edition, 49*, 712–715.
85. Jodlowski, A. D., Yepez, A., Luque, R., Camacho, L., & de Miguel, G. (2016). Benign-by-design solventless mechanochemical synthesis of three-, two-, and one-dimensional hybrid perovskites. *Angewandte Chemie International Edition, 55*, 14972–14977.
86. Park, C.-M., & Sohn, H.-J. (2007). Black phosphorus and its composite for lithium rechargeable batteries. *Advanced Materials, 19*, 2465–2468.
87. Sheldon, R. A. (2005). Green solvents for sustainable organic synthesis: State of the art. *Green Chemistry, 7*, 267–278.
88. Buzaglo, M., Bar, I. P., Varenik, M., Shunak, L., Pevzner, S., & Regev, O. (2017). Graphite-to-graphene: Total conversion. *Advanced Materials, 29*, 1603528.
89. Lakowicz, J. R. (1983). *Principles of fluoresence spectrosecopy*. New York: Plenum Press.
90. Yusoff, A. R. B. M., & Nazeeruddin, M. K. (2018). Low-dimensional perovskites: From synthesis to stability in perovskite Solar Cells. *Advanced Energy Materials, 8*, 1702073.
91. Li, J., Xu, L., Wang, T., Song, J., Chen, J., Xue, J., Dong, Y., Cai, B., Shan, Q., Han, B., & Zeng, H. (2017). 50-fold EQE improvement up to 6.27% of solution-processed all-inorganic perovskite $CsPbBr_3$ QLEDs via surface ligand density control. *Advanced Materials, 29*, 1603885.
92. Schmidt, T., Lischka, K., & Zulehner, W. (1992). Excitation-power dependence of the near-band-edge photoluminescence of semiconductors. *Physical Review B: Condensed Matter and Materials Physics, 45*, 8989–8994.
93. Gao, L. F., Wen, T., Xu, J. Y., Zhai, X. P., Zhao, M., Hu, G. W., Chen, P., Wang, Q., & Zhang, H. L. (2016). Iron-doped carbon nitride-type polymers as homogeneous organocatalysts for visible light-driven hydrogen evolution. *ACS Appl. Mater. Interfaces, 8*, 617–624.
94. Strauss, V., Margraf, J. T., Dolle, C., Butz, B., Nacken, T. J., Walter, J., Bauer, W., Peukert, W., Spiecker, E., Clark, T., & Guldi, D. M. (2014). Carbon nanodots: Toward a comprehensive understanding of their photoluminescence. *Journal of the American Chemical Society, 136*, 17308–17316.
95. Lakowicz, J. R. (1999). *Principles of fluorescence spectroscopy* (2nd ed.). New York: Kluwer Academic/Plenum Publishers.

96. Wu, K., Liang, G., Shang, Q., Ren, Y., Kong, D., & Lian, T. (2015). Ultrafast interfacial electron and hole transfer from $CsPbBr_3$ perovskite quantum dots. *Journal of the American Chemical Society, 137*, 12792–12795.
97. Tang, B., Dong, H., Sun, L., Zheng, W., Wang, Q., Sun, F., Jiang, X., Pan, A., & Zhang, L. (2017). Single-mode lasers based on cesium lead halide perovskite submicron spheres. *ACS Nano, 11*, 10681–10688.
98. Zhou, H., Yuan, S., Wang, X., Xu, T., Wang, X., Li, H., Zheng, W., Fan, P., Li, Y., Sun, L., & Pan, A. (2017). Vapor growth and tunable lasing of band gap engineered cesium lead lalide perovskite micro/nanorods with triangular cross section. *ACS Nano, 11*, 1189–1195.
99. Yang, B., Zhang, F., Chen, J., Yang, S., Xia, X., Pullerits, T., Deng, W., & Han, K. (2017). Ultrasensitive and fast all-inorganic perovskite-based photodetector via fast carrier diffusion. *Advanced Materials, 29*, 1703758.
100. Liu, Q., Wang, Y., Sui, N., Wang, Y., Chi, X., Wang, Q., Chen, Y., Ji, W., Zou, L., & Zhang, H. (2016). Exciton relaxation dynamics in photo-excited $CsPbI_3$ perovskite nanocrystals. *Scientific Reports, 6*, 29442.
101. Mondal, N., & Samanta, A. (2017). Complete ultrafast charge carrier dynamics in photo-excited all-inorganic perovskite nanocrystals ($CsPbX_3$). *Nanoscale, 9*, 1878–1885.
102. Yang, D., Li, X., Wu, Y., Wei, C., Qin, Z., Zhang, C., Sun, Z., Li, Y., Wang, Y., & Zeng, H. (2019). Surface halogen compensation for robust performance enhancements of $CsPbX_3$ perovskite quantum dots. *Advanced Optical Materials, 7*, 1900276.
103. Begum, R., Parida, M. R., Abdelhady, A. L., Murali, B., Alyami, N. M., Ahmed, G. H., Hedhili, M. N., Bakr, O. M., & Mohammed, O. F. (2017). Engineering interfacial charge transfer in $CsPbBr_3$ perovskite nanocrystals by heterovalent doping. *Journal of the American Chemical Society, 139*, 731–737.
104. Scheidt, R. A., Samu, G. F., Janáky, C., & Kamat, P. V. (2018). Modulation of charge recombination in $CsPbBr_3$ perovskite films with electrochemical bias. *Journal of the American Chemical Society, 140*, 86–89.
105. Queisser, H. J., & Haller, E. E. (1998). Defects in semiconductors: Some fatal, some vital. *Science, 281*, 945–950.
106. Wang, H.-C., Bao, Z., Tsai, H.-Y., Tang, A.-C., & Liu, R.-S. (2018). Perovskite quantum dots and their application in light-emitting diodes. *Small, 14*, 1702433.
107. Di Stasio, F., Christodoulou, S., Huo, N., & Konstantatos, G. (2017). Near-unity photoluminescence quantum yield in $CsPbBr_3$ nanocrystal solid-state films via postsynthesis treatment with lead bromide. *Chemistry of Materials, 29*, 7663–7667.
108. Kovalenko, M. V., Protesescu, L., & Bodnarchuk, M. I. (2017). Properties and potential optoelectronic applications of lead halide perovskite nanocrystals. *Science, 358*, 745–750.
109. Gan, J., He, J., Hoye, R. L. Z., Mavlonov, A., Raziq, F., MacManus-Driscoll, J. L., Wu, X., Li, S., Zu, X., Zhan, Y., Zhang, X., & Qiao, L. (2019). α-$CsPbI_3$ colloidal quantum dots: Synthesis, photodynamics, and photovoltaic applications. *ACS Energy Letters, 4*, 1308–1320.
110. Sanehira, E. M., Marshall, A. R., Christians, J. A., Harvey, S. P., Wheeler, L. M., Ciesielski, P. N., Schulz, P., Lin, L. Y., Luther, J. M., & Beard, M. C. (2017). Enhanced mobility $CsPbI_3$ quantum dot arrays for record-efficiency, high-voltage photovoltaic cells. *Science Advances, 3*, eaao4204.
111. Aharonovich, I., Englund, D., & Toth, M. (2016). Solid-state single-photon emitters. *Nature Photonics, 10*, 631.
112. Lounis, B., & Orrit, M. (2005). Single-photon sources. *Reports on Progress in Physics, 68*, 1129–1179.
113. Utzat, H., Sun, W., Kaplan, A. E. K., Krieg, F., Ginterseder, M., Spokoyny, B., Klein, N. D., Shulenberger, K. E., Perkinson, C. F., Kovalenko, M. V., & Bawendi, M. G. (2019). Coherent single-photon emission from colloidal lead halide perovskite quantum dots. *Science, 363*, 1068–1072.
114. Hu, F. R., Zhang, H. C., Sun, C., Yin, C. Y., Lv, B. H., Zhang, C. F., Yu, W. W., Wang, X. Y., Zhang, Y., & Xiao, M. (2015). Superior optical properties of perovskite nanocrystals as single photon emitters. *ACS Nano, 9*, 12410–12416.

115. Raino, G., Nedelcu, G., Protesescu, L., Bodnarchuk, M. I., Kovalenko, M. V., Mahrt, R. F., & Stoferle, T. (2016). Single cesium lead halide perovskite nanocrystals at low temperature: Fast single-photon emission, reduced blinking, and exciton fine structure. *ACS Nano, 10*, 2485–2490.
116. Park, Y. S., Guo, S., Makarov, N. S., & Klimov, V. I. (2015). Room temperature single-photon emission from individual perovskite quantum dots. *ACS Nano, 9*, 10386–10393.
117. Krishnakanth, K. N., Seth, S., Samanta, A., & Rao, S. V. (2018). Broadband femtosecond nonlinear optical properties of $CsPbBr_3$ perovskite nanocrystals. *Optics Letters, 43*, 603–606.
118. Han, Q., Wu, W., Liu, W., Yang, Q., & Yang, Y. (2018). Two-photon absorption and upconversion luminescence of colloidal $CsPbX_3$ quantum dots. *Optical Materials, 75*, 880–886.
119. Li, J., Ren, C., Qiu, X., Lin, X., Chen, R., Yin, C., & He, T. (2018). Ultrafast optical nonlinearity of blue-emitting perovskite nanocrystals. *Photonics Research, 6*, 554–559.
120. Ferrando, A., Pastor, J. P. M., & Suarez, I. (2018). Toward metal halide perovskite nonlinear photonics. *Journal of Physical Chemistry Letters, 9*, 5612–5623.
121. Wang, Y., Li, X., Zhao, X., Xiao, L., Zeng, H., & Sun, H. (2016). Nonlinear absorption and low-threshold multiphoton pumped stimulated emission from all-inorganic perovskite nanocrystals. *Nano Letters, 16*, 448–453.
122. Xing, G., Mathews, N., Sun, S., Lim, S. S., Lam, Y. M., Gratzel, M., Mhaisalkar, S., & Sum, T. C. (2013). Long-range balanced electron- and hole-transport lengths in organic-inorganic $CH_3NH_3PbI_3$. *Science, 342*, 344–347.
123. Stranks, S. D., Eperon, G. E., Grancini, G., Menelaou, C., Alcocer, M. J., Leijtens, T., Herz, L. M., Petrozza, A., & Snaith, H. J. (2013). Electron-hole diffusion lengths exceeding 1 micrometer in an organometal trihalide perovskite absorber. *Science, 342*, 341–344.
124. Kulbak, M., Cahen, D., & Hodes, G. (2015). How important is the organic part of lead halide perovskite photovoltaic cells? Efficient $CsPbBr_3$ cells. *Journal of Physical Chemistry Letters, 6*, 2452–2456.
125. Yettapu, G. R., Talukdar, D., Sarkar, S., Swarnkar, A., Nag, A., Ghosh, P., & Mandal, P. (2016). Terahertz conductivity within colloidal CsPbBr3 perovskite nanocrystals: Remarkably high carrier mobilities and large diffusion lengths. *Nano Letters, 16*, 4838–4848.
126. Chiba, T., Hoshi, K., Pu, Y.-J., Takeda, Y., Hayashi, Y., Ohisa, S., Kawata, S., & Kido, J. (2017). High-efficiency perovskite quantum-dot light-emitting devices by effective washing process and interfacial energy level alignment. *ACS Applied Materials & Interfaces, 9*, 18054–18060.
127. Chiba, T., Hayashi, Y., Ebe, H., Hoshi, K., Sato, J., Sato, S., Pu, Y.-J., Ohisa, S., & Kido, J. (2018). Anion-exchange red perovskite quantum dots with ammonium iodine salts for highly efficient light-emitting devices. *Nature Photonics, 12*, 681–687.
128. Swarnkar, A., Marshall, A. R., Sanehira, E. M., Chernomordik, B. D., Moore, D. T., Christians, J. A., Chakrabarti, T., & Luther, J. M. (2016). Quantum dot–induced phase stabilization of a-$CsPbI_3$ perovskite for high-efficiency photovoltaics. *Science, 354*, 92–95.
129. Ramasamy, P., Lim, D. H., Kim, B., Lee, S. H., Lee, M. S., & Lee, J. S. (2016). All-inorganic cesium lead halide perovskite nanocrystals for photodetector applications. *Chemical Communications, 52*, 2067–2070.
130. Li, X., Yu, D., Cao, F., Gu, Y., Wei, Y., Wu, Y., Song, J., & Zeng, H. (2016). Healing all-inorganic perovskite films via recyclable dissolution-recyrstallization for compact and smooth carrier channels of optoelectronic devices with high stability. *Advanced Functional Materials, 26*, 5903–5912.
131. Dong, Y., Gu, Y., Zou, Y., Song, J., Xu, L., Li, J., Xue, J., Li, X., & Zeng, H. (2016). Improving all-inorganic perovskite photodetectors by preferred orientation and plasmonic effect. *Small, 12*, 5622–5632.
132. Zhou, L., Yu, K., Yang, F., Zheng, J., Zuo, Y., Li, C., Cheng, B., & Wang, Q. (2017). All-inorganic perovskite quantum dot/mesoporous TiO2 composite-based photodetectors with enhanced performance. *Dalton Transactions, 46*, 1766–1769.
133. Gong, M., Sakidja, R., Goul, R., Ewing, D., Casper, M., Stramel, A., Elliot, A., & Wu, J. Z. (2019). High-performance all-inorganic $CsPbCl_3$ perovskite nanocrystal photodetectors with superior stability. *ACS Nano, 13*, 1772–1783.

134. Liu, X., Yu, D., Song, X., & Zeng, H. (2018). Metal halide perovskites: Synthesis, ion migration, and application in field-effect transistors. *Small, 14*, 1801460.
135. He, X., Liu, P., Zhang, H., Liao, Q., Yao, J., & Fu, H. (2017). Patterning multicolored microdisk laser arrays of cesium lead halide perovskite. *Advanced Materials, 29*, 1604510.
136. Huang, L., Gao, Q., Sun, L. D., Dong, H., Shi, S., Cai, T., Liao, Q., & Yan, C. H. (2018). Composition-graded cesium lead halide perovskite nanowires with tunable dual-color lasing performance. *Advanced Materials, 30*, e1800596.
137. Yang, S., Wang, Y., & Sun, H. (2015). Advances and prospects for whispering gallery mode microcavities. *Advanced Optical Materials, 3*, 1136–1162.
138. Eaton, S. W., Lai, M., Gibson, N. A., Wong, A. B., Dou, L., Ma, J., Wang, L. W., Leone, S. R., & Yang, P. (2016). Lasing in robust cesium lead halide perovskite nanowires. *Proceedings of the National Academy of Sciences of the United States of America, 113*, 1993–1998.
139. Wang, Y., Li, X., Nalla, V., Zeng, H., & Sun, H. (2017). Solution-processed low threshold vertical cavity surface emitting lasers from all-inorganic perovskite nanocrystals. *Advanced Functional Materials, 27*, 1605088.
140. Sheik-Bahae, M., Said, A. A., & Van Stryland, E. W. (1989). High-sensitivity, single-beam n_2 measurements. *Optics Letters, 14*, 955–957.
141. Wei, T.-C., Mokkapati, S., Li, T.-Y., Lin, C.-H., Lin, G.-R., Jagadish, C., & He, J.-H. (2018). Nonlinear absorption applications of $CH_3NH_3PbBr_3$ perovskite crystals. *Advanced Functional Materials, 28*, 1707175.
142. Bai, T., Dong, N., Cheng, H., Cheng, Q., Wang, J., & Chen, Y. (2017). $CH_3NH_3PbI_3$ perovskite: Poly(N-vinylcarbazole) blends for broadband optical limiting. *RSC Advances, 7*, 1809–1813.
143. Zhang, H., Lu, S. B., Zheng, J., Du, J., Wen, S. C., Tang, D. Y., & Loh, K. P. (2014). Molybdenum disulfide (MoS_2) as a broadband saturable absorber for ultra-fast photonics. *Optics Express, 22*, 7249–7260.
144. Chen, Y., Jiang, G., Chen, S., Guo, Z., Yu, X., Zhao, C., Zhang, H., Bao, Q., Wen, S., Tang, D., & Fan, D. (2015). Mechanically exfoliated black phosphorus as a new saturable absorber for both Q-switching and mode-locking laser operation. *Optics Express, 23*, 12823–12833.
145. Bao, Q., Zhang, H., Wang, Y., Ni, Z., Yan, Y., Shen, Z. X., Loh, K. P., & Tang, D. Y. (2009). Atomic-layer graphene as a saturable absorber for ultrafast pulsed lasers. *Advanced Functional Materials, 19*, 3077–3083.
146. Zhou, Y., Hu, Z., Li, Y., Xu, J., Tang, X., & Tang, Y. (2016). $CsPbBr_3$ nanocrystal saturable absorber for mode-locking ytterbium fiber laser. *Applied Physics Letters, 108*, 261108.
147. Li, J., Dong, H., Xu, B., Zhang, S., Cai, Z., Wang, J., & Zhang, L. (2017). $CsPbBr_3$ perovskite quantum dots: Saturable absorption properties and passively Q-switched visible lasers. *Photonics Research, 5*, 457–460.
148. Xu, Y.-F., Yang, M.-Z., Chen, B.-X., Wang, X.-D., Chen, H.-Y., Kuang, D.-B., & Su, C.-Y. (2017). A $CsPbBr_3$ perovskite quantum dot/graphene oxide composite for photocatalytic CO_2 reduction. *Journal of the American Chemical Society, 139*, 5660–5663.
149. Hou, J., Cao, S., Wu, Y., Gao, Z., Liang, F., Sun, Y., Lin, Z., & Sun, L. (2017). Inorganic colloidal perovskite quantum dots for robust solar CO_2 reduction. *Chemistry – A European Journal, 23*, 9481–9485.
150. Zhu, X., Lin, Y., Sun, Y., Beard, M. C., & Yan, Y. (2019). Lead-halide perovskites for photocatalytic α-alkylation of aldehydes. *Journal of the American Chemical Society, 141*, 733–738.
151. Yakunin, S., Sytnyk, M., Kriegner, D., Shrestha, S., Richter, M., Matt, G. J., Azimi, H., Brabec, C. J., Stangl, J., Kovalenko, M. V., & Heiss, W. (2015). Detection of X-ray photons by solution-processed lead halide perovskites. *Nature Photonics, 9*, 444–450.
152. Birowosuto, M. D., Cortecchia, D., Drozdowski, W., Brylew, K., Lachmanski, W., Bruno, A., & Soci, C. (2016). X-ray scintillation in lead halide perovskite crystals. *Scientific Reports, 6*, 37254.
153. Chen, Q., Wu, J., Ou, X., Huang, B., Almutlaq, J., Zhumekenov, A. A., Guan, X., Han, S., Liang, L., Yi, Z., Li, J., Xie, X., Wang, Y., et al. (2018). All-inorganic perovskite nanocrystal scintillators. *Nature, 561*, 88–93.

154. Becker, M. A., Vaxenburg, R., Nedelcu, G., Sercel, P. C., Shabaev, A., Mehl, M. J., Michopoulos, J. G., Lambrakos, S. G., Bernstein, N., Lyons, J. L., Stöferle, T., Mahrt, R. F., Kovalenko, M. V., et al. (2018). Bright triplet excitons in caesium lead halide perovskites. *Nature, 553*, 189.
155. Huo, C., Cai, B., Yuan, Z., Ma, B., & Zeng, H. (2017). Two-dimensional metal halide perovskites: Theory, synthesis, and optoelectronics. *Small Methods, 1*, 1600018.
156. Chen, D., Fang, G., & Chen, X. (2017). Silica-coated Mn-doped $CsPb(Cl/Br)_3$ inorganic perovskite quantum dots: Exciton-to-Mn energy transfer and blue-excitable solid-state lighting. *ACS Applied Materials & Interfaces, 9*, 40477–40487.
157. Leng, M., Yang, Y., Zeng, K., Chen, Z., Tan, Z., Li, S., Li, J., Xu, B., Li, D., Hautzinger, M. P., Fu, Y., Zhai, T., Xu, L., et al. (2018). All-inorganic bismuth-based perovskite quantum dots with bright blue photoluminescence and excellent stability. *Advanced Functional Materials, 28*, 1704446.
158. Leng, M., Chen, Z., Yang, Y., Li, Z., Zeng, K., Li, K., Niu, G., He, Y., Zhou, Q., & Tang, J. (2016). Lead-free, blue emitting bismuth halide perovskite quantum dots. *Angewandte Chemie International Edition, 55*, 15012–15016.
159. Zhang, J., Yang, Y., Deng, H., Farooq, U., Yang, X., Khan, J., Tang, J., & Song, H. (2017). High quantum yield blue emission from lead-free inorganic antimony halide perovskite colloidal quantum dots. *ACS Nano, 11*, 9294–9302.
160. Hong, W. L., Huang, Y. C., Chang, C. Y., Zhang, Z. C., Tsai, H. R., Chang, N. Y., & Chao, Y. C. (2016). Efficient low-temperature solution-processed lead-free perovskite infrared light-emitting diodes. *Advanced Materials, 28*, 8029–8036.
161. Zhang, H. H., Wang, X., Liao, Q., Xu, Z. Z., Li, H. Y., Zheng, L. M., & Fu, H. B. (2017). Embedding perovskite nanocrystals into a polymer matrix for tunable luminescence probes in cell imaging. *Advanced Functional Materials, 27*, 1604382.
162. Chen, Z., Wang, J. J., Ren, Y., Yu, C., & Shum, K. (2012). Schottky solar cells based on $CsSnI_3$ thin-films. *Applied Physics Letters, 101*, 093901.
163. Kumar, M. H., Dharani, S., Leong, W. L., Boix, P. P., Prabhakar, R. R., Baikie, T., Shi, C., Ding, H., Ramesh, R., Asta, M., Graetzel, M., Mhaisalkar, S. G., & Mathews, N. (2014). Lead-free halide perovskite solar cells with high photocurrents realized through vacancy modulation. *Advanced Materials, 26*, 7122–7127.
164. Dai, X., Zhang, Z., Jin, Y., Niu, Y., Cao, H., Liang, X., Chen, L., Wang, J., & Peng, X. (2014). Solution-processed, high-performance light-emitting diodes based on quantum dots. *Nature, 515*, 96.
165. Dai, X., Deng, Y., Peng, X., & Jin, Y. (2017). Quantum-dot light-emitting diodes for large-area displays: Towards the dawn of commercialization. *Advanced Materials, 29*, 1607022.
166. Zhang, Y., Saidaminov, M. I., Dursun, I., Yang, H., Murali, B., Alarousu, E., Yengel, E., Alshankiti, B. A., Bakr, O. M., & Mohammed, O. F. (2017). Zero-dimensional Cs_4PbBr_6 perovskite nanocrystals. *Journal of Physical Chemistry Letters, 8*, 961–965.
167. Saidaminov, M. I., Almutlaq, J., Sarmah, S., Dursun, I., Zhumekenov, A. A., Begum, R., Pan, J., Cho, N., Mohammed, O. F., & Bakr, O. M. (2016). Pure Cs_4PbBr_6: Highly luminescent zero dimensional perovskite solids. *ACS Energy Letters, 1*, 840–845.
168. Liu, Z., Bekenstein, Y., Ye, X., Nguyen, S. C., Swabeck, J., Zhang, D., Lee, S.-T., Yang, P., Ma, W., & Alivisatos, A. P. (2017). Ligand mediated transformation of cesium lead bromide perovskite nanocrystals to lead depleted Cs_4PbBr_6 nanocrystals. *Journal of the American Chemical Society, 139*, 5309–5312.
169. Odenthal, P., Talmadge, W., Gundlach, N., Wang, R., Zhang, C., Sun, D., Yu, Z.-G., Valy Vardeny, Z., & Li, Y. S. (2017). Spin-polarized exciton quantum beating in hybrid organic–inorganic perovskites. *Nature Physics, 13*, 894.
170. Long, G., Jiang, C., Sabatini, R., Yang, Z., Wei, M., Quan, L. N., Liang, Q., Rasmita, A., Askerka, M., Walters, G., Gong, X., Xing, J., Wen, X., et al. (2018). Spin control in reduced-dimensional chiral perovskites. *Nature Photonics, 12*, 528–533.

Chapter 7
Quantum Dot Materials Toward High-Speed and Ultrafast Laser Applications

Xu Wang, Jiqiang Ning, Changcheng Zheng, and Ziyang Zhang

Abstract Self-assembled In(Ga)As/GaAs quantum dots (QDs) have attracted much attention for both high-speed and ultrafast laser applications because of their fascinating optical and electronic properties. Here, we will review recent development of InAs/GaAs quantum dots and their applications to high-speed lasers and ultrafast lasers. The chapter includes two main sections, one is focusing on developing high-quality 1310 nm InAs/GaAs quantum dot structures and fabricating high-performance lasers including ultrashort cavity Fabry-Pérot (F-P) and distributed feedback (DFB) lasers. We will discuss effects of the modulation p-doping on optical properties of 1310 nm InAs/GaAs QDs and share our latest results on ultrashort cavity F-P and DFB lasers. The other is about the recent works on the development of 1550 nm InAs/GaAs quantum dot semiconductor saturable absorber mirrors (QD-SESAMs) and the realization of a high repetition rate diode-pumped solid-state and *Q*-switched Er-doped fiber laser mode-locked by the utilization of 1550 nm QD-SESAM.

7.1 Introduction

7.1.1 1310 nm InAs/GaAs QD for High-Speed Lasers

Owing to their superior properties including ultralow and temperature-insensitive threshold currents, and low relative intensity noise accompanied by a low linewidth enhancement factor, self-assembled InAs/GaAs quantum dot (QD) lasers operating at 1.3 μm have attracted great interest as promising light sources for

X. Wang · J. Ning · Z. Zhang (✉)
Key Lab of Nanodevices and Applications, Suzhou Institute of Nano-Tech and Nano-Bionics, Chinese Academy of Sciences, Suzhou, People's Republic of China
e-mail: zyzhang2014@sinano.ac.cn

C. Zheng
Division of Natural and Applied Sciences, Duke Kunshan University, Kunshan, People's Republic of China

P. Yu, Z. M. Wang (eds.), *Quantum Dot Optoelectronic Devices*, Lecture Notes in Nanoscale Science and Technology 27,
https://doi.org/10.1007/978-3-030-35813-6_7

high-speed device applications in short-haul local area network and metropolitan area network communication systems [1–6]. The high-speed characteristics of InAs/GaAs QD lasers highly depend on both optical and electronic properties of self-assembled QD structures. Recently, great progress has been made to investigate and optimize the properties of InAs/GaAs QD by using modulation p-doping methods [7–10]. It has been demonstrated that holes in the p-doped barrier or spacer layer are transferred into the hole ground state (GS) in the InAs QD layer, which results in the compensation of the thermal broadening of the hole distribution [11–14]. Therefore, modulation p-doping of QD lasers shows significant improvement in the optical gain, which is very favorable for high-speed device applications. Additionally, it has also been found that modulation p-doping can effectively improve the QD laser modulation response and reduce the linewidth enhancement factor [15]. The induced intermixing by thermal annealing is also attracting considerable attention as a way of adjusting and controlling the optical properties of InAs QDs [16, 17].

The thermal intermixing effect is expected to play an important role in QD structures due to the large surface-area-to-volume ratio compared to bulk or quantum well (QW). Generally, the sharper emission peak induced by rapid thermal annealing (RTA) process indicates improvement in optical quality. However, the investigations on the influence of the RTA process on the elementary composition, structure, and optical properties of InAs QDs are mostly based on solo QD structures whose properties are greatly different from that of the final practical laser device [18–21]. In the practical InAs/GaAs QD laser structure, to construct the p-i-n configuration, a very thick p-type AlGaAs top cladding layer is prepared on the InAs/GaAs QD active layer. The growth temperature of high-quality p-type AlGaAs cladding layer is usually higher than that of the InAs/GaAs QD active layer. The high-temperature growth process of p-type AlGaAs cladding layer is equivalent to a RTA treatment to the InAs/GaAs QD layer, which will strongly impact properties of the InAs/GaAs QD layer.

For high-speed lasers, another important aspect is to find an appropriate cavity length, since a very short cavity will increase the response time due to the easily saturated gain, while a long cavity slows the response owing to the increased photon lifetime [22–25]. Modulation p-doping in the InAs/GaAs QD structures has been proven to be an effective way to optimize the device performances. Introducing p-doping allows the cavity length to be shortened, because the large built-in hole concentration can reduce the effect of gain saturation and therefore allow GS lasing in shorter cavities with higher current densities. So far, it has been demonstrated that the modulation p-doping 1.3 μm InAs/GaAs QD Fabry-Pérot (F-P) laser with p-i-n configuration exhibits 15 Gb/s high-speed data transmission [26], and InAs/GaAs QD-based distributed feedback (DFB) lasers, which provide longitudinal single-mode emission with a narrow line width, are indispensable for dense wavelength division multiplexing (WDM) systems.

7.1.2 1550 nm InAs/GaAs QD-SESAM for Ultrafast Lasers

Self-assembled InAs/GaAs quantum dots (QDs), as a mature structure, have been successfully employed in high-performance laser diodes, photodetectors, and superluminescent light-emitting diodes due to precisely customized parameters and great reproducibility [27–29]. InAs/GaAs QD-based SESAMs offer many potential advantages for the design and development of ultrashort pulse laser because of extremely broad gain bandwidths and fast carrier dynamics in comparison with commercial QW-based SESAMs [30, 31]. Furthermore, in high repetition rate SESAM mode-locked lasers, the pulse energy is very low requiring relatively tight focusing onto the QW-SESAM for achieving saturation. The tight focusing can be realized by using a highly curved intracavity mirror in a very compact laser cavity structure, which limits the design freedom of laser cavity [32]. This issue can be resolved by using QD-based SESAMs due to their lower saturation fluence in comparison with QW-based SESAMs. For QD-based SESAMs, both saturation fluence and modulation depth are highly dependent on the areal QD density, which provides the additional degree of freedom for the design of laser cavity [33].

The operating wavelength of the InAs/GaAs QD-based SESAMs is mainly in the range of ~1–1.3 μm [34–36], while it is still a huge challenge to fabricate a high-performance QD-SESAMs working at the wavelength of 1550 nm. Though InAs QDs grown on InP dominate the 1550 nm region [37–41], InAs/InP QD technology is still under developing compared with that of InAs/GaAs QD technology due to the difficulty to obtain the high-quality distributed Bragg reflectors (DBRs) on InP substrate materials. Moreover, in comparison with InP-based substrate materials, GaAs-based substrate materials exhibit many advantages such as higher thermal conductivity, larger band offsets, lower cost of substrates, and higher lattice match with high-quality DBRs [42]. Following the introduction of a strained InGaAs cap layer to reach 1310 nm for InAs/GaAs QDs, the expected In content in QD cap layer should be higher for realizing 1550 nm InAs/GaAs QDs. However, high In content in the QDs and the surrounding matrix very readily leads to non-radiative recombination centers which rapidly degrades the crystal and optical quality of the QDs [43], making the QD device unsuitable for application. To solve this issue, different materials were used including quaternary alloys like InGaNAs and InGaAsSb which made the epitaxial growth process more complex and reduced the optical and material qualities of InAs QDs [44, 45]. In addition, 1550 nm InAs/GaAs QDs has also been grown on thick metamorphic InGaAs buffer layers or virtual substrates [46]. However, it is restrained by the difficulty in growing thick lattice-matched InGaAs alloys or DBRs. Therefore, it is highly desirable to explore new techniques to grow 1550 nm InAs/GaAs QDs structures with improved quality.

Ultrafast lasers mode-locked by using 1550 nm SESAMs exhibit a lot of advantages such as low timing jitter, high individual optical spectral mode signal-to-noise ratio, and high pulse-to-pulse phase coherence. These characters are highly desired for ultrahigh-speed data transmission systems, optical clock, frequency metrology, continuum wave generation, and coherent optical communications [47–49].

However, as the issue mentioned above, so far, SESAMs based on 1550 nm InAs/GaAs QDs are still researched in the lab, with only a few reports on 1550 nm InAs/GaAs QDs mode-locked laser for ultrafast applications [30, 31].

7.2 Modulation p-Doping 1310 nm InAs/GaAs QD Ultrashort Cavity F-P and DFB Lasers

This section deals with the epitaxial growth of 1310 nm InAs/GaAs QD materials and the fabrication of 1310 nm InAs/GaAs QD-based ultrashort cavity F-P and DFB lasers. This section also deals with the effect of modulation p-doping on optical properties of 1310 nm InAs/GaAs QD materials and the effect of cavity length on performances of ultrashort cavity F-P and DFB lasers. In one early work underlying 1310 nm InAs/GaAs QD materials, Li et al. successfully grew the 1310 nm InAs/GaAs QD laser structure and studied the effect of modulation p-doping and RTA on optical properties of 1310 nm InAs/GaAs QD materials [50]. In the same work, Li et al. also fabricated the 1310 nm InAs/GaAs QD-based ultrashort cavity F-P and DFB lasers and investigated performances of ultrashort cavity F-P and DFB lasers with the cavity length ranging from 350 to 500 μm [50]. Arsenijević et al. reported a 15 Gb/s high-speed 1310 nm modulation p-doped QD laser with a 500-μm-long F-P QD laser [26].

7.2.1 Epitaxial Growth of 1310 nm InAs/GaAs QD Materials and Fabrication of Ultrashort Cavity F-P and DFB Lasers

Here, we give a brief description of the fabrication process of a F-P and DFB laser with ultrashort cavity. The schematic diagram of the InAs/GaAs QD-based laser diode structure, in a typical p-i-n configuration, is presented in Fig. 7.1. All the InAs/GaAs QD laser structures were grown by a molecular-beam epitaxy (MBE) system on Si-doped GaAs (100) substrates. The QD active region of these laser structures are eight stacks of QD layers which are separated by GaAs barriers with a thickness of 33 nm as shown in the inset of Fig. 7.1. The QD layer consists of 2.7 nm monolayer (ML) InAs covered with 6 nm InGaAs stain reducing layer. The QD active region is sandwiched by an-type $Al_{0.3}Ga_{0.7}As$ cladding layer with a thickness of 2800 nm bellow and a p-type $Al_{0.3}Ga_{0.7}As$ cladding layer with a thickness of 1800 nm above to form the laser structure. The p-doped as grown QDs (labeled QDP) sample was sequentially grown with identical structures. The modulation p-doping with Be was conducted in a 6 nm layer located in the GaAs spacer layer 10 nm beneath each InAs/GaAs layer to obtain a concentration of 25 acceptors per dot. The RTA process was performed at the temperature of 700 °C in a nitrogen ambient for 5 min. A GaAs layer was employed to protect the QD samples during the RTA process. In order to obtain the effective light excitation and PL signal collection, we etched away the upper p-type $Al_{0.3}Ga_{0.7}As$ cladding layer by using wet etching above the QD active region (Fig. 7.2).

As the first step in fabricating the DFB lasers, the ridge waveguide lasers were fabricated using optical lithography and inductively coupled plasma (ICP) dry etching. The width of these ridge waveguide was set to be 3.5 μm, and the length of laser cavities was set to be 350, 400, and 500 μm, respectively. For the fabrication of DFB lasers, first-order Bragg gratings fabricated alongside the ridge waveguide using electron beam lithography (EBL) were etched by ICP. The etching stopped at ~150 nm above the QD active range in order to form a good coupling with light.

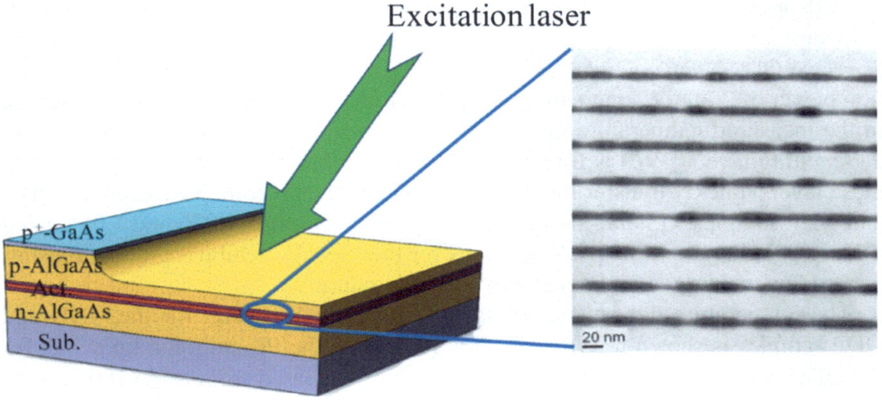

Fig. 7.1 Schematic structure of the QD sample (PL experiments were carried out on the samples corroded out of p^+-AlGaAs layers). Inset: cross-sectional TEM image of the QD active layer structure. (Reproduced with permission from Ref. [50] (copyright 2018 ACS))

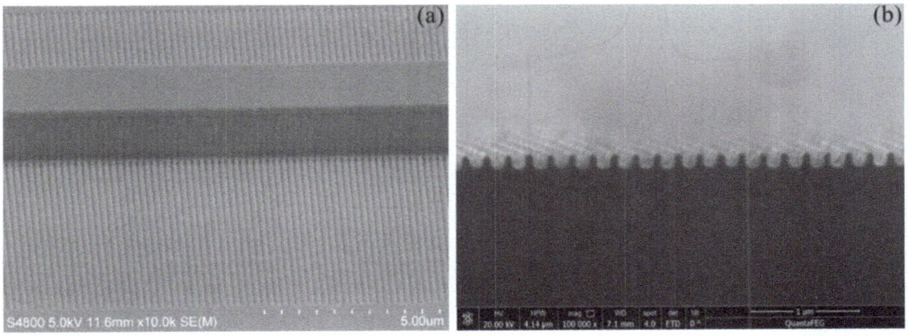

Fig. 7.2 (**a**) Top view of the SEM image of LC-DFB laser structure with first-order grating. (**b**) Cross-sectional SEM image of grating after dry etching. (Reproduced with permission from Ref. [50] (copyright 2018 ACS))

Au/Ge/Ni/Au and Ti/Au were deposited to form bottom and top Ohmic contacts, respectively. The length of laser cavities was set to be 500 μm and cavity facets were not coated. In order to minimize self-heating effects, the substrates were thinned to around 80 μm. All the laser bars which were mounted on a copper heat sink were measured at room temperature (RT) under continuous wave (CW) operation.

7.2.2 Optical Properties of 1310 nm InAs/GaAs QD Materials

Temperature-dependent PL spectra were measured at the temperature range from 4 to 300 K with the 532 nm line of an Ar^+ laser as excitation source and detected with an InGaAs detector. Before the PL measurements, each temperature point was kept for 20 min to avoid temperature fluctuations during spectrum acquisition. Temperature-dependent PL spectra of undoped as grown QDs (QDU) and p-doped as grown QDs (QDP) samples are shown in Fig. 7.3a, b, respectively. Under excitation power of 200 mW, one emission peak from ground state (GS) can be observed in the measured temperature range. No emission from the InGaAs capping layers and wetting layers (WL) is found, which suggests that photo-generated carriers in these layers quickly transfer into the InAs QD and then recombine there. For QDU and QDP samples, GS emission peaks are located at around 1.026 and 1.008 eV at 4 K, respectively, and the peak positions for these peaks exhibit a gradual redshift toward lower energy with the increase of temperature.

The temperature dependence of the PL emission linewidth for the QDU and QDP samples is shown in Fig. 7.4a. It is found that the PL emission linewidth for the QDU sample decreases with the temperature increased from 80 to 180 K, sug-

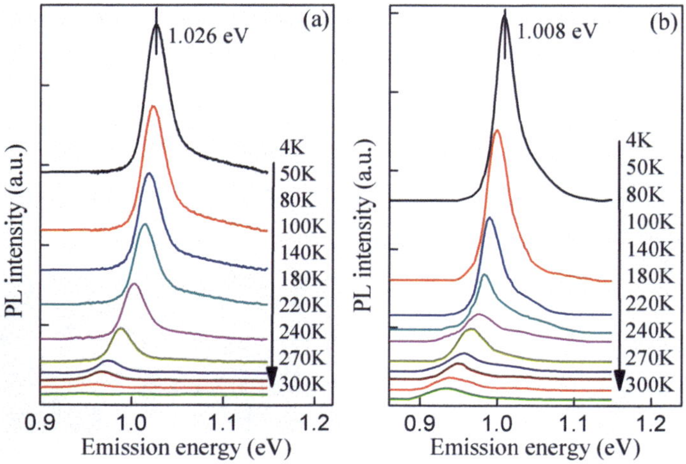

Fig. 7.3 PL spectra measured at 4–300 K from (**a**) the undoped and (**b**) p-doped eight-layer InAs/GaAs QD laser structures. (Reproduced with permission from Ref. [50] (copyright 2018 ACS))

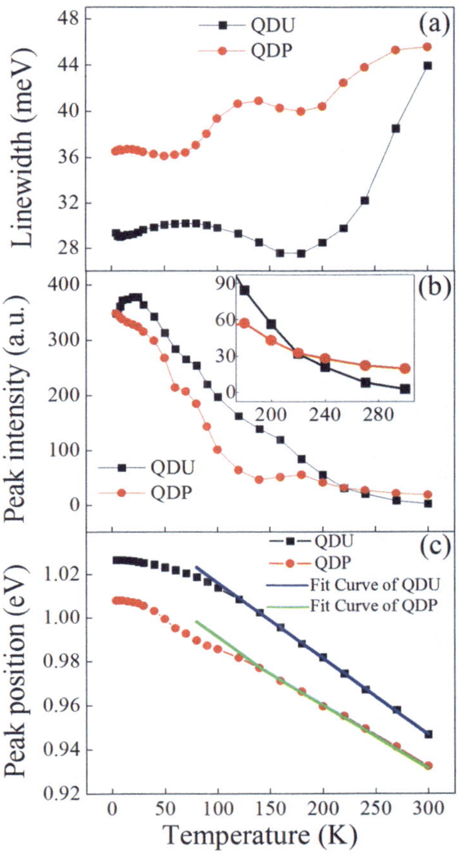

Fig. 7.4 (**a**) PL linewidth, (**b**) PL peak intensity, and (**c**) PL peak position as functions of temperature. Black squares and red solid circles are for the QDU and QDP samples, respectively. The inset in (**b**) shows the plots in the high-temperature range with an enlarged vertical scale. The blue and green lines in (**c**) show the linear fitting results from 140 to 300 K for QDU and QDP samples, respectively. (Reproduced with permission from Ref. [50] (copyright 2018 ACS))

gesting that thermally activated carriers are redistributed among small and large QDs [51]. When the temperature is increased beyond 180 K, the emission linewidth rapidly increases because of the increased electron-phonon scattering. After the modulation p-doping, the QDP sample exhibits a broader linewidth than that of the QDU sample, which can be mainly attributed to the state filling effect induced by modulation p-doping [52].

Figure 7.4b shows the temperature dependence of the normalized PL intensity for the QDU and QDP samples. With the temperature increased from 4 to 25 K, an increase of PL intensity for the QDU sample can be clearly observed. It is well-known that the WL can work as a carrier reservoir, which increases the PL intensity of QDs by enhancing multiphonon-assisted relaxation process. However, no increased peak intensity at the temperature range from 4 to 25 K is found for QDP sample indicating that fast carrier capture and relaxation processes are dominant owing to the enhanced carrier-carrier scattering [53]. When the temperature is further increased, the PL peak intensity decreases for both QDU and QDP samples, which implies the activated carriers escape from the GS to barriers or states of non-radiative recombination centers. Moreover, it is clearly observed that the QDP sample has the less PL intensity than that of QDU sample in the temperature range of 4–200 K, which is ascribed to the increased density of dopant-related trap states in the p-doped GaAs [54]. We also note that the PL intensity for QDP sample is still very strong at 300 K as present in the inset of Fig. 7.4b. The value of PL intensity for QDP sample at 300 K has about a factor of sixteen decrease in comparison with that at 4 K, while the decrease is about one hundred and fifty times for QDU sample. The decreased quenching effect of PL emission in QDP sample can be explained by the strong compensation of the holes induced by modulation doping, which inhibits thermal broadening of holes from GS with increasing temperature, resulting in a higher GS gain in the QDP sample.

The PL peak positions as a function of temperature are presented in Fig. 7.4c. Peak positions of both QDU and QDP samples show a redshift with the increase of temperature. However, the variation of the PL peak position with temperature cannot be well fitted by Varshini's formula in terms of band gap shrinkage suggesting that the carriers transfer from the small QDs over barriers to the large QDs when increasing the temperature [55, 56]. The PL peak energy of the QDP sample is smaller than that of QDU sample in the measured temperature range, which is induced by the higher temperature for the growth process of AlGaAs cladding layer. The growth process of the AlGaAs cladding layer at high temperature is equivalent to a RTA process, in which a strong In/Ga element intermixing between QDs and surrounding barrier layers occurred in QDU sample. For the QDP sample, the Ga vacancy propagation is inhibited due to the modulation p-doping, which results in the reduced interdiffusion and reduced intermixing effect [57]. As a result, the QDP sample experienced a less change in the emission behavior against temperature. The PL peak position exhibits a fast redshift in the range from 140 to 300 K, and it can be linearly fitted. Fitting lines have slopes of 0.35 and 0.29 meV/K corresponding to QDU and QDP samples, respectively. The smaller slope of the QDP sample suggests that PL emission properties are effectively modified, making its PL peak position less temperature sensitive.

As discussed above, the emission properties of the self-assembled InAs/GaAs QD system are significantly determined by the intermixing effect. In fact, it has been demonstrated that well-controlled element intermixing in QD materials can effectively adjust their emission wavelength, making it an attractive technique for the manufacturing of multiple-section photonic integrated circuits [58]. Though a large blueshift of the emission wavelength can be realized, an undesirable decrease of PL intensity was also obtained during the intermixing process [59]. Moreover, the growth of high-quality AlGaAs cladding layer usually needs a much higher temperature than that for the growth of QD active layer. This high growth temperature can be equivalent to a relatively strong RTA treatment for the InAs QD active region. The RTA treatment is often used to get rid of the point defects which are induced during the epitaxy growth process. Therefore, it is necessary to study the effect of the RTA treatment on the optical properties of QDU and QDP samples. In order to have a clear sight on the different situations induced by RTA treatment for QDU and QDP samples, the structural and compositional details of these samples are shown in Fig. 7.5, with the Ga vacancies and doped Be atoms included.

Temperature-dependent PL spectra of the annealed undoped QDs (QDAU) and the annealed p-doped QDs (QDAP) samples were also measured with the same excitation power of 200 mW as in measuring the unannealed samples. Figure 7.6a, b presents the temperature-dependent PL spectra of QDAU and

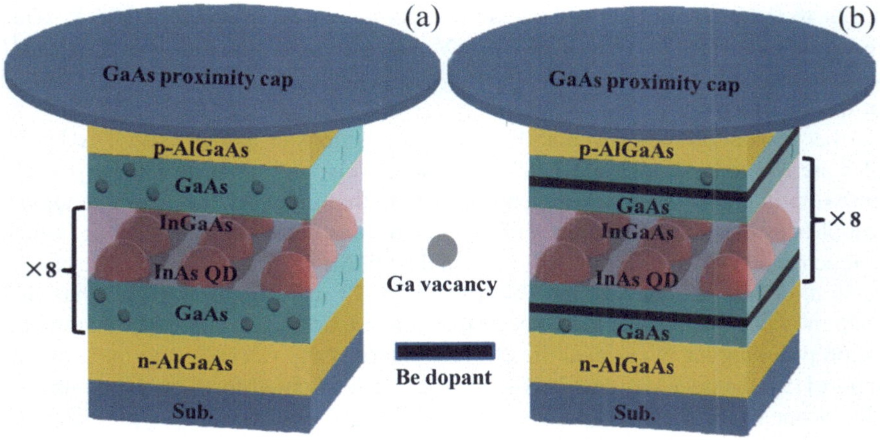

Fig. 7.5 Schematic illustration of the RTA process of (**a**) undoped and (**b**) p-doped eight-layer InAs/InGaAs/GaAs QD structures using a GaAs proximity cap, indicating the lower concentrations of Ga vacancies in p-doped samples due to the Be dopants. (Reproduced with permission from Ref. [50] (copyright 2018 ACS))

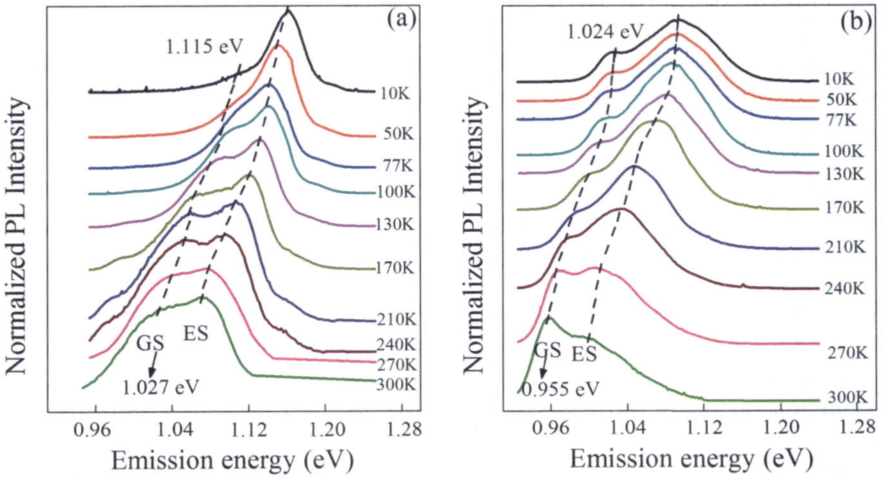

Fig. 7.6 Normalized PL spectra measured at 10–300 K from (**a**) the QDAU and (**b**) QDAP samples, respectively. The dashed lines show the tendency of temperature dependence of GS and ES emission peaks. (Reproduced with permission from Ref. [50] (copyright 2018 ACS))

QDAP samples, respectively. At the temperature range from 10 to 300 K, a distinctive double-peaked feature can be clearly observed corresponding to the GS and excited state (ES) emission of InAs QDs, respectively. The observed ES peak can be attributed to interdiffusion mechanism of In and Ga atoms, which induces a shallower confining profile and consequently a reduced GS gain. Moreover, the GS emission peak of the QDAP sample exhibits a redshift of ~68.9 meV (from 1.024 to 0.955 eV) when the temperature is increased from 10 to 300 K. This redshift is smaller than that of ~88.0 meV (from 1.115 to 1.027 eV) for the QDAU sample indicating the QDAP sample has the less temperature-dependent behavior of the PL emission.

Further inspection indicates that the GS peak positions of QDAP and QDAU samples are located at 0.955 and 1.027 eV, respectively, and the peak intensity of GS emission in QDAP sample is much higher than that in QDAU sample. The QDAP sample shows a ~22.3 meV blueshift of GS emission energy upon annealing, which is much smaller than that of ~80.4 meV in the QDAU sample. Moreover, the energy separation between the GS and ES of QDAP (~71 meV) is much larger than that (~45 meV) of QDAU. All these spectral characteristics result in the fact that modulation p-doping has weakened the effect of intermixing. According to the previous report [19], the degree of intermixing effect is mainly determined by the number of Ga vacancies. The concentrations of Ga vacancies are lower in p-type materials, which leads to the reduction of In-Ga intermixing. This is well agreed with the smaller blueshift in the peak position and the lower intensity in the p-doped samples. The effect of group III intermixing on QDs increases the lateral size, rather than the vertical height of QDs, and the intersubband energy is considered to be mainly governed by the lateral dimension of QDs. This explains the observed energy separation difference between GS and ES in QDAP and QDAU samples, and these results discussed above demonstrate that the bandgap energy and energy separation between quantized states in InAs/GaAs QDs with modulation p-doping can be effectively tailored by RTA treatment. Additionally, effects of modulation p-doping on carrier dynamics are also found by comparing temperature-dependent spectral evolutions in Fig. 7.6a, b. As shown in Fig. 7.6a, the GS and ES emissions are present in all the spectral lines of the QDAU sample from 10 to 300 K, but their relative intensity changes remarkably, i.e., the intensity of the GS emission largely increases relative to the ES intensity. The same intensity evolution is also observed for the QDAP sample as shown in Fig. 7.6b except that GS gets much stronger than the ES band at RT, which suggests that we have a consideration on the carrier relaxation process in understanding the spectral evolution. It is clearly observed that the relative increase of the GS intensity along with the increased temperature causes a decrease of intensity of the ES emission indicating the evolution of the carrier distribution between GS and ES bands with temperature. The thermal distribution of carriers is governed by the carrier relaxation mechanism which induces more carriers relaxing from the ES band to the GS band with the increase of temperature. The observed relative intensity evolution revealed by Fig. 7.6 can be well explained by the well-known phonon-assisted carrier relaxation. Nevertheless, in comparison with the QDAU sample, the QDAP sample has a much stronger GS emission than the ES emission. This result is attributed to the carrier scattering-induced relaxation effect due to the modulation p-doping that further relaxes more carriers from the ES to the GS band at increased temperatures in addition to the phonon-assisted relaxation. Therefore, we should note that, when we observe a stronger emission of the GS band at RT with the InAs/GaAs QDs, the carrier relaxation from the ES band as well as the state density of the GS band should be considered.

7.2.3 Performances of Ultrashort Cavity F-P and DFB Lasers toward High-Speed Application

High-speed lasers are fascinating for their applications in high-speed data computation and communication, and shortening the laser cavity length (*L*) is an effective method to realize the high-speed performance due to the reduced photon lifetime. However, the shortening of the laser cavity imposes strict requirements on active materials, demanding higher gain characteristics. As presented in Fig. 7.7, the very short cavity laser based on InAs/GaAs QDs shows a superior lasing performance. Power-current (*P-I*) characteristics of QDU lasers with very short *L* of 500, 400, and 350 μm are shown in Fig. 7.7a, c, e, respectively. Two thresholds can be clearly observed in Fig. 7.7a. The first is at 23 mA (1.31 kA/cm^2) corresponding to the GS lasing at around 1318 nm, followed by the GS saturation profile starting from 35 to 50 mA, and the second is at 50 mA (2.86 kA/cm^2) corresponding to the ES lasing at around 1223 nm. The slop efficiency of GS lasing is calculated to be around 0.24 W/A, while that of ES lasing is as high as 0.41 W/A due to the double degeneracy and high gain of the ES band. The inset of Fig. 7.7a presents the detailed lasing profile of the QDU laser with the *L* of 500 μm, which indicates the transition from GS to ES lasing: the only GS lasing at low injection current, the simultaneous GS and ES lasing at increased current, and finally the only ES lasing at a further increased current. The observed results of the saturation of the GS gain and increased population of the ES are due to Pauli-blocking. As presented in Fig. 7.7c, e, for QDU lasers with the *L* of 400 and 350 μm, threshold currents for ES lasing are 53 (3.79 kA/cm^2) and 57 mA (4.65 kA/cm^2), respectively. In comparison with the 500-μm-long device, the increased threshold current can be attributed to the increased mirror loss owing to the shortened *L*. Furthermore, no GS lasing can be observed in the insets of Fig. 7.7c, e. The inhibition of the GS lasing can be explained by the shortened cavity length *L* and the reduced GS density in undoped QD materials. As we have seen in the spectral analysis mentioned above, the In-Ga intermixing effect is remarkable in the undoped QD samples, which greatly decreased the number of confined states in the GS band. As a result, the reduced state density in the GS band with the cavity length shortened to 400 and 350 μm cannot fulfill the gain requirement for lasing. On the contrary, due to the reduced intermixing effect, the high state density of the GS band in the modulation p-doped QD sample can be well maintained. Therefore, the GS lasing is expected to be observed in QDP laser with a shorter cavity.

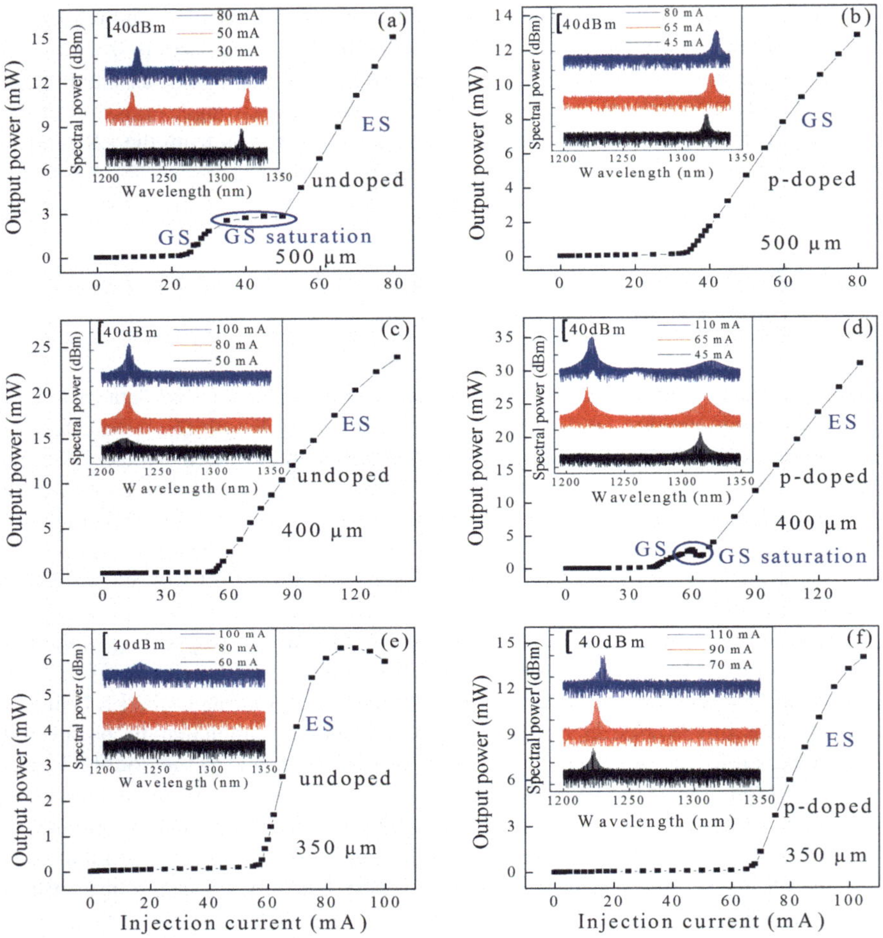

Fig. 7.7 *P-I* characteristics of the QD lasers with different *L*, measured under CW operation at RT; (**a**), (**c**), and (**e**) are for QDU lasers with 500, 400, and 350 μm; (**b**), (**d**), and (**f**) are for QDP lasers with 500, 400, and 350 μm, respectively. Inset: the corresponding lasing spectra of the device under various injection currents. (Reproduced with permission from Ref. [50] (copyright 2018 ACS))

We present *P-I* characteristics of QDP lasers with different *L* in Fig. 7.7b, d, f. As shown in Fig. 7.6b, for QDP laser with a cavity length *L* of 500 μm, the GS lasing is observed with a threshold current of 34 mA (1.94 kA/cm^2) and a slope efficiency of 0.30 W/A. In comparison with QDU laser with the same *L*, the higher threshold current density and higher slope efficiency of the GS lasing can be attributed to an increase in non-radiative recombination [60] and improved gain characteristics of the QD active layer caused by modulation p-doping. Three lasing spectra measured at different injection currents exhibit only the GS lasing at around 1330 nm and emission intensity larger than 52 dB by subtracting maximum power of the background noise as shown in the inset of Fig. 7.7b, which suggests the modulation p-doping weakens the effect of gain saturation and results in a significant inhibition of the ES lasing. The QDP laser with the very short cavity of 400 μm still maintains an excellent GS lasing at around 1315 nm at the injection current of 45 mA (3.21 kA/cm^2) as shown in Fig. 7.7d. With increased injection current up to 64 mA (4.57 kA/cm^2), simultaneous two-state lasing behavior is observed. We note that the GS emission still maintains a large intensity with further increased injection current, even at >110 mA.

It is well-known that the suppressed GS lasing of QD lasers can be attributed to the strong competition for the holes between the GS and ES electrons. The modulation p-doping can significantly increase the hole concentration in a QD structure, which effectively reduces this competition and consequently facilitates the GS lasing. In addition, it is also found that the increased hole capture rate is favorable for GS lasing in the two-state lasing regime. The holes introduced by modulation p-doping play an important role in the dominant carrier relaxation from the GaAs barriers to the GS of dots because of the enhanced Auger scattering. GS lasing characteristics of a 1.3 μm p-doped InAs/GaAs QD laser with a very short *L* of 500 μm were also studied by Arsenijević et al. [26]. In addition, the comprehensive experimental comparison between the modulation bandwidths of GS and ES emissions was reported. By depositing dichroic mirrors with different reflectivity on each cavity facet of two 1.3 μm p-doped InAs/GaAs QD lasers, one laser exhibits only the GS lasing, while the one emits only at ES. Large-signal measurements are performed in back-to-back configuration with these two lasers. The lasers are directly modulated in the on-off keying scheme with a non-return-to-zero pseudorandom binary sequence, which have an amplitude of 0.32 $V_{p\text{-}p}$ and a word length of 2^7–1 bits. The eye diagrams are measured by employing an 80 GSa/s real-time oscilloscope and a 50 GHz photodetector. Figure 7.8 presents bit error ratio versus received power plots for GS and ES lasers. As shown in the inset of Fig. 7.8a, the eye diagram of the GS laser exhibits a *Q* factor of 4.1 at a data rate of 15 Gb/s and the maximum

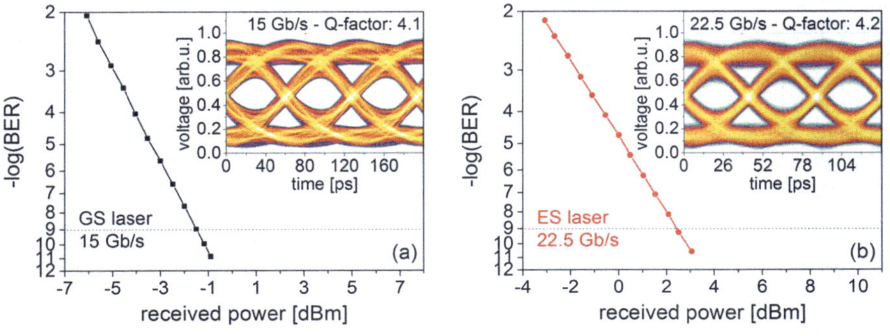

Fig. 7.8 Bit error ratio versus received power plots and eye diagrams at maximum received power as insets for the GS (**a**) and ES laser (**b**) at bit rates of 15 Gb/s and 22.5 Gb/s, respectively. (Reproduced with permission from Ref. [26] (copyright 2014 AIP))

received power. For the ES laser, the bit rate is 22.5 Gb/s as presented in Fig. 7.8b, corresponding to the *Q* factor of 4.2; while the ES laser was operated at 15 Gb/s, the *Q* factor of 7.8 was obtained. Error-free operation (BER < 10^{-9}) is realized and no error floor is detected. The 3.8 dB power difference at the BER of 10^{-9} mainly causes the difference in bit rate as well as the lower conversion efficiency and a large noise at the ES wavelength. The ES therefore offers an increase of digital bandwidth about 7.5 Gb/s in comparison with the GS laser.

Owing to their narrow emission spectrum, DFB lasers are key devices for modern optical data communication, which is indispensable for long-distance optical transmission and WDM systems. DFB lasers with the *L* of 500 μm were fabricated by employing the superior material properties and device performances which have been observed in the modulation p-doped QD system. *P-I* characteristics of the p-doped QD DFB laser obtained under CW operation at RT are shown in Fig. 7.9. The threshold current is determined to be 30 mA corresponding to the threshold current density of 1.71 kA/cm^2. With the injection current increased up to 150 mA, the output power of the DFB laser is measured to be 23 mW without the appearance of obvious heating saturation phenomena. An output spectrum of the DFB laser working at RT under CW operation is shown in the inset of Fig. 7.9. This DFB laser exhibits a single longitudinal mode operation and a SMSR as high as 51 dB at a pump current of 55 mA. In recent reports, Stubenrauch et al. realized a 10 Gb/s data transmission across 30 km of single mode fiber by using a 800-μm-long p-doped InGaAs QD DFB laser. In comparison, our DFB laser has a much shorter cavity of 500 μm and as-cleaved facets. The high-speed performance of the DFB laser can be further improved by shortening *L* due to the decreasing of the cavity's photon lifetime, and superior gain materials are, therefore, highly demanded. As we have demonstrated here, the modulation p-doped InAs/GaAs QDs possess a great potential for further improving high-speed performance of DFB lasers.

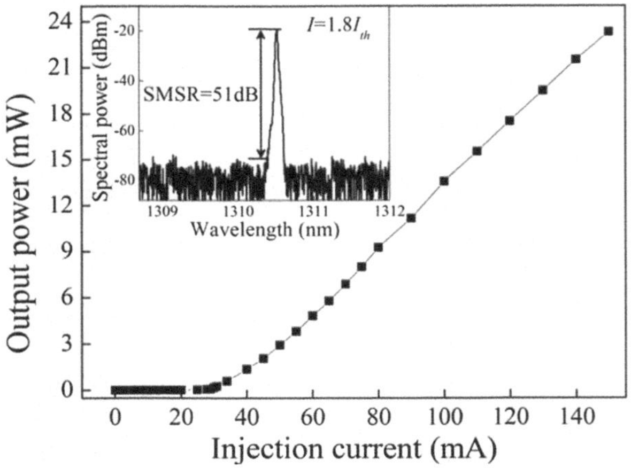

Fig. 7.9 *P-I* characteristic of p-doped QD DFB laser under CW operation at RT. Inset: lasing spectrum of p-doped QD DFB laser measured at 1.8I_{th}. (Reproduced with permission from Ref. [50] (copyright 2018 ACS))

7.3 1550 nm Ultrafast Lasers Based on InAs/GaAs QD-SESAMs

Ultrafast pulse lasers, which have high repetition rate and ultrashort pulse duration, have widespread applications in optical communication, ultrafast optics, and nonlinear optics due to their high peak power, low thermal effect, and high pulse energy [61]. The saturable absorber (SA) is the critical optical component of these ultrashort pulse lasers, and the saturable absorption materials used for SAs should have wide optical bandwidth, fast response time, and low loss properties. This section describes the development of InAs/GaAs QDs working at 1550 nm range and the fabrication of a high-performance 1550 nm QD-SESAM [30]. By using this 1550 nm QD-SESAM, a mode-locked Er:Yb:glass laser with the high repetition rate of 10 GHz and a *Q*-switched EDF laser with the picosecond pulse duration are successfully constructed.

7.3.1 Epitaxial Growth 1550 nm InAs/GaAs QD and the Fabrication of 1550 QD-SESAMs

Five InGaAs/GaAs QD test structures were prepared on semi-insulating GaAs (100) substrates by applying molecular-beam epitaxy (MBE) technique. Each QD test structure contains three dot-in-well (DWELL) layers. The DWELL layer consists of a 2.9 nm monolayer InAs QDs and different InGaAs QW pairs which include a 1 nm buffer layer (BL) and a 6 nm InGaAs CL. Samples A, B, and C have a symmetric DWELL structure where both BL and CL were grown with the same In content of 31%, 35%, and 39%, respectively. The band structures of these symmetric DWELL structures are presented schematically in band diagram a in Fig. 7.10a. Samples D and E have an asymmetric DWELL structures as shown in band diagram b in Fig. 7.10a. For sample D, the InAs QDs were directly grown on a GaAs BL and covered by a 6 nm $In_{0.31}Ga_{0.69}As$ CL, and sample E was grown by employing a 1 nm $In_{0.18}Ga_{0.82}As$ BL and a 6 nm $In_{0.31}Ga_{0.69}As$ CL. The growth temperatures of GaAs and InGaAs layers were set to be 565 and 530 °C, respectively.

In order to understand optical properties of five QD samples, PL measurements were performed at room temperature with a 532 nm solid-state laser as excitation source. Figure 7.10b shows RT-PL spectra obtained from the structures. The emission peak position exhibits a red shift from 1552 nm for sample A to 1592 nm for sample B with the increase of In content in the InGaAs CL as well as a reduction in PL peak intensity. Sample C has a very weak PL signal, which can be attributed to many non-radiative recombination centers induced by even higher In content in InGaAs CL. The emission at around 1540 nm is also observed in sample D, but the PL intensity is much weaker than that of sample A. For sample E, the emission is located at 1565 nm and the optical quality is improved, e.g., the PL intensity is enhanced by 2.5 times and PL emission linewidth is narrowed by 10%. As we have

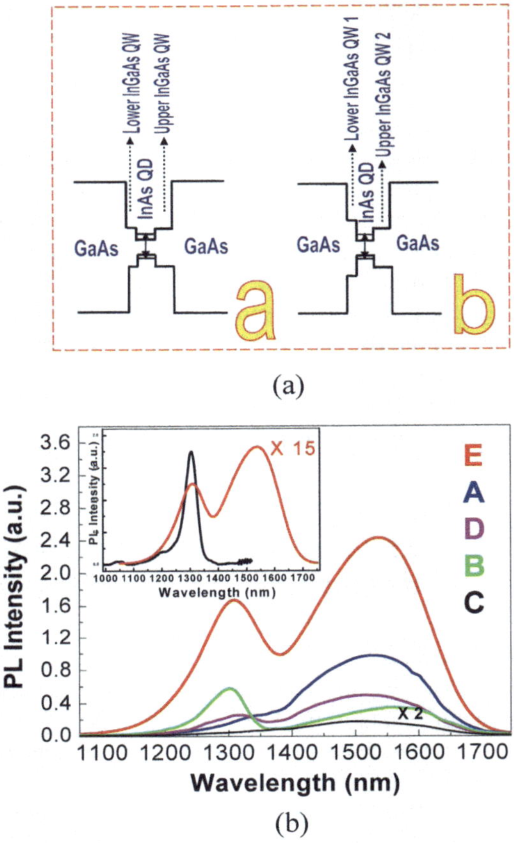

Fig. 7.10 (**a**) Schematic band diagram for InAs QD inserting in a symmetric and an asymmetric InGaAs QW pairs; (**b**) room temperature PL spectra of different 1550 nm InAs QD structures under identical excitation power of 120 mW. Inset: a comparison of room temperature PL characteristics between a 1310 nm QD sample and a 1550 nm QD sample. (Reproduced with permission from Ref. [30] (copyright 2012 Nature))

seen from the inset of Fig. 7.10b, the RT-PL peak/integrated intensity was found to be only ~15/4 times lower in sample E compared to a 1310 nm QD test sample which has a DWELL structure with 1 nm $In_{0.18}Ga_{0.82}As$ BL and a 6 nm $In_{0.18}Ga_{0.82}As$ CL. The main reasons for the improved optical quality are a high areal QD density of about $4 \times 10^{10}/cm^2$, more homogeneous dot size distribution, and small number of coalesced QDs, which was confirmed by using atomic force microscope measurements [30]. This highlights the high crystal perfection and high material quality for this growth technique, significantly better than other previous reports [62], in which the PL peak intensity of the QDs became much weaker when the emission wavelength was shifted from 1.3 to 1.5 μm.

Based on InAs/GaAs QD of sample E, a 1550 nm InAs/GaAs QD-SESAM has been fabricated. The QD-SESAM contains one DWELL layer prepared on a bottom distributed Bragg reflector (DBR) mirror, which includes 31 pairs of un-doped GaAs (115 nm) and $Al_{0.98}Ga_{0.02}As$ (134 nm) layers. The cross-sectional transmission electron microscopy (TEM) image of the QD-SESAM is shown in Fig. 7.11. This QD-SESAM is measured with a precision saturation fluence measurement system

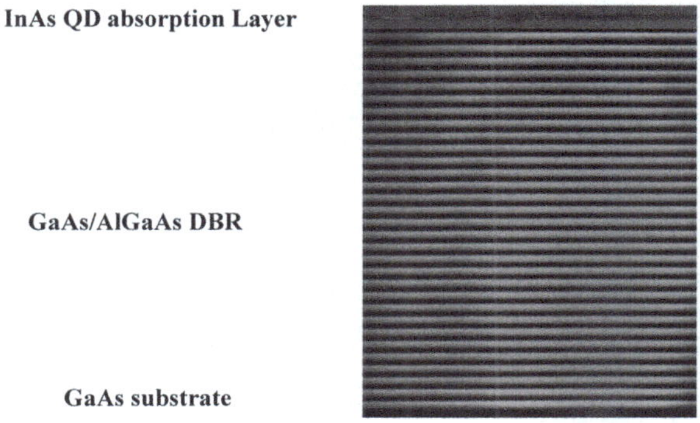

Fig. 7.11 Cross-sectional TEM image of the 1.55 μm QD-SESAM. (Reproduced with permission from Ref. [30] (copyright 2012 Nature))

and exhibits a saturation fluence of 9 μJ/cm³ and a modulation depth of 0.4%. In comparison, the standard QW-SESAM was characterized with the same precision system and shows a saturation fluence of 15 μJ/cm³, a modulation depth of 0.5%, and nonsaturable losses of 0.1%. When operated with the QD-SESAM, the laser exhibits a lower lasing threshold than that of a standard QW-SESAM, indicating that the QD-SESAM has nonsaturable losses of 0.1% or lower.

7.3.2 High Repetition Rate Er:Yb:Glass Laser Based on 1550 nm QD-SESAM

The generation of ultrashort pulses with high repetition rate is very important for high-bit-rate optical communication. By using the 1550 nm QD-SESAM mentioned above, a high pulse repetition rate (≥10 GHz) diode-pumped solid-state laser has been realized. The Er-doped glass oscillator (ERGO) laser is shown in Fig. 7.12. The laser cavity consists of 1-mm-thick Er:Yb-doped glass plate as the gain medium, a 1550 nm InAs/GaAs QD-SESAM, a fold high reflection mirror, and an output coupler with 0.5% transmission. Both the fold mirror and the output coupler are strongly curved with radius of curvature <4 mm. The pump laser source is a single-mode fiber-pigtailed diode with the center wavelength of 976 nm and the maximum output power of 600 mW. The laser is fundamentally mode-locked, suggesting that only one pulse is circulating within the laser cavity. The whole length of the laser cavity is slightly less than 15 mm. Passive fundamental mode-locking with the QD-SESAM allows ultralow pulse timing jitter and pulse-to-pulse optical phase coherence, which is essential for high data rate communication systems with coherent modulation formats.

In our experiment, the 1550 nm InAs/GaAs QD-SESAM is inserted into the laser cavity shown in Fig. 7.12, and the obtained output characteristic of this mode-locked ERGO laser is present in Fig. 7.13. The *Q*-switching threshold is at 70 mW of pump power. The average output power of the mode-locked laser is 8 mW when the output power of pump laser is set to be 130 mW. The mode-locked ERGO laser

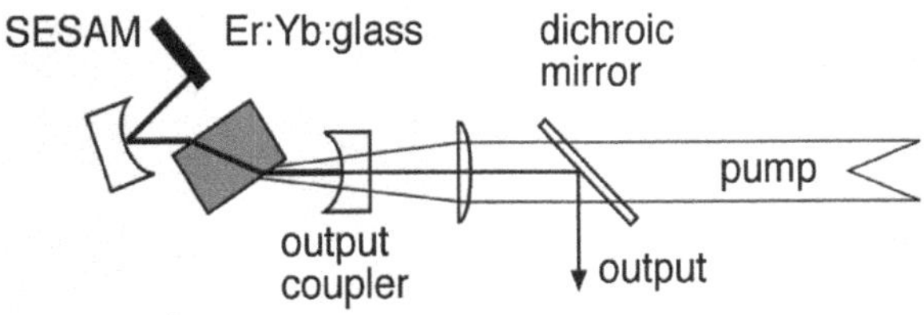

Fig. 7.12 10 GHz ERGO laser layout. The gain medium is Er:Yb-doped glass and QD-SESAM is placed at the end of the cavity. (Reproduced with permission from Ref. [30] (copyright 2012 Nature))

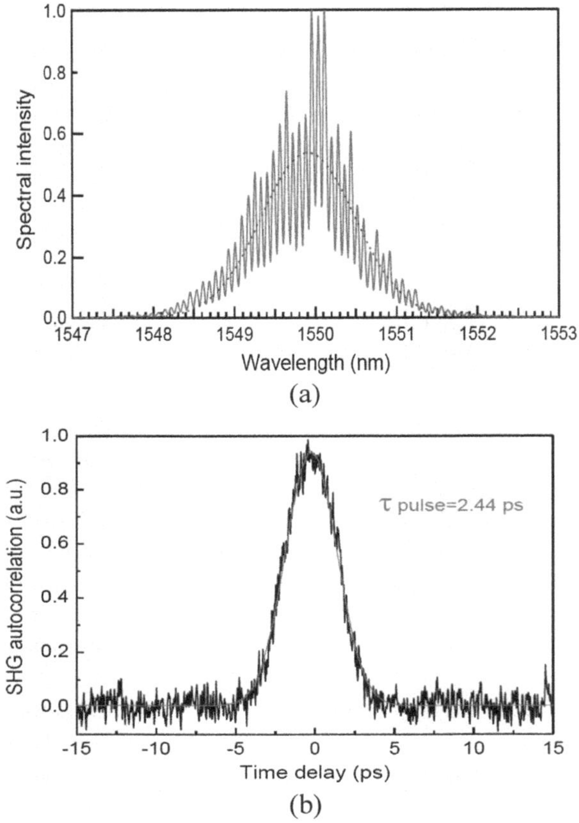

Fig. 7.13 10 GHz ERGO laser output: (**a**) optical power spectrum and (**b**) second harmonic generation (SHG) autocorrelation trace. The spectrum is fitted to a Gaussian curve with FWHM = 1.4 nm. The SHG autocorrelation trace is fitted to $sech^2$ pulse autocorrelation. (Reproduced with permission from Ref. [30] (copyright 2012 Nature))

exhibits a very high repetition rate of 10 GHz due to the much shorter laser cavity. The output spectrum has a full width at half maximum (FWHM) of 1.4 nm, and the pulse duration is 2.4 ps obtained by assuming $sech^2$ pulse shape. The pulse duration and the output power are in the same range (1.9 ps, 12 mW, and 2.4 ps, 8 mW, for QW and QD, respectively) for the QW and QD-SESAM mode-locked 10 GHz

ERGO lasers. Critically, using the QD-SESAM, the measured *Q*-switching threshold is halved. This opens the route for the development of higher-frequency mode-locked lasers.

7.3.3 Ultrashort Pulse EDF Laser Q-Switched with 1550 nm QD-SESAM

Passive *Q*-switching and mode-locking are still the most prominent techniques to produce all kinds of ultrashort laser pulses, which can be realized by inserting the SA into a fiber laser cavity. As we have mentioned previously, in comparison with the standard QW-SESAM, the 1550 nm InAs/GaAs QD-SESAM does not only exhibit the lower saturation fluence and nonsaturable losses, but it also has a broader operation range. Therefore, a *Q*-switched fiber laser with the picosecond pulse duration was successfully constructed by using the InAs/GaAs QD-SESAM. Figure 7.14a shows the schematic setup of the *Q*-switched fiber laser. The fiber laser is composed of several fiber components, including a 0.75-m-long Er-doped fiber (EDF) with the dispersion parameter of −36 ps/nm/km at 1550 nm, 15.5 m standard single mode fiber, 980/1550 wavelength division multiplexing (WDM) coupler, a polarization insensitive isolator (PI-ISO), one polarization controller (PC), a 90:10 optical coupler (OC), and a InAs/GaAs QD-SESAM. A 976 nm laser diode (LD) serves as the pump source.

In this experiment, as shown in Fig. 7.14b, the *Q*-switching is observed when the injection current is over the threshold of ~199 mA with the PC at an appropriate state. With further increased injection current, the output power increases linearly and the maximum power output is observed to be 10.5 mW. As shown in Fig. 7.14c, the center wavelength is at 1558 nm and ~3 dB spectral bandwidth is about 0.8 nm. Figure 7.14d demonstrates that the round-trip time of the cavity is ~27 μs. Figure 7.14e presents the single-pulse train at the range of 20 μs. The pulse width of this *Q*-switched fiber laser is about 3.1 μs by fitting the curve line with asech^2 pulse function. These results indicate that InAs/GaAs QD-SESAM have great potential in many applications, such as optical fiber communications, optical logical gate, laser materials processing, etc.

7.4 Conclusions and Outlook

In this chapter, we have discussed effects of the modulation p-doping on the optical properties of 1310 nm InAs/GaAs QDs. The comparative analysis of the PL characteristics of four 1310 nm InAs/GaAs QD samples reveals that the modulation p-doped QD sample is much more temperature-stable and has notable insensitivity to intermixing. For approaching ultrahigh-speed QD laser, CW GS lasing has been

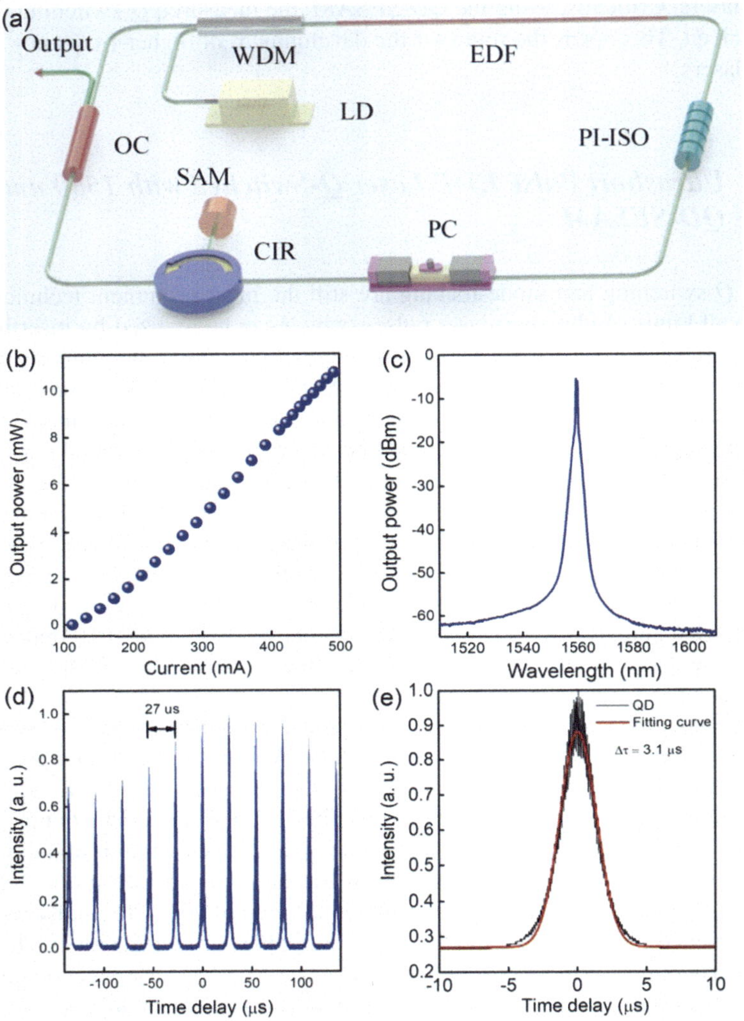

Fig. 7.14 (**a**) Experimental setup of the fiber laser with InAs/GaAs QD-SESAM. Experimental result: (**b**) changes of output power versus injection current; (**c**) output optical spectra; output pulse train at ranges of (**d**) 300 μs and (**e**) 20 μs

achieved in a 400-μm-long narrow ridge F-P laser, and a 500-μm-long DFB laser has been fabricated by using the p-doped QD materials. Moreover, we have also described the development of an asymmetric InAs/GaAs QD structure with high optical quality in the 1550 nm range. The fabricated 1550 nm SESAM based on this InAs/GaAs QD structure exhibits significant improvement in nonlinear characteristic, which allows us to develop a high repetition rate Er:Yb:glass laser and a picosecond pulse duration EDF laser.

In recent years, many works have been reported on the realization of high-speed devices based on InAs/GaAs QDs. Modulation p-doping InAs/GaAs QD has been regarded as an effective method to realize high-speed application for long-distance optical transmission and WDM systems. An optimized cavity length is very important for high-speed laser but difficult to trace. Therefore, how to solve this issue will be a main development direction for high-speed laser based on InAs/GaAs QD materials. For ultrafast applications, the SA that is made of various materials has emerged, such as carbon nanotubes, graphene, and black phosphorus, but the poor stability and repeatability severely limit their development. SESAMs will still dominate in commerce due to their high repeatability and excellent stability. Future development of QD-SESAM will aim to reduce the saturation fluence, decrease nonsaturable losses, and increase modulation depth for the ultrafast laser with high repetition rate and narrow pulse duration.

References

1. Bimberg, D., Kirstaedter, N., Ledentsov, N. N., Alferov, Z. I., Kop'ev, P. S., & Ustinov, V. M. (1997). InGaAs-GaAs quantum-dot lasers. *IEEE Journal of Selected Topics in Quantum Electronics, 3*, 196–205.
2. Shchekin, O. B., & Deppe, D. G. (2002). 1.3 μm InAs quantum dot laser with T_0=161 K from 0 to 80°C. *Applied Physics Letters, 80*, 3277–3279.
3. Kim, J., & Chuang, S. L. (2006). Theoretical and experimental study of optical gain, refractive index change, and linewidth enhancement factor of p-doped quantum-dot lasers. *IEEE Journal of Quantum Electronics, 42*, 942–952.
4. Capua, A., Rozenfeld, L., Mikhelashvili, V., Eisenstein, G., Kuntz, M., & Laeemmlin, M. (2007). Direct correlation between a highly damped modulation response and ultra low relative intensity noise in an InAs/GaAs quantum dot laser. *Optics Express, 15*, 5388–5393.
5. Kami, O., Capua, A., Eisenstein, G., Franke, D., Kreissl, J., Kuenzel, H., Arsenijević, D., Schmeckebier, H., Stubenrauch, M., Kleinert, M., Bimberg, D., Gilfert, C., & Reithmaier, J. P. (2013). Nonlinear pulse propagation in a quantum dot laser. *Optics Express, 21*, 5715–5736.
6. Hantschmann, C., Vasil'ev, P. P., Chen, S., Liao, M., Seeds, A. J., Liu, H., Penty, R. V., & White, I. H. (2018). Gain switching of monolithic 1.3 μm InAs/GaAs quantum dot lasers on silicon. *Journal of Lightwave Technology, 36*, 3837–3842.
7. Korenev, V. V., Savelyev, A. V., Zhukov, A. E., Omelchenko, A. V., & Maximov, M. V. (2014). The analytical approach to the multi-state lasing phenomenon in undoped and p-doped InAs/InGaAs semiconductor quantum dot lasers. *Proceedings of SPIE, 9134*, 913406.
8. Liu, Q. L., Hou, C. C., Chen, H. M., Ning, J. Q., Li, Q. Z., Huang, Y. Q., Zhao, Z. Y., Wang, Z. G., Jin, P., & Zhang, Z. Y. (2018). Effects of modulation P-doping on thermal stability of InAs/GaAs quantum dot superluminescent diodes. *Journal of Nanoscience and Nanotechnology, 6*, 7536–7541.
9. Chia, C. K., Chua, S. J., Wang, Y. J., Yong, A. M., & Chow, S. Y. (2007). Impurity free vacancy disordering of InAs/GaAs quantum dot and InAs/InGaAs dot-in-a-well structures. *Thin Solid Films, 515*, 3927–3931.
10. Saha, J., Panda, D., Tongbram, B., Das, D., Chavan, V., & Chakrabarti, S. (2019). Higher performance optoelectronic devices with $In_{0.21}Al_{0.21}Ga_{0.58}As/In_{0.15}Ga_{0.85}As$ capping of III-V quantum dots. *Journal of Luminescence, 210*, 75–82.

11. Liu, H. Y., Sellers, I. R., Badcock, I. R., Mowbray, D. J., & Skolnick, M. S. (2004). Improved performance of 1.3 μm multilayer InAs quantum-dot lasers using a high-growth-temperature GaAs spacer layer. *Applied Physics Letters, 85*, 704–706.
12. Mi, Z., Bhattacharya, P., & Fathpour, S. (2005). High-speed 1.3 μm tunnel injection quantum-dot lasers. *Applied Physics Letters, 96*, 153109.
13. Yu, H. C., Wang, J. S., Su, Y. K., Chang, S. J., Lai, F. I., Chang, Y. H., Kuo, H. C., Sung, C. P., Yang, P. D., Lin, K. F., Wang, J. M., Chi, J. Y., Hsiao, R. S., & Mikhrin, S. (2006). 1.3-μm InAs-InGaAs quantum-dot vertical-cavity surface-emitting laser with fully doped DBRs grown by MBE. *IEEE Photonics Technology Letters, 18*, 418–420.
14. Liu, C. Y., Yoon, S. F., Cao, Q., Tong, C. Z., & Li, H. F. (2007). Low transparency current density and high temperature operation from ten-layer p-doped 1.3 μm InAs/InGaAs/GaAs quantum dot lasers. *Applied Physics Letters, 90*, 041103.
15. Tong, C., Xu, D., & Yoon, S. F. (2009). Carrier relaxation and modulation response of 1.3-μm InAs-GaAs quantum dot lasers. *Journal of Lightwave Technology, 27*, 5442–5450.
16. Liu, S. W., Liu, R. Y., & Lin, H. C. (2017). Tailoring energy band alignment of vertically aligned InGaAs quantum dots capped with GaAs(Sb)/AlGaAsSb composite structure after thermal annealing treatment. *ACS Photonics, 4*, 242–250.
17. Abdollahinia, A., Banyoudeh, S., Rippien, A., Schnabel, F., Eyal, O., Cestier, I., Kalifa, I., Mentovichm, E., Eisenstein, G., & Reithmaier, J. P. (2018). Temperature stability of static and dynamic properties of 1.55 μm quantum dot lasers. *Optics Express, 26*, 6056–6066.
18. Dai, Y., Fan, J., Chen, Y., Lin, R., Lee, S., & Lin, H. (1997). Temperature dependence of photoluminescence spectra in InAs/GaAs quantum dot superlattices with large thicknesses. *Journal of Applied Physics, 82*, 4489–4492.
19. Cao, Q., Yoon, S., Liu, C., & Tong, C. (2008). Effects of rapid thermal annealing on optical properties of p-doped and undoped InAs/InGaAs dots-in-a-well structures. *Journal of Applied Physics, 104*, 033522.
20. Liu, H. Y., Dai, Q. F., Wu, L. J., Lan, S., Trofimov, V., & Varentsova, S. (2012). Effects of p-type doping on the optical properties of InAs/GaAs quantum dots. *Solid State Communications, 152*, 435–439.
21. Chaabani, W., Melliti, A., Maaref, M., Testelin, C., & Lemaitre, A. (2016). Rapid thermal annealing and modulation-doping effects on InAs/GaAs quantum dots photoluminescence dependence on excitation power. *Physics B, 493*, 53–57.
22. Nakahara, K., Tsuchiya, T., Kitatani, T., Shinoda, K., Taniguchi, T., Kikawa, T., Aoki, M., & Mukaikubo, M. (2007). 40-Gb/s direct modulation with high extinction ratio operation of 1.3-μm InGaAlAs multiquantum well ridge waveguide distributed feedback lasers. *IEEE Photonics Technology Letters, 19*, 1436–1438.
23. Zhang, C., Srinivasan, S., Tang, Y., Heck, M. J., Davenport, M. L., & Bowers, J. E. (2014). Low threshold and high speed short cavity distributed feedback hybrid silicon lasers. *Optics Express, 22*, 10202–10209.
24. Jhang, Y. H., Mochida, R., Tanabe, K., Takemasa, K., Sugawara, M., Iwamoto, S., & Arakawa, Y. (2016). Direct modulation of 1.3 μm quantum dot lasers on silicon at 60 °C. *Optics Express, 24*, 18428–18435.
25. Banyoudeh, S., Abdollahinia, A., Eual, O., Schkovskyi, V., Eisenstein, G., & Reithmaier, J. P. (2016). Temperature-insensitive high-speed directly modulated 1.55 μm quantum dot lasers. *IEEE Photonics Technology Letters, 28*, 2451–2454.
26. Arsenijević, D., Schliwa, A., Schmeckebier, H., Stubenrauch, M., Spiegelberg, M., Bimberg, D., Mikhelashvili, V., & Eisenstein, G. (2014). Comparison of dynamic properties of ground- and excited-state emission in p-doped InAs/GaAs quantum-dot lasers. *Applied Physics Letters, 104*, 181101.
27. Deppe, D. G., Shavritranuruk, K., Ozgur, G., Chen, H., & Freisem, S. (2009). Quantum dot laser diode with low threshold and low internal loss. *Electronics Letters, 45*, 54–55.
28. Wu, J., Shao, D., Dorogan, V. G., Li, A. Z., Li, S., DeCuir, E. A., Manasreh, M. O., Wang, Z. M., Mazur, Y. I., & Salamo, G. J. (2010). Intersublevel infrared photodetector with

strain-free GaAs quantum dot pairs grown by high-temperature droplet epitaxy. *Nano Letters, 10*, 1512–1516.
29. Zhang, Z., Hogg, R. A., Lv, X. Q., & Wang, Z. G. (2010). Self-assembled quantum-dot superluminescent light-emitting diodes. *Advances in Optics and Photonics, 2*, 201–228.
30. Zhang, Z., Oehler, A. E. H., Resan, B., Kurmulis, S., Zhou, K. J., Wang, Q., Mangold, M., Suedmeyer, T., Keller, U., Weigarten, K. J., & Hogg, R. A. (2012). 1.55 μm InAs/GaAs quantum dots and high repetition rate quantum dot SESAM mode-locked laser. *Scientific Reports, 2*, 477.
31. Resan, B., Kurmulis, S., Zhang, Z., Oehler, A. E. H., Markovic, V., Mangold, M., Sudmeyer, T., Keller, U., Hogg, R. A., & Weigarten, K. J. (2016). 10 GHz pulse repetition rate Er:Yb:glass laser modelocked with quantum dot semiconductor saturable absorber mirror. *Applied Optics, 55*, 3776–3780.
32. Oehler, A. E. H., Suedmeter, T., Weingarten, K. J., & Keller, U. (2008). 100 GHz passively mode-locked Er:Yb:glass laser at 1.5 μm with 1.6-ps pulses. *Optics Express, 16*, 21930–21935.
33. Maas, D. J. H. C., Bellancourt, A. R., Hoffmann, M., Rudin, B., Barbarin, Y., Golling, M., Sudmeyer, T., & Keller, U. (2008). Growth parameter optimization for fast quantum dot SESAMs. *Optics Express, 16*, 18646–18656.
34. White, S. E., & Cataluna, M. A. (2015). Unlocking spectral versatility from broadly-tunable quantum-dot lasers. *Photonics, 2*, 719–744.
35. Gorodetsky, A., Yadav, A., Avrutin, E., Fedorova, K. A., & Rafailov, E. U. (2018). Photoelectric properties of InAs/GaAs quantum dot photoconductive antenna wafers. *IEEE Journal of Selected Topics in Quantum Electronics, 24*, 1900105.
36. Salhi, A., Alshaibani, S., Alaskar, Y., Albadri, A., Alyamani, A., & Missous, M. (2018). Tuning the optical properties of InAs QDs by means of digitally-alloyed GaAsSb strain reducing layers. *Applied Physics Letters, 113*, 103101.
37. Paranthoen, C., Bertru, N., Dehaese, O., Corre, A. L., Loualiche, S., & Lambert, B. (2001). Height dispersion control of InAs/InP quantum dots emitting at 1.55 μm. *Applied Physics Letters, 78*, 1751–1753.
38. Miyazawa, T., Takemoto, K., Sakuma, Y., Hirose, S., Usuki, T., Yokoyama, N., Takatsu, M., & Arakawa, Y. (2005). Single-photon generation in the 1.55-μm optical-fiber band from an InAs/InP quantum dot. *Japanese Journal of Applied Physics, 44*, 20–23.
39. Zilkie, A. J., Meier, J., & Smith, P. W. E. (2006). InAs/InGaAsP quantum dot amplifier operating at 1.55 μm. *Optics Express, 14*, 11453–11459.
40. Lelarge, F., Dagens, B., Renaudier, J., Brenot, R., Accard, A., Dijk, F. V., & Make, D. (2007). Recent advances on InAs/InP quantum dash based semiconductor lasers and optical amplifiers operating at 1.55 μm. *IEEE Journal of Selected Topics in Quantum Electronics, 13*, 111–124.
41. Rosales, R., Merghem, K., Martinez, A., Akrout, A., Tourrenc, J. P., Accard, A., Lelarge, F., & Ramdane, A. (2011). InAs/InP quantum-dot passively mode-locked lasers for 1.55-μm applications. *IEEE Journal of Selected Topics in Quantum Electronics, 17*, 1292–1301.
42. Mikhrin, V. S., Vasil'ev, A. P., & Semenova, E. S. (2006). InAs/InGaNAs/GaNAs QW and QD heterostructures emitting at 1.4-1.8 μm. *Semiconductors, 40*, 342–345.
43. Seravalli, L., Frigeri, P., Trevisi, G., & Franchi, S. (2008). 1.59 μm room temperature emission from metamorphic InAs/InGaAs quantum dots grown on GaAs substrates. *Applied Physics Letters, 92*, 213104.
44. Ripalda, J. M., Granados, D., & Gonzalez, Y. (2005). Room temperature emission at 1.6 mu m from InGaAs quantum dots capped with GaAsSb. *Applied Physics Letters, 87*, 202108.
45. Richter, M., Damilano, B., Massies, J., Duboz, J. Y., Wieck, A. D., et al. (2006). *Progress in Semiconductor Materials V-Novel Materials and Electronic and Optoelectronic Applications, 891*, 185–190.
46. Ledentsov, N. N., Kovsh, A. R., Zhukov, A. E., Maleev, N. A., Mikhrin, S. S., Vasil'ev, A. P., Semenova, E. S., Maximov, M. V., Shernyakov, Y. M., Kryzhanovskaya, N. V., Ustinov, V. M., & Bimberg, D. (2003). High performance quantum dot lasers on GaAs substrates operating in 1.5 μm range. *Electronics Letters, 39*, 1126–1128.

47. Bakopoilos, P. (2007). Multi-wavelength laser source for dense wavelength division multiplexing networks. *Proceedings of Optical Fiber Communications Conference paper OWJ2.*
48. Bartels, A., Heinecke, D., & Diddams, S. A. (2009). 10-GHz self-referenced optical frequency comb. *Science, 326*, 681–682.
49. Hillerkuss, D., Schmogrow, R., Schellinger, T., et al. (2011). 26 Tbits/s line-rate super-channel transmission utilizing alloptical fast Fourier transform processing. *Nature Photonics, 5*, 364–371.
50. Li, Q., Wang, X., Chen, H. M., Huangm, Y., Hou, C., Wang, J., Zhang, R., Ning, J., Min, J., Zheng, C. C., & Zhang, Z. (2018). Development of modulation p-doped 1310 nm InAs/GaAs quantum dot laser materials and ultrashort cavity Fabry–Perot and distributed-feedback laser diodes. *ACS Photonics, 5*, 1084–1093.
51. Mazur, Y. I., Liang, B., Wang, Z. M., Tarasov, G., Guzun, D., & Salamo, G. (2007). Development of continuum states in photoluminescence of self-assembled InGaAs/GaAs quantum dots. *Journal of Applied Physics, 101*, 014301.
52. Kumagai, N., Watanabe, K., Nakata, Y., & Arakawa, Y. (2007). Optical properties of p-type modulation-doped InAs quantum dot structures grown by molecular beam epitaxy. *Journal of Crystal Growth, 301*, 805–808.
53. Berg, T. W., & Mørk, J. (2003). Quantum dot amplifiers with high output power and low noise. *Applied Physics Letters, 82*, 3083–3085.
54. Harbord, E., Spencer, P., Clarke, E., & Murray, R. (2009). Radiative lifetimes in undoped and p-doped InAs/GaAs quantum dots. *Physical Review B: Condensed Matter and Materials Physics, 80*, 195312.
55. Paul, S., Roy, J., & Basu, P. (1991). Empirical expressions for the alloy composition and temperature dependence of the band gap and intrinsic carrier density in $Ga_xIn_{1-x}As$. *Journal of Applied Physics, 69*, 827–829.
56. Su, L., Liang, B., Wang, Y., Guo, Q., Li, X., Wang, S., Fu, G., Mazur, Y. I., Ware, M. E., & Salamo, G. J. (2016). The continuum state in photoluminescence of type-II $In_{0.46}Al_{0.54}As/Al_{0.54}Ga_{0.46}As$ quantum dots. *Applied Physics Letters, 109*, 183103.
57. Babinski, A., & Jasinski, J. (2002). Post-growth thermal treatment of self-assembled InAs/GaAs quantum dots. *Thin Solid Films, 412*, 84–88.
58. Gunawan, O., Djie, H., & Ooi, B. (2005). Electronics states of interdiffused quantum dots. *Physical Review B, 71*, 205319.
59. Zhao, H., Yoon, S., Ngo, C., Wang, R., Cao, Q., & Liu, C. (2011). Effects of annealing and p-doping on the two-state competition in 1.3 μm InAs/GaAs quantum-dot lasers. *IEEE Transactions on Nanotechnology, 10*, 1211–1213.
60. Smowton, P. M., Sandall, I. C., Mowbray, D. J., Liu, H. Y., & Hopkinson, M. (2007). Temperature-dependent gain and threshold in p-doped quantum dot lasers. *IEEE Journal of Selected Topics in Quantum Electronics, 13*, 1261–1266.
61. Gao, F., Luo, S., Ji, H., Liu, S. T., Xu, F., Lv, Z., Lu, D., Ji, C., & Yang, T. (2016). Ultrashort pulse and high power mode-locked laser with chirped InAs/InP quantum dot active layers. *IEEE Photonics Technology Letters, 28*, 1481–1484.
62. Sopanen, M., Xin, H. P., & Tu, C. W. (2000). Self-assembled GaInNAs quantum dots for 1.3 and 1.55 μm emission on GaAs. *Applied Physics Letters, 76*, 994–996.

Chapter 8
Bioresource-Derived Graphene Quantum Dots: A Tale of Sustainable Materials and Their Applications

Sankarapillai Mahesh and Kizhisseri Devi Renuka

8.1 Introduction

Carbon is an elite element when coming to the material research field [1]. Carbon has delivered numerous useful nanostructures including carbon nanotube [2], graphene, etc. [3]. The history of carbon nanomaterials began with the synthesis of graphene oxide from graphite, and it was named as 'graphon' in the older time [4]. Then a large number modified synthetic procedures were developed for the production of graphene oxide in bulk quantity [5]. Along with these developments, the invention of fullerene, an important allotrope of carbon, occurred [6], and it was followed by the discovery of one-dimensional carbon nanotubes [7]. At the same time, the existence of single sheet of graphite was a hot topic of debate, and its existence was only theoretical until in 2011 the monolayer graphene through scotch-tape technique was synthesized [8]. The other important carbon nanomaterials invented include nano-horns [9], nano-onions [10], etc. Thus, the carbon nanomaterial family included members in various dimensions ranging from zero to three.

The graphene materials proved as potential candidates for material synthesis with different applications. But the disadvantages associated with them like zero-band gap encouraged the scientists to go for structural modification. This eventually led to the development of quantum dots. The quantum dots based on carbon is divided into two different classes: first one is carbon quantum dot [11] and the other is graphene quantum dot [12]. The main difference between them is that basically, carbon quantum dots possess a quasi-spherical structure with size less than 10 nm,

S. Mahesh (✉)
Polymers and Special Chemicals Division, Vikram Sarabhai Space Centre, Indian Space Research Organisation, Trivandrum, India
e-mail: mahesh@vssc.gov.in

K. D. Renuka
Electrocatalysis Division, CSIR-CECRI Madras Unit, Chennai, India

P. Yu, Z. M. Wang (eds.), *Quantum Dot Optoelectronic Devices*,
Lecture Notes in Nanoscale Science and Technology 27,
https://doi.org/10.1007/978-3-030-35813-6_8

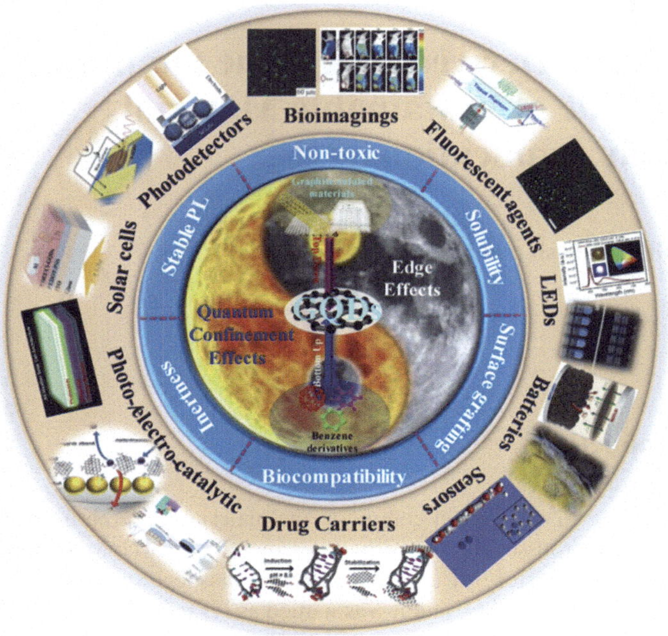

Fig. 8.1 The synthetic routes, properties and applications of graphene quantum dots (GQDs). (Adapted from Ref. [25], open-access article)

and on the other hand, the graphene quantum dots contain graphene lattice of size 100 nm with layer thickness less than 10 nm.

Right from their discovery, graphene quantum dots (GQDs) continue their journey in the research world with scope of exploring further in each step [13]. The GQDs are zero-dimensional in nature and have strong fluorescence properties owing to the effects of quantum confinement [14] and edge effects [15]. The GQDs are capable of absorbing in a wide spectral range including ultraviolet, infrared and visible [16, 17]. The GQDs possess fine band gap tunable over the spectral range [18]. This makes them excellent choice in the optoelectronic applications. Thus, the GQDs possess different physical and chemical properties from the other carbon nanomaterial. The common methods of synthesis of GQDs are classified into two: bottom-up [19] and top-down [20] approaches. In the bottom-up approach, the GQDs are synthesized from different polycyclic aromatic hydrocarbons like glucose and hexabenzocoronene with structure similar to graphene through stepwise chemical reactions. In the top-down approach, the direct cutting of graphene-based materials like graphite powder and graphene oxide results in GQDs. But the synthesis of GQDs with specific energy gap is done through two methods. The first one is the size separation which employs chromatography [21] or acid treatment of coal and carbon fibres [22]. The second method is chemical functionalization with modifiers having either strong electron-donating or electron-withdrawing capacity [23, 24]. The graphene quantum dots, no matter through whatever way synthesized are now widely getting explored for various optoelectronic and biological applications (Fig. 8.1).

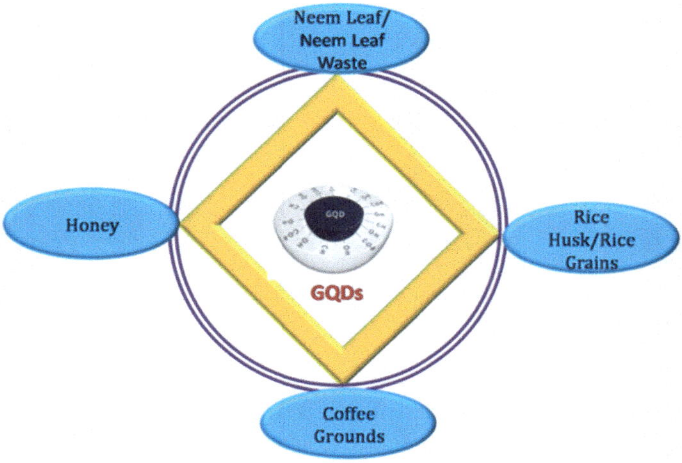

Fig. 8.2 The different bioresource materials for GQD synthesis. (Adapted with permission from Ref. [21], Copyright (2019) Royal Society of Chemistry)

The GQDs, however in pure form, possess limitations when coming to the application side. This results in the functionalization of GQDs. The functionalization is done either through doping with heteroatoms [26] or by forming composite materials with inorganic materials [27], organic molecules [28], polymers [29], etc. The functionalized GQDs showed tremendous applications including drug delivery, bioimaging, batteries, sensors, etc. [30]. Recently the development of green chemistry and sustainable chemistry has influenced the synthesis of graphene quantum dots from various bioresources [31]. This is mainly due to the fact that the usual synthetic procedures include tedious synthetic processes at drastic conditions of temperature and pressure. Also, such synthetic methods utilize toxic reagents and require extensive purification treatment. The utilization of bioresources for the GQD synthesis resulted in simple and cost-effective methods. The present study therefore deals with the synthesis of GQDs from various bioresource materials including neem leaf waste, neem leaf, rice husk, rice grain, coffee grounds and honey (Fig. 8.2).

The GQDs are synthesized from biomass waste by Suryawanshi and his group [32]. They used dead neem leaves for the same. Here amine-terminated GQDs are developed by the pyrolytic treatment of dead neem leaves followed by hydrothermal treatment. This amine-terminated GQDs (Am-GQDs) can act as fluorescent sensors for metal ions. Am-GQDs are characterized by various analytical techniques including X-ray photoelectron spectroscopy (XPS), Fourier transform infrared (FTIR) spectroscopy and transmission electron microscopy (TEM). In the UV-visible absorption spectrum Am-GQDs & GQDs exhibits an absorption band at 300 nm due to the n–$\pi*$ transition and a peak below 300 nm due to the π–$\pi*$ transition associated with aromatic domains. These Am-GQDs selectively detected the presence of Ag+ ion in solution media in combination with the L-cysteine amino acid. The mechanism of the detection is based on Pearson's HSAB theory. According to this concept, soft acids form strong bond formation with soft bases quickly and hard acids

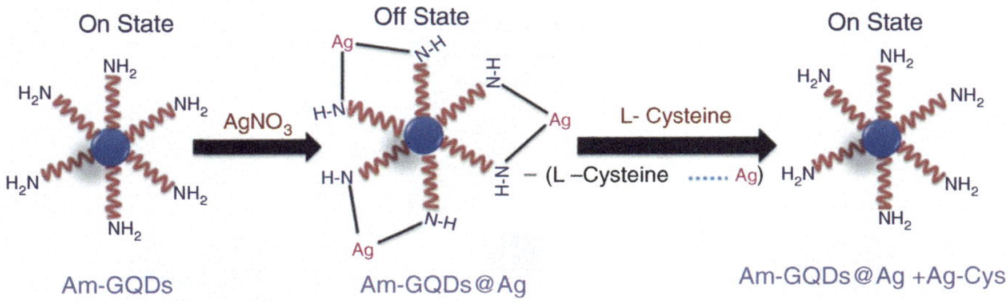

Fig. 8.3 The Am-GQDs selectively sensing Ag+ ions. (Adapted with permission from Ref. [33] Copyright (2019) Royal Society of Chemistry)

with hard bases. Since Ag is a soft acid and S (from L-cysteine) is a soft base, they form stronger bond and lead to the selective sensing of Ag+ ions (Fig. 8.3).

In another work bioresource utilized for synthesizing GQD is plant extracts of neem and fenugreek [33]. In 2014 Roy et al. synthesized GQDs through hydrothermal synthetic routes from both neem (N-GQD) and fenugreek (F-GQD) leaves. These GQDs exhibited good optical properties. The GQDs show strong absorption band around 300 nm. The photoluminescence studies (PL) revealed their green fluorescence emission. The quantum yields (QY) of N-GQDs and F-GQDs are 41.2% and 38.9%. The chlorophylls were extracted by soaking neem leaves in ethanol. The chlorophyll molecules are absorbed at 430 nm under violet light and 662 nm under red light, which proved that the type of chlorophylls present is chlorophyll a. These GQDs showed greater stability under acidic conditions. There was no change in PL intensity even under the continuous excitation for 6 h under xenon lamp. This proved the photostability of GQDs.

Altogether this makes them suitable candidates for optoelectronic applications. The N-GQDs are then developed into light converter for the production of white light. This is done by mixing solutions of N-GQDs with quinine sulphate (QS) and chlorophyll (CPY) derived from the same neem extract in polymethyl methacrylate (PMMA) in appropriate ratios (Fig. 8.4).

The same group later utilized N-GQDs for the in vivo imaging of apoptic cells. The GQDs are first modified with an antibody annexin V (Ab_{A5}). The modified GQDs are absorbed at 265 nm in the UV-visible spectra. The GQDs showed greater photostability. The biotoxicity assay is carried out in zebrafish model. The antibody-modified Ab_{A5}-GQDs are then used to image the apoptic cells in zebrafish. These GQDs exhibited excellent biocompatibility in the zebrafish embryo. The apoptosis is a mechanism which helps in tissue modelling and homeostatic regulation during the embryo development. The utilized annexin-5 (A5) fused enhanced green fluorescent protein (EGFP) (A5-EGFP) incorporated transgenic zebrafish and Ab_{A5}-GQDs for the in vivo imaging of apoptic cells in zebrafish. The GQDs due to their low toxicity were employed for the imaging of various cells like HeLa cells for human cervical cancer cell line, MCF-7 cells for human breast adenocarcinoma cell line and MCF-10A cells for normal human mammary epithelial cell line [34] (Fig. 8.5). These GQDs are thus excellent candidates for biomedical applications.

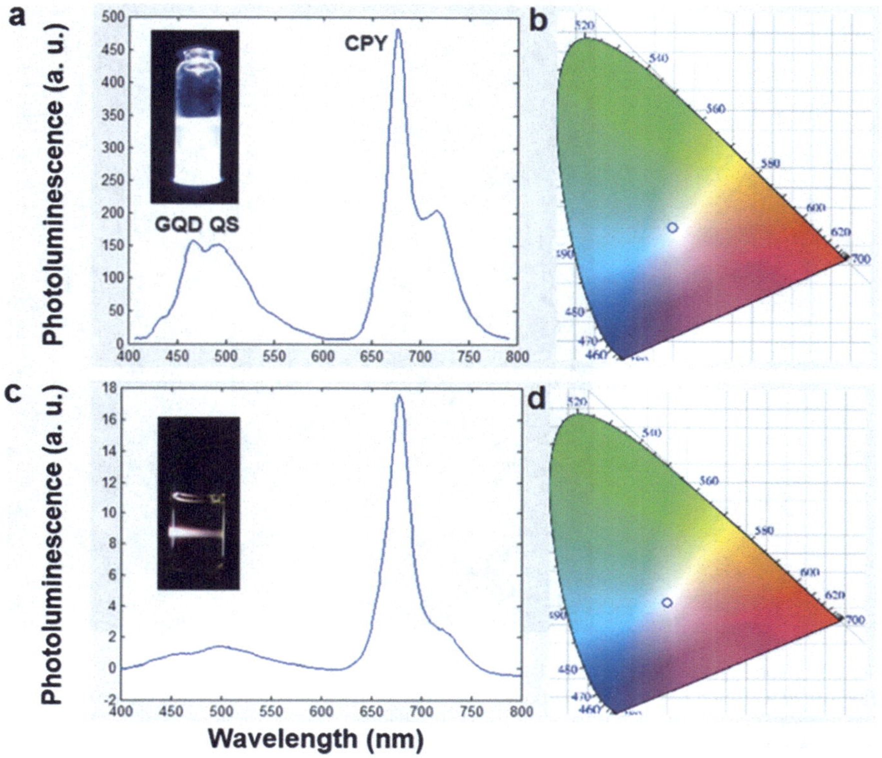

Fig. 8.4 (**a**) PL spectrum of N-GQD-QS-CPY coated over polyethylene terephthalate (PET) cap with inset showing the white-light emission under excitation with near UV-LED light. (**b**) CIE spectrum under near UV-LED excitation. (**c**) PL spectrum of N-GQD-QS-CPY solution under excitation with 325 nm. (**d**) the CIE spectrum under 325 nm excitation. (Adapted with permission from Ref. [29], Copyright (2019) Royal Society of Chemistry)

Another interesting bioresource material employed for the synthesis of GQDs is rice grains. The rice grains are first made into rice powder. The rice powder is then heated on a hot plate which led to the formation of GQDs with percentage yield of 56%. The GQDs are then treated with oleylamine to yield amine-functionalized GQDs. By controlling the calcination time from 3 to 10 min, it is observed that the size of GQDs is increasing from 2 to 6.5 nm. The as-synthesized GQDs in the UV-visible absorption spectra exhibit two bands. The π–$\pi*$ band appears between 200 and 270 nm, and the n–$\pi*$ band appears in region beyond 260 nm. The increase in size of GQDs from 2 to 6.5 nm results in a red shift in the absorption peak due to the quantum confinement effect. The PL measurements clearly showed the red shift in the emission spectra. The colour of the GQD solution changes from blue to cyan with the size change from 2 to 6.5 nm. The quantum yield (QY) also varies from 16% to 24%.

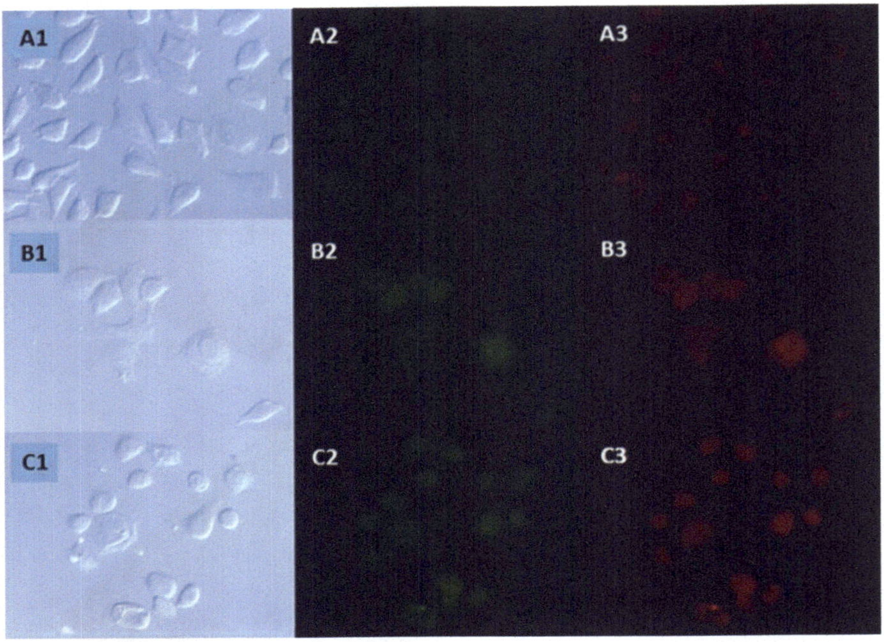

Fig. 8.5 Representative images of (**A1–A3**) HeLa, (**B1–B3**) MCF-7 and (**C1–C3**) MCF-10A cells after treated with GQDs (2 mg mL^{-1}) in cell 70 medium for 4 h at 37 °C. **A1**, **B1** and **C1** are the bright-field images of those cells. **A2**, **B2** and **C2** are the PL images using a green filter (530 nm, bandwidth 55 nm). **A3**, **B3** and **C3** are the PL images using a red filter (607 nm, bandwidth 36 nm). Excitation wavelength that is 488 nm (argon laser) was used to record the green and red PL images. (Adapted with permission from Ref. [30], Copyright (2019) Royal Society of Chemistry)

The GQDs are then amine-functionalized, and their quantum yield increases to 125% due to the electron-donating ability of amine groups. The GQDs are then employed for the bioimaging of HeLa cells. The GQDs exhibit excellent biocompatibility with HeLa cells. The HeLa cells after treating with GQDs are examined for bioimaging capacity. The studies are conducted under laser of three different wavelengths 375 nm, 488 nm and 561 nm. The luminescence from the normal cells is then put into comparison with that from GQD-treated HeLa cells. The bright-field images of GQD-treated HeLa cells are stronger than the same of normal cells. The result is consistent under three different laser wavelengths 375 nm, 488 nm and 561 nm (Fig. 8.6). The GQDs are thus excellent candidates for bioimaging [35].

The GQDs are then synthesized from rice husk biomass (RH). First carbon flakes with sp^2 hybridization is produced from the RH through bottom-up approach. Here rice husk is heated under nitrogen atmosphere to product rice husk ash (RHA) containing carbon and silica. The RHA is then treated with sodium hydroxide solution to produce a mixture of rice husk carbon (RHC) and sodium silicate. The RHC is separated and subjected to top-down approach involving a hydrothermal reaction. The process resulted in rice husk GQDs (RH-GQDs) with yield 15%. The obtained GQDs have their size ranging from 3 to 6 nm. Upon exposure to UV light of wavelength 365 nm, the RH-GQDs exhibited blue luminescence. The RH-GQDs have a

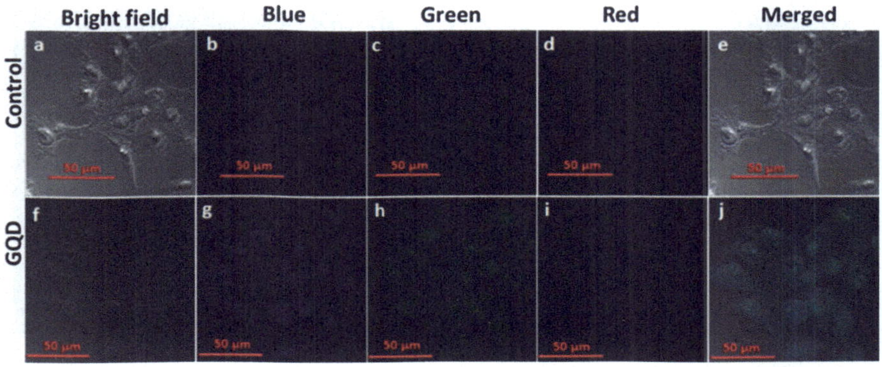

Fig. 8.6 Cellular imaging of GQDs using HeLa cells (**a–j**) of both untreated control and GQD-treated row shows confocal fluorescence image: (**a** and **f**) bright-field images of untreated control and GQD-treated cells, (**b** and **g**) blue (pseudocolour) at 375 nm excitation, (**c** and **h**) green (pseudocolour) at 488 nm (**d** and **i**), red (pseudocolour) at 561 nm and (**e** and **j**) merged images. All scale bar, 50 μm (Adapted with permission from Ref. [9], Copyright (2019) Royal Society of Chemistry)

quantum yield of 8.1%. Two PL bands are observed in the PL mapping which clearly indicate the presence of two luminescent centres in the RH-GQDs. The reason behind the appearance of two PL bands is the presence of defects and intrinsic states arising from size/edge effects.

The RH-GQDs are water-dispersible and thus suitable candidates for biomedical applications. The RH-GQDs showed excellent biocompatibility with HeLa cells. Therefore, the HeLa cells treated with RH-GQDs are subjected for in vitro bioimaging. Stable fluorescence is exhibited by the cells over repeated measurements (Fig. 8.7). Thus RH-GQDs have greater applicability as fluorescent bio-probes [25].

Then Wang and his group developed GQDs from coffee grounds through a hydrothermal procedure. The GQDs are further functionalized with polyethyleneimine (PEI) molecule. From the atomic force microscopic analysis, the average heights of GQDs and PEI-GQDs are 2.23 ± 0.54 nm and 3.18 ± 0.74 nm, respectively. This clearly indicates the existence of multiple layers in both types of GQDs. The optical properties of GQDs and PEI-GQDs are observed to be interesting. The GQDs exhibited absorption at 370 nm and, upon excitation on the same wavelength, emit blue fluorescence at 460 nm with a quantum yield of 8%. The PEI-GQD absorbs at 390 nm and emits cyan fluorescence at 480 nm with a quantum yield of 24%. The emission of both GQDs and PEI-GQDs is observed to be excitation-dependent. The GQDs exhibited a lifetime of $\tau = 1.81$ ns, and PEI-GQDs have the same around $\tau = 4.06$ ns.

Both GQDs and PEI-GQDs are therefore highly useful for biomedical imaging. They are employed for the bioimaging of HeLa cells. Due to the higher quantum yield and low cytotoxicity, PEI-GQDs served as better bioimaging agent. The GQDs and PEI-GQDs are also employed as fluorescent probes. Both of them selectively detected Fe^{3+} and Cu^{2+} ions. The presence of nitrogen-containing functional groups in the GQDs and PEI-GQDs is the reason behind the strong binding affinity and fast chelating kinetics of Fe^{3+} and Cu^{2+} ions with them (Fig. 8.8) [38].

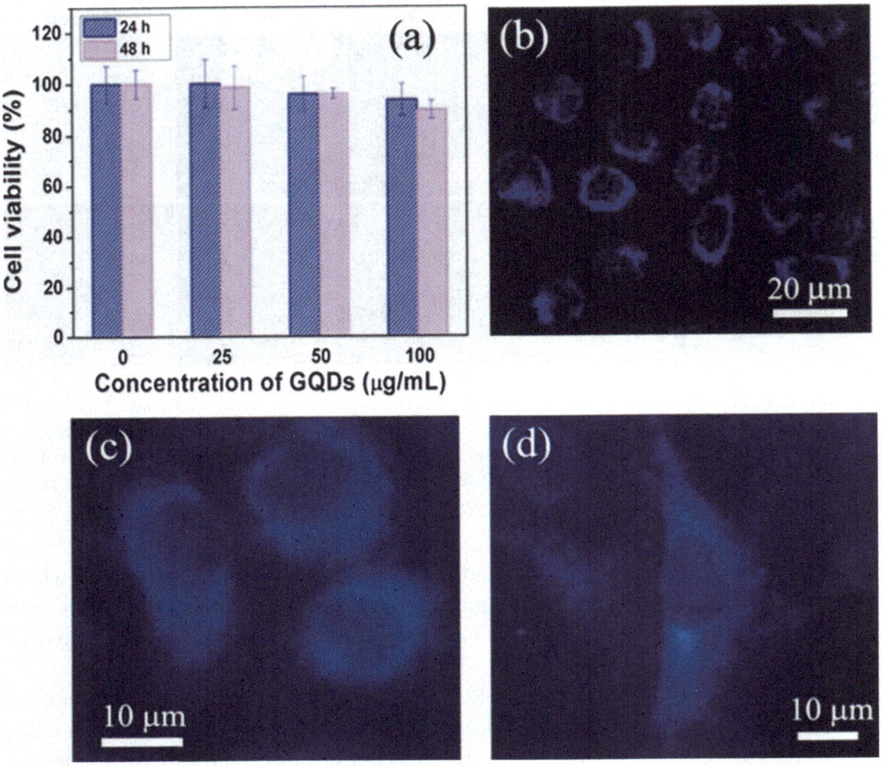

Fig. 8.7 (**a**) Viability of Hela cells combined with different concentrations of RH-GQDs synthesized at 200 °C after 24 and 48 h; cell imaging of Hela cells combined with 50 μg/mL RH-GQDs excited by (**b**) 335 nm in a confocal microscope and (**c**, **d**) 365 nm in a conventional fluorescent microscope. (Adapted with permission from Ref. [36] Copyright (2019), American Chemical Society)

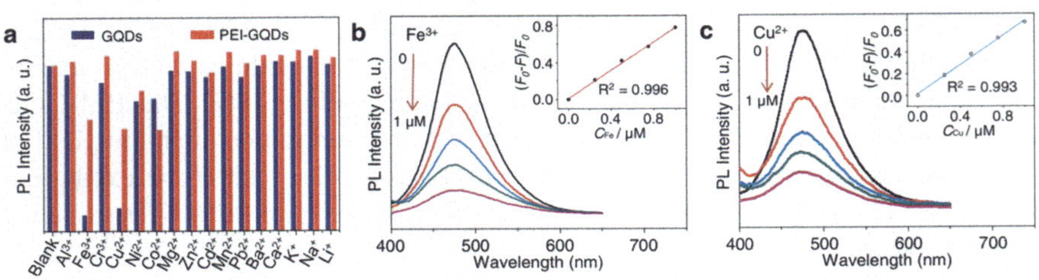

Fig. 8.8 (**a**) The difference in relative PL intensities of PEI-GQDs between the blank and solutions containing different metal ions. PL intensity changes of PEI-GQDs for different concentrations of Fe^{3+} (**b**) and Cu^{2+} (**c**) (Adapted with permission from Ref. [37]. Copyright (2019) Elsevier)

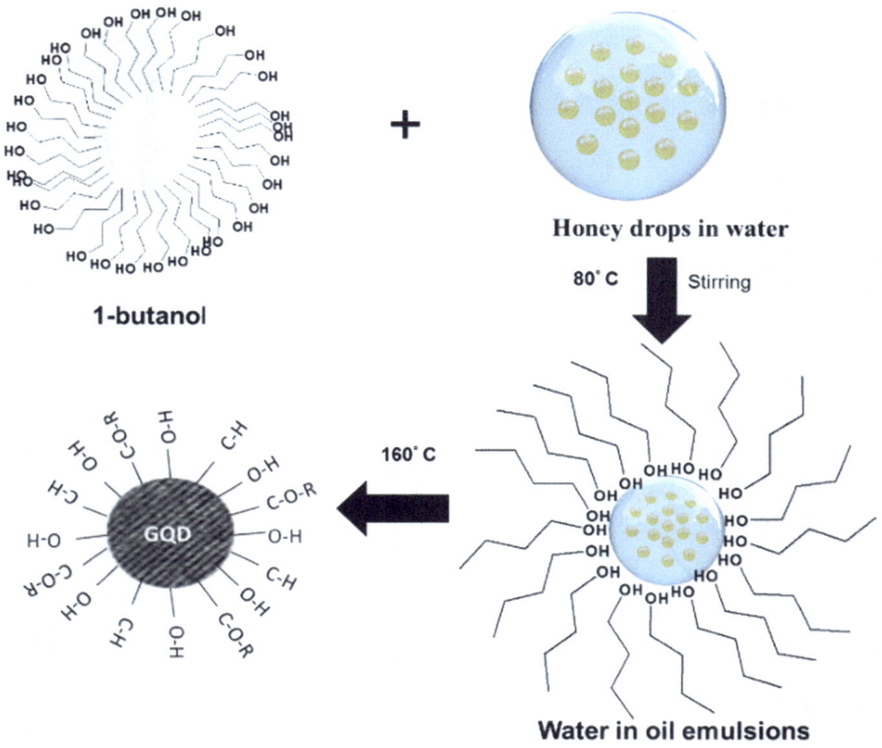

Fig. 8.9 Synthesis of GQDs from honey. (Adapted with permission from Ref. [22]. Copyright (2019) John Wiley and Sons)

The GQDs are then developed from honey, which is extracted by bees from the nectars of flowers, and this is an easily available bioresource [39]. The GQDs are prepared through the procedure called emulsion template-assisted carbonization. This synthetic method yielded nearly monodispersed GQDs. In this method, aqueous solutions of honey are emulsified in 1-butanol. The hydroxyl functional group stabilizes the emulsion against coalescence. Since each emulsion contains a limited amount of honey molecule, the formation of undesirable bulk structures gets prevented. This emulsion acts as microreactors. The nearly monodispersed GQDs were synthesized via carbonization of honey molecules in water-in-oil emulsions (Fig. 8.9). To form the emulsions, a 10 wt % honey solution (aq.) was mixed with 1-butanol, and the mixture was aged at 80 ° C under vigorous stirring for 1 h. To the mixture a catalytic amount of hexadecylamine was then added, and the temperature was raised to 160 °C under argon to initiate the carbonization. The GQDs thus prepared were purified by centrifuging. The GQDs are then dissolved in methanol [37].

The GQDs are then characterized by various spectroscopic and microscopic techniques. From the transmission electron microscopic analysis (TEM), the GQDs have size ranging from 2.4 nm. Again, the GQDs are further characterized by IR, XRD, Raman, etc. The UV-visible absorption studies showed the presence of two

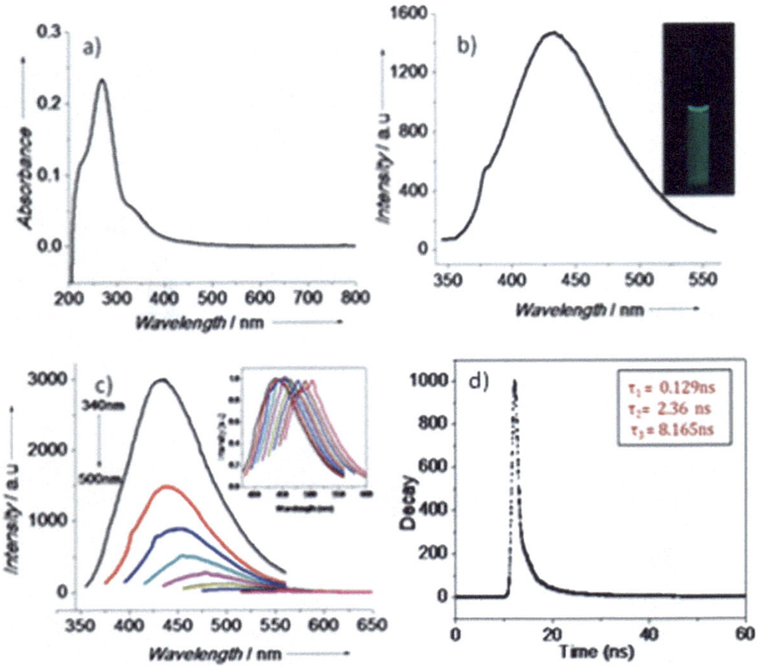

Fig. 8.10 Optical properties of the GQDs. (**a**) UV-visible spectra of GQD. Inset is the photograph of the corresponding GQD under UV light. (**b**) The PL (at 340 nm excitation) spectrum of GQD. (**c**) The excitation-dependent PL behaviour of GQDs; GQDs were excited at wavelengths ranging from 340 to 500 nm (10 nm interval) inset; the normalized PL emission spectra. (**d**) Decay profile of GQD. (Adapted with permission from Ref. [22]. Copyright (2019) John Wiley and Sons)

peaks in GQDs. The absorption peak at 267 nm is due to π–$\pi*$ transition of C=C bond, and the absorption at 340 nm corresponds to π–$\pi*$ transition of C=O bond (Fig. 8.10a). It is observed that under UV light, the GQDs have a strong green emission. The photoluminescence studies are conducted with a fluorometer. Upon excitation at 340 nm, the GQDs exhibited a broad peak around 435 nm along with a strong green fluorescence under UV light (Fig. 8.10b). The GQDs also have an excitation-dependent fluorescence behaviour. Upon changing excitation wavelength from 340 to 500 nm, the emission spectra shift from 435 to 520 nm (Fig. 8.10c). It is observed that the emission peaks are red-shifted and intensity is decreased. The formation of electronic conjugated structures, which act as emissive traps, is responsible for this phenomenon. The fluorescent emission is due to the presence of functional groups on the surface of the GQDs. The photoluminescence excitation spectrum (PLE) of the GQDs is recorded by monitoring the emission at 420 nm. The quantum yield of the GQDs is 3.6%. The lifetime measurements showed the presence of three components: one fast component at (τ_1 = 0.129 ns) and two slow components (τ_2 = 2.36 ns and τ_3 = 8.165 ns) (Fig. 8.10d).

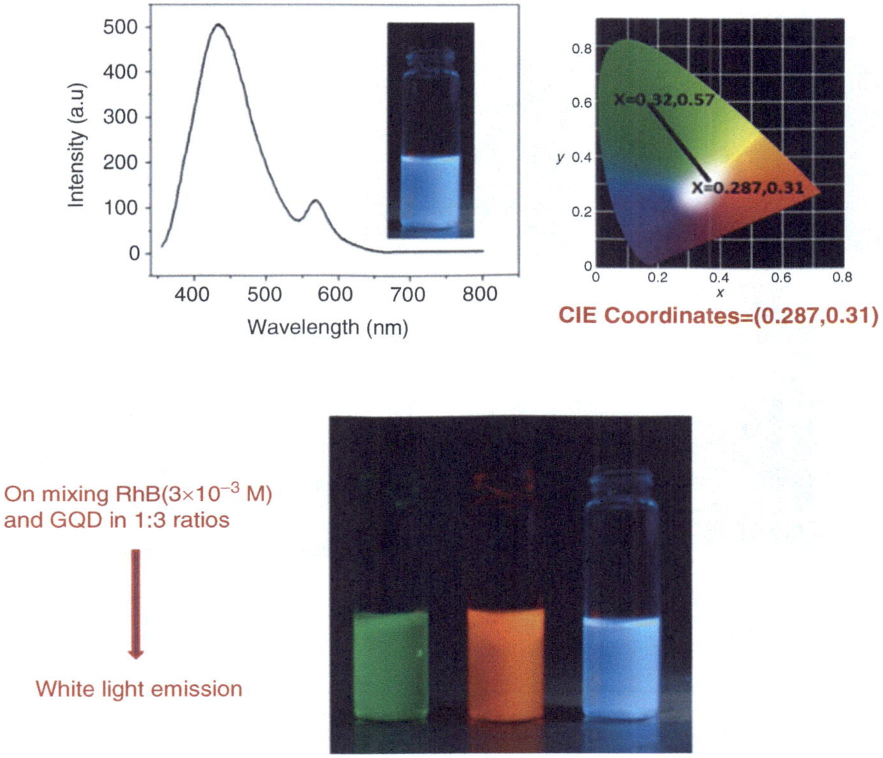

Fig. 8.11 Application of GQD as component of white-light emission. (Adapted with permission from Ref. [22]. Copyright (2019) John Wiley and Sons)

The GQDs are then employed as a component of white-light emission. A mixture is prepared with GQDs ($\lambda_{max} = 280$ nm), with pure green fluorescence ($\lambda_{max} = 435$ nm) as the donor chromophore and rhodamine B (RhB) with orange red fluorescence ($\lambda_{max} = 544$ nm) as the acceptor. On excitation with 340 nm light, the GQDs-RhB combination exhibited a white-light emission. The Commission International de l'Eclairage (CIE) 1931 chromaticity coordinates ($x = 0.287$, $y = 0.310$) for GQDs after mixing suggest that a highly pure white-light emission was achieved (Fig. 8.11). The GQDs have a stable and high intense fluorescence emission. Hence, they are utilized as transparent security ink for security spot marking on currency or any other paper. This can be of use in hidden writing or anti-counterfeiting measures (Fig. 8.12).

The honey-derived GQDs are then coupled with photochromic azobenzene system derived from cardanol. The binary system is then employed as molecular fluorescent probe for the selective screening of harmful pesticide carbofuran. This fluorescence switch is then developed into molecular logic gate and molecular keypad lock [40]. The chemical inputs for both logic gate and keypad lock are the same photochromic azobenzene system DOAZOC1 and harmful carbamate pesticide carbofuran. The output is the variation in fluorescence intensity of GQDs.

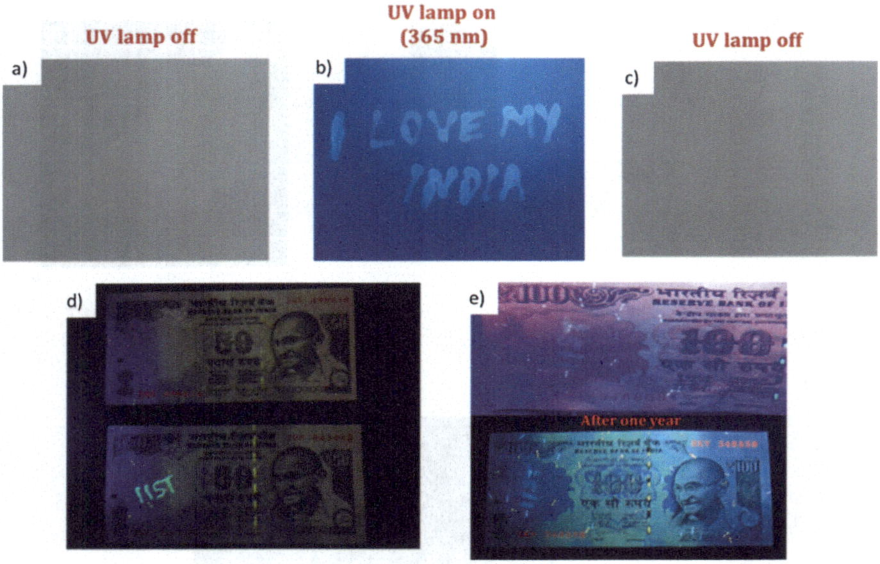

Fig. 8.12 (**a–c**) GQDs as a security ink. (**d**, **e**) GQD solution as counterfeit purpose. (Adapted with permission from Ref. [22]. Copyright (2019) John Wiley and Sons)

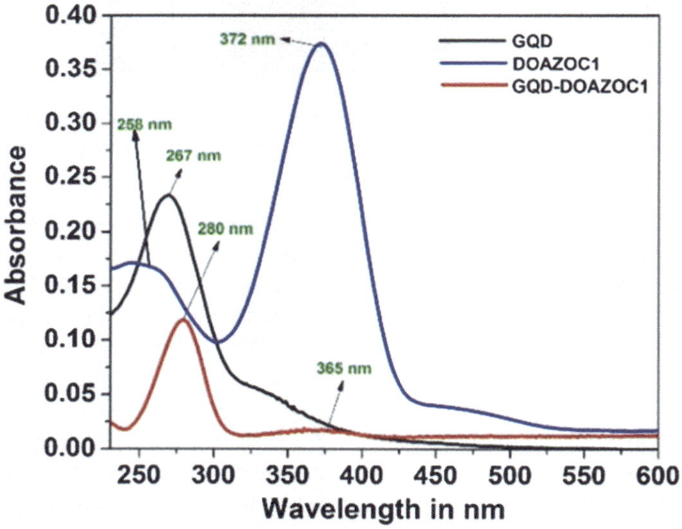

Fig. 8.13 UV-visible absorption spectra of GQDs, DOAZOC1 and complex GQD-DOAZOC1. (Adapted with permission from Ref. [28]. Copyright (2019) John Wiley and Sons)

The UV-visible spectrum of DOAZOC1 (Fig. 8.13) shows an absorption maximum corresponding to π–π∗ band at 372 nm and n–π∗ band at 258 nm. The π–π∗ band of DOAZOC1 shows a blue shift from 372 nm to 365 nm upon complex formation with GQDs. Similarly, the peak of GQDs at 266 nm shows a red shift to 280 nm, and n–π∗ of DOAZOC1 at 258 nm is shifted to 232 nm.

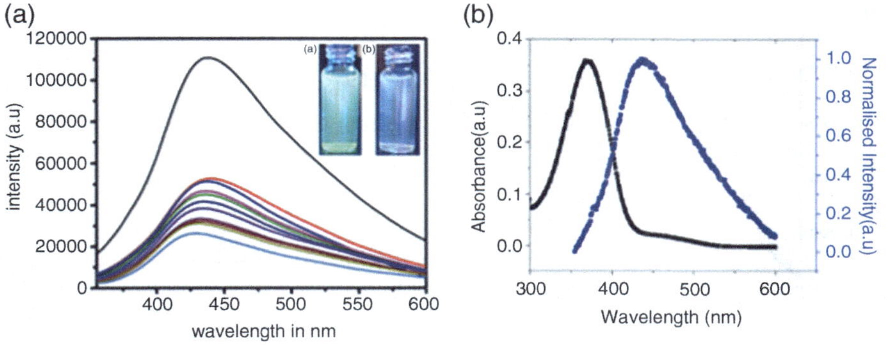

Fig. 8.14 (**a**) Fluorescence spectra of GQDs and complex GQD-DOAZOC1.Inset is the visual fluorescence of (1) GQDs (2) GQD-DOAZOC1 complex. (**b**) Comparison of decay profile of GQD and GQD-DOAZOC1 complex. (Adapted with permission from Ref. [28]. Copyright (2019) John Wiley and Sons)

When a solution of DOAZOC1 in methanol at a concentration of 10^{-4} M is added to GQD solution, a fluorescence quenching of GQDs occurs (Fig. 8.14a). It is observed that there is no significant overlap between the absorption spectra of DOAZOC1 and the emission peak of the GQDs (Fig. 8.14b). The possible reason for the fluorescence quenching of the GQDs by DOAZOC1 may be the photoinduced electron transfer (PET) and the hydrogen bonding between GQDs and DOAZOC1. Generally, the GQDs are good electron acceptors [41], whereas DOAZOC1 with two nitrogen atoms are good electron donor [36].

The average lifetimes of the GQDs and GQD-DOAZOC1 are found to be 6.2 and 4.3 ns, respectively. The reduction in the average lifetime of GQD-DOAZOC1 confirms the occurrence of photoinduced electron transfer (PET) in the system. Hydrogen bonding also plays a crucial role in the quenching of fluorescence of GQDs [42, 43]. The GQDs are derived from honey which is rich in sucrose. As such there are a large number of hydroxyl functional groups on the surface of GQDs. This makes a strong hydrogen bonding with DOAZOC1 and thereby decreases the fluorescence intensity (Fig. 8.16a).

The fluorescence property of GQD-DOAZOC1 system is used for the detection of poisonous carbofuran pesticide. Different concentrations of carbofuran solution are added to the GQD-DOAZOC1. The enhancement in fluorescent intensity with increasing concentration of carbofuran is observed (Fig. 8.15a). The time-resolved fluorescence analysis also confirms the enhancement in the fluorescence intensity. After the addition of carbofuran, the average lifetime of GQD-DOAZOC1 system is increased from 4.3 to 5.2 ns. This confirms the dissociation of DOAZOC1 from the complex (Fig. 8.15b). This fluorescent enhancement is due to the dissociation of GQDs from DOAZOC1 after the introduction of carbofuran. This is because of the fact that DOAZOC1 shows higher affinity to carbofuran than the GQD.

When carbofuran approaches the GQD-DOAZOC1 complex, there occurs a competition in hydrogen bonding between GQDs and DOAZOC1 with carbofuran. Since the hydrogen bonding between the amide group (carbofuran) and carboxylic

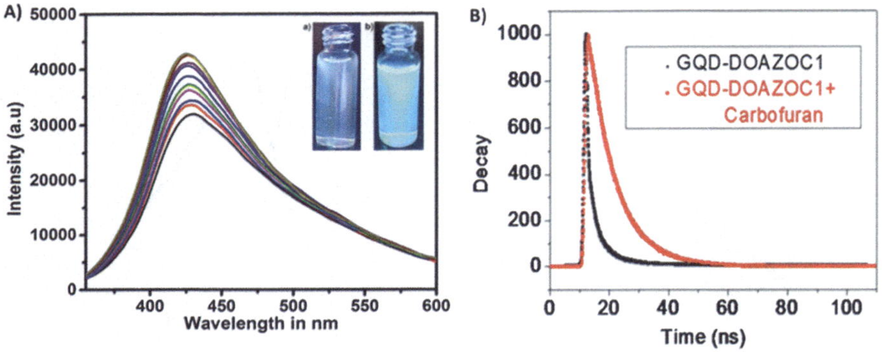

Fig. 8.15 (**a**) Fluorescence spectra of GQD-DOAZOC1-carbofuran complex. Inset is the visual fluorescence of (1) GQD-DOAZOC1 (2) GQD-DOAZOC1-carbofuran complex. (**b**) Comparison of decay profile of GOD-DOAZOC1 complex and that in presence of carbofuran. (Adapted with permission from Ref. [28]. Copyright (2019) John Wiley and Sons)

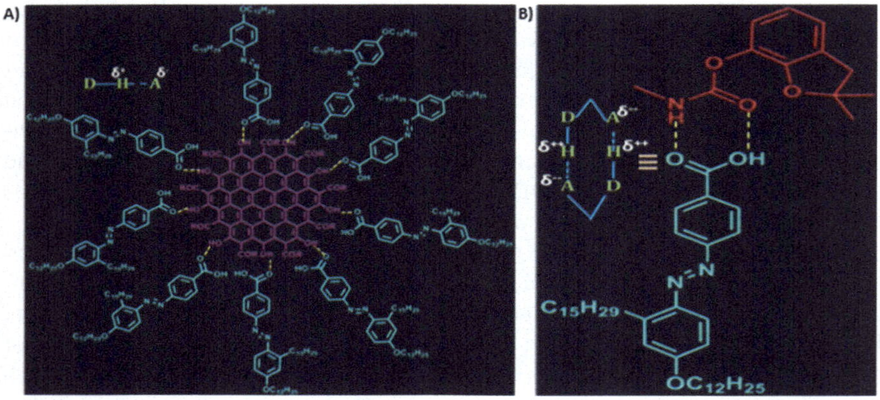

Fig. 8.16 (**a**) Pattern of hydrogen bonding between GQDs and DOAZOC1. (**b**) Pattern of hydrogen bonding between carbofuran and DOAZOC1. (Adapted with permission from Ref. [28]. Copyright (2019) John Wiley and Sons)

acid group (DOAZOC1) is strong compared with that of hydroxyl group (GQD) (Fig. 8.16b), addition of carbofuran enhances the fluorescent intensity by dissociating the GQD-DOAZOC1 complex. The typical enthalpy of the hydrogen bond between the hydroxyl donor and oxygen acceptor is 21 kJ/mol, whereas that between amide donor and oxygen acceptor is 8 kJ/mol. So, the complex between carbofuran and DOAZOC1 possesses a total enthalpy of 29 kJ/mol compared to 21 kJ/mol between GQD-DOAZOC1. This additional enthalpy factor favours the dissociation of GQD-DOAZOC1 upon addition of carbofuran [44, 45].

A molecular logic gate is developed by observing the change in the fluorescence intensity of the GQD-DOAZOC1 by the addition of carbofuran. The excitation wavelength for all the logic functions was set at 340 nm, and the emission wavelength is taken as 435 nm. DOAZOC1-carbofuran-induced fluorescence of the

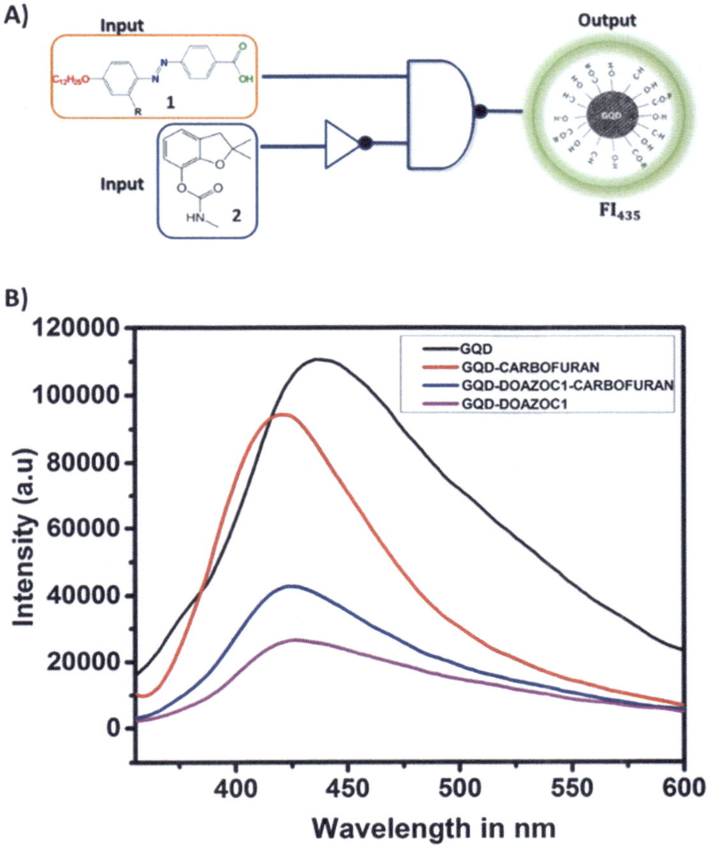

Fig. 8.17 (**a**) IMPLICATION logic gate designed with DOAZOC1 and carbofuran as inputs. The output is the measurement of fluorescence intensity of GQD. (**b**) Comparison of emission spectra of GQDs, GQDs-DOAZOC1, GQDs-carbofuran and GQDs-DOAZOC1-carbofuran complex. (Adapted with permission from Ref. [28]. Copyright (2019) John Wiley and Sons)

GQDs enabled the design of 'IMPLICATION' logic system based on Boolean operations (Fig. 8.17a). For input, the presence of DOAZOC1 or carbofuran is defined as '1' and the absence as '0'. The fluorescence intensity at a wavelength of 435 nm (FI_{435}) is defined as the output (0 or 1) for the logic gate. Only with the input of DOAZOC1 and without carbofuran (1/0), the fluorescent intensity is low, and the output is '0' which implies off state (curve pink in Fig. 8.17b). With all other possible input combinations [(0/0), (0/1), (1/1)], the output signal is '1' which corresponds the on state. With no input or with the input carbofuran alone, there is no change in the fluorescent intensity. The output signal remains at '1' (Fig. 8.17b).

When the system is subjected to the two inputs together, there is more interaction between DOAZOC1 and carbofuran. This results in the release of DOAZOC1 from the GQDs. The output signal is 1 (Fig. 8.17b). The truth table (Table 8.1) contains the details of all combinations. The symbol of the IMPLICATION logic gate is presented in Fig. 8.17a.

Table 8.1 Truth table

Input		Output
DOAZOC1 (1)	Carbofuran (2)	FI_{435}
0	0	1
1	0	0
0	1	1
1	1	1

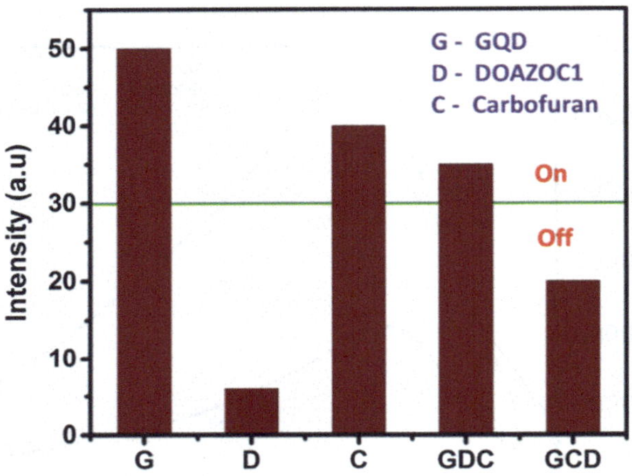

Fig. 8.18 The impact of DOAZOC1 and carbofuran on the fluorescence emission of GQDs at 435 nm upon distinct input sequences. The letters G, D and C stand for GQDs, DOAZOC1 (1×10^{-4} M) and carbofuran (1×10^{-4} M), respectively. GDC and GCD are the two sequences of addition. The output is designated as '0' or '1' depending on whether the intensity of fluorescence emission is higher or lower than 30. (Adapted with permission from Ref. [28]. Copyright (2019) John Wiley and Sons)

The system is then employed as a security device to store the information at the molecular level. The important criterion that must be satisfied for the system to act as a molecular keypad lock is that it should get unlocked only by a particular sequence of chemical inputs. The system is then examined to find out whether the emission of GQD-DOAZOC1 complex is sequence-dependent or not. The emission intensity at 340 nm is monitored for the system. If the fluorescence intensity $I > 30$, the output is '1', and if $I < 30$ it is '0'. The emission intensity of GQDs at 340 nm is dependent on the sequence of addition of inputs of DOAZOC1 and carbofuran. From Fig. 8.18 it is clear that in the sequence where DOAZOC1 is first added to GQDs followed by carbofuran, the output is '1'. On the other hand, a reversal in the sequence produces an output of '0'. The probable reason for the change in output in the sequence GCD is that the initial bonding between GQDs and carbofuran may cause a reduction in number of the carbofuran units available for the coordination with DOAZOC1, thereby decreasing the fluorescence recovery of GQDs.

Then the effect of isomeric form *trans or cis* of molecule DOAZOC1 over the fluorescence quenching of GQDs is monitored. The quenching of fluorescence intensity

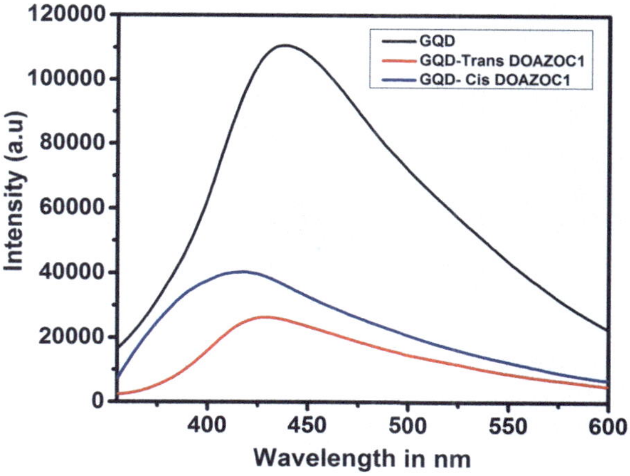

Fig. 8.19 The quenching of fluorescence intensity of GQD by trans-DOAZOC1 and cis-DOAZOC1. (Adapted with permission from Ref. [28]. Copyright (2019) John Wiley and Sons)

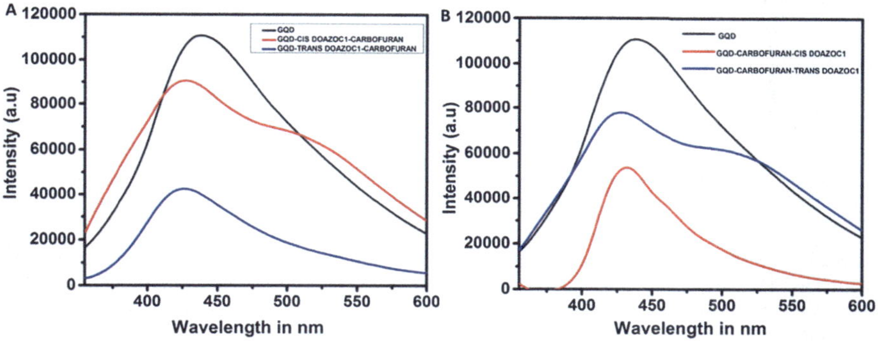

Fig. 8.20 (**a**) The sequential addition 'DC' with both trans and cis isomers of DOAZOC1. (**b**) The sequential addition 'CD' with both trans and cis isomers of DOAZOC1. (Adapted with permission from Ref. [28]. Copyright (2019) John Wiley and Sons)

of GQDs is more for *trans*-DOAZOC1 (76.4%) compared to the *cis*-DOAZOC1 (63.5%) (Fig. 8.19) [46]. But the fluorescence recovery is more for cis when the carbofuran is added. It is observed that the turn on sequence 'DC' is unaffected by the isomeric change of DOAZOC1 from trans to cis (Fig. 8.20a). This is true for the turn off state also (Fig. 8.20b). Thus, it is established that the system cannot be unlocked creating a new password which involves only a change in the isomeric form.

A molecular keypad lock is constructed to visualize this sequence-dependent addition. It appears as a crossword puzzle which can be solved by the correct password only. As given in Fig. 8.21, the chemical inputs DOAZOC1 and carbofuran are represented by the letters 'D' and 'C'. When the sequential input 'D' is executed first followed by 'C', the emission at 340 nm becomes switched ON. Therefore, the password 'DC' can unlock the system. In contrast when the input 'C' is executed first, by 'D', the system is OFF due to low emission at 340 nm. So, the system remains locked. The keypad lock can be unlocked only by the input of the correct

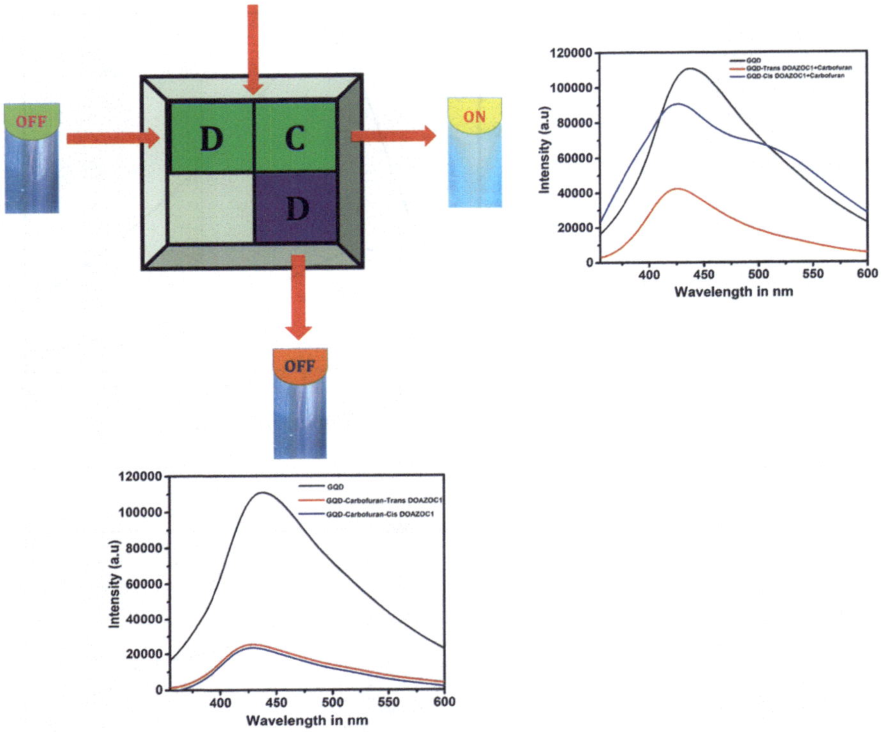

Fig. 8.21 Schematic representation of molecular keypad lock with corresponding fluorescence curves. The letters 'D' and 'C' represent the chemical inputs DOAZOC1 and carbofuran, respectively. The keys 'I' and 'O' depict the fluorescence on (unlock) and fluorescence off (lock) states respectively. (Adapted with permission from Ref. [28]. Copyright (2019) John Wiley and Sons)

password 'DC'. The wrong password 'CD' fails to unlock the system. More importantly the fluorescence outputs at 435 nm created by the sequential inputs of DC and CD are found to be stable even after 1 week. The password remains the same irrespective of the cis/trans isomeric form of azobenzene. This logic gate and keypad lock are the first ones to be constructed from components of sustainable origin.

Sometimes the GQDs are mixed with bioresource materials for developing materials with applications. In one of such works, graphene oxide (GO) is exfoliated to nitrogen-doped graphene quantum dots (N-GQDs) through a hydrothermal process. The synthesis also resulted in the formation of large graphene sheets. The N-GQDs formed composites with the graphene sheets by distributing over the latter. The composites are then coated over the surface of paper fibres. The paper is nothing other than cellulose material, a common bioresource product. The entire composite paper containing N-GQDs, graphene sheets (G) and paper (C) (N-GQDs/G/C) exhibited a strong fluorescence peak around 475 nm in the PL spectra. The cellulose paper otherwise non-fluorescent exhibited strong fluorescence behaviour after the composite formation. It is clearly evident from the confocal laser analysis (CLSM) images (Fig. 8.22).

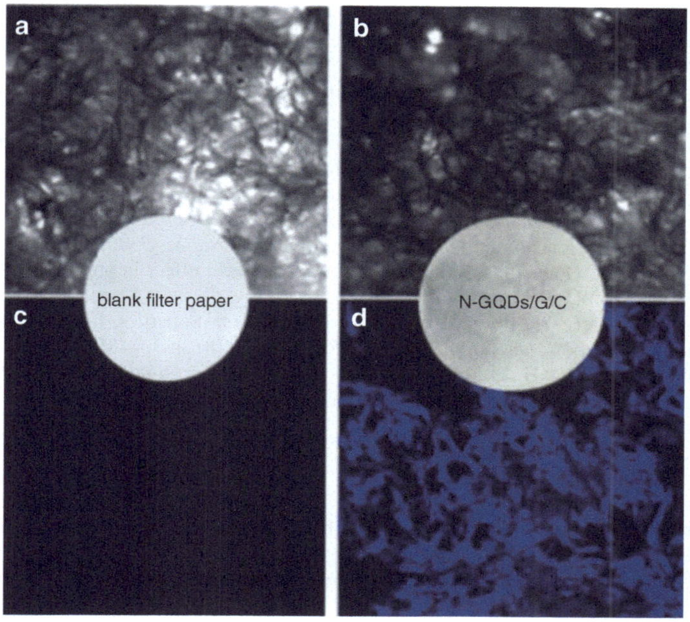

Fig. 8.22 CLSM images of blank paper in (**a**) bright field and (**c**) dark field; composite N-GQDs/G/C in (**b**) bright field and (**d**) dark field. (Adapted from open-access article, DOI: 10.15376/biores.11.3.6299-6308)

8.2 Conclusion

The graphene quantum dots (GQDs) are an important member of carbon nanomaterial family. These are zero-dimensional materials with excellent properties. The GQDs exhibit strong fluorescence properties due the quantum confinement and edge effects. This makes them suitable candidates for various applications. The prominence attained by green chemistry and sustainable chemistry made the scientists to shift their focus from the conventional methods of synthesis of GQDs to their derivation from bioresources. Thus, the GQDs are synthesized from different bioresources including neem leaves, rice grains, coffee grounds, honey, etc. All the bioresource-derived GQDs showed greater potential to be used either in optoelectronic or biological applications.

References

1. dos Santos, M. C., Maynart, M. C., et al. (2017). Carbon-based materials: Recent advances, challenges, and perspectives. *Reference module in materials science and materials engineering*. Elsevier.
2. Schnorr, J. M., & Swager, T. M. (2011). Emerging applications of carbon nanotubes. *Chemistry of Materials, 23*(3), 646–657.

3. Mohan, V. B., Lau, K.-t., et al. (2018). Graphene-based materials and their composites: A review on production, applications and product limitations. *Composites Part B: Engineering, 142*, 200–220.
4. Brodie, B. C. (1859). XIII. On the atomic weight of graphite. *Philosophical Transactions of the Royal Society of London, 149*, 249–259.
5. Chen, D., Feng, H., et al. (2012). Graphene oxide: Preparation, functionalization, and electrochemical applications. *Chemical Reviews, 112*(11), 6027–6053.
6. Kroto, H., Heath, J. R., O'Brien, S. C., Curl, R. F., & Smalley, R. E. (1987). Long Carbon Chain molecules in circular stellar shells, *The Astrophysical Journal, 314*, 352–355.
7. Lijima, S. (1991). Helical microtubes of graphitic carbon. *Nature, 354*, 56–58.
8. Novoselov, K. S., Geim, A. K., et al. (2004). Electric field effect in atomically thin carbon films. *Science, 306*(5696), 666–669.
9. Karousis, N., Suarez-Martinez, I., et al. (2016). Structure, properties, functionalization, and applications of carbon Nanohorns. *Chemical Reviews, 116*(8), 4850–4883.
10. Mykhailiv, O., Zubyk, H., et al. (2017). Carbon nano-onions: Unique carbon nanostructures with fascinating properties and their potential applications. *Inorganica Chimica Acta, 468*, 49–66.
11. Xu, X., Ray, R., et al. (2004). Electrophoretic analysis and purification of fluorescent single-walled carbon nanotube fragments. *Journal of the American Chemical Society, 126*(40), 12736–12737.
12. Ponomarenko, L., Schedin, F., et al. (2008). Chaotic Dirac billiard in graphene quantum dots. *Science, 320*(5874), 356–358.
13. Tian, P., Tang, L., et al. (2018). Graphene quantum dots from chemistry to applications. *Materials Today Chemistry, 10*, 221–258.
14. Li, L.-s., & Yan, X. (2010). Colloidal graphene quantum dots. *The Journal of Physical Chemistry Letters, 1*(17), 2572–2576.
15. Zhu, S., Zhang, J., et al. (2011). Strongly green-photoluminescent graphene quantum dots for bioimaging applications. *Chemical Communications, 47*(24), 6858–6860.
16. Tan, X., Li, Y., et al. (2015). Electrochemical synthesis of small-sized red fluorescent graphene quantum dots as a bioimaging platform. *Chemical Communications, 51*(13), 2544–2546.
17. Tang, L., Ji, R., et al. (2012). Deep ultraviolet photoluminescence of water-soluble self-passivated graphene quantum dots. *ACS Nano, 6*(6), 5102–5110.
18. Eda, G., Lin, Y. Y., et al. (2010). Blue photoluminescence from chemically derived graphene oxide. *Advanced Materials, 22*(4), 505–509.
19. Van Tam, T., Kang, S. G., et al. (2017). Synthesis of B-doped graphene quantum dots as a metal-free electrocatalyst for the oxygen reduction reaction. *Journal of Materials Chemistry A, 5*(21), 10537–10543.
20. Tayyebi, A., Akhavan, O., et al. (2018). Supercritical water in top-down formation of tunable-sized graphene quantum dots applicable in effective photothermal treatments of tissues. *Carbon, 130*, 267–272.
21. Shen, J., Zhu, Y., et al. (2011). Facile preparation and upconversion luminescence of graphene quantum dots. *Chemical Communications, 47*(9), 2580–2582.
22. Kwon, W., Kim, Y.-H., et al. (2014). Electroluminescence from graphene quantum dots prepared by amidative cutting of tattered graphite. *Nano Letters, 14*(3), 1306–1311.
23. Luo, P., Ji, Z., et al. (2013). Aryl-modified graphene quantum dots with enhanced photoluminescence and improved pH tolerance. *Nanoscale, 5*(16), 7361–7367.
24. Zhu, S., Zhang, J., et al. (2012). Surface chemistry routes to modulate the photoluminescence of graphene quantum dots: From fluorescence mechanism to up-conversion bioimaging applications. *Advanced Functional Materials, 22*, 4732–4740.
25. Wang, Z., Yu, J., et al. (2016b). Large-scale and controllable synthesis of graphene quantum dots from rice husk biomass: A comprehensive utilization strategy. *ACS Applied Materials & Interfaces, 8*(2), 1434–1439.
26. Kundu, S., Yadav, R. M., et al. (2015). Synthesis of N, F and S co-doped graphene quantum dots. *Nanoscale, 7*(27), 11515–11519.

27. Dhar, S., Majumder, T., et al. (2017). Phenomenal improvement of external quantum efficiency, detectivity and responsivity of nitrogen doped graphene quantum dot decorated zinc oxide nanorod/polymer schottky junction UV detector. *Materials Research Bulletin, 95*, 198–203.
28. Wen, T., Yang, B., et al. (2014). Organosilane-functionalized graphene quantum dots and their encapsulation into bi-layer hollow silica spheres for bioimaging applications. *Physical Chemistry Chemical Physics, 16*(42), 23188–23195.
29. Kovalchuk, A., Huang, K., et al. (2015). Luminescent polymer composite films containing coal-derived graphene quantum dots. *ACS Applied Materials & Interfaces, 7*(47), 26063–26068.
30. Zheng, X. T., Ananthanarayanan, A., et al. (2015). Glowing graphene quantum dots and carbon dots: Properties, syntheses, and biological applications. *Small, 11*(14), 1620–1636.
31. Mahesh, S., Lekshmi, C. L., et al. (2017). New paradigms for the synthesis of graphene quantum dots from sustainable bioresources. *New Journal of Chemistry, 41*(17), 8706–8710.
32. Suryawanshi, A., Biswal, M., et al. (2014). Large scale synthesis of graphene quantum dots (GQDs) from waste biomass and their use as an efficient and selective photoluminescence on–off–on probe for Ag+ ions. *Nanoscale, 6*(20), 11664–11670.
33. Roy, P., Periasamy, A. P., et al. (2014). Plant leaf-derived graphene quantum dots and applications for white LEDs. *New Journal of Chemistry, 38*(10), 4946–4951.
34. Roy, P., Periasamy, A. P., et al. (2015). Photoluminescent graphene quantum dots for in vivo imaging of apoptotic cells. *Nanoscale, 7*(6), 2504–2510.
35. Kalita, H., Mohapatra, J., et al. (2016). Efficient synthesis of rice based graphene quantum dots and their fluorescent properties. *RSC Advances, 6*(28), 23518–23524.
36. Ghosh, T., Chatterjee, S., et al. (2015). Photoinduced electron transfer from various aniline derivatives to graphene quantum dots. *The Journal of Physical Chemistry A, 119*(49), 11783–11790.
37. Kwon, W., Do, S., et al. (2012). Formation of highly luminescent nearly monodisperse carbon quantum dots via emulsion-templated carbonization of carbohydrates. *RSC Advances, 2*(30), 11223–11226.
38. Wang, L., Li, W., et al. (2016a). Facile synthesis of fluorescent graphene quantum dots from coffee grounds for bioimaging and sensing. *Chemical Engineering Journal, 300*, 75–82.
39. Mahesh, S., Lekshmi, C. L., et al. (2016). Simple and cost-effective synthesis of fluorescent graphene quantum dots from honey: Application as stable security ink and white-light emission. *Particle & Particle Systems Characterization, 33*(2), 70–74.
40. Renuka, K. D., Lekshmi, C. L., et al. (2017). Sustainable bioresource-derived components for molecular keypad lock and IMPLICATION logic gate construction. *Chemistry Select, 2*(35), 11615–11619.
41. Li, Y., Hu, Y., et al. (2011). An electrochemical avenue to green-luminescent graphene quantum dots as potential electron-acceptors for photovoltaics. *Advanced Materials, 23*(6), 776–780.
42. Hwang, E., Hwang, H. M., et al. (2016). Chemically modulated graphene quantum dot for tuning the photoluminescence as novel sensory probe. *Scientific Reports, 6*, 39448.
43. Sun, H., Wu, L., et al. (2013). Recent advances in graphene quantum dots for sensing. *Materials Today, 16*(11), 433–442.
44. Emsley, J. (1980). Very strong hydrogen bonding. *Chemical Society Reviews, 9*(1), 91–124.
45. Larson, J., & McMahon, T. (1984). Gas-phase bihalide and pseudobihalide ions. An ion cyclotron resonance determination of hydrogen bond energies in XHY-species (X, Y= F, Cl, Br, CN). *Inorganic Chemistry, 23*(14), 2029–2033.
46. Van Amersfoort, E. S., & Van Strijp, J. A. G. (1994). Evaluation of a flow cytometric fluorescence quenching assay of phagocytosis of sensitized sheep erythrocytes by polymorphonuclear leukocytes. *Cytometry, 17*(4), 294–301.

Chapter 9
Quantum Dot Interfaces for Memristor

Sajeeda Shaikh, Rafiq Mulla, M. N. Kalasad, and Mohammad Hussain K. Rabinal

Abstract Memristor, a two-terminal device, whose resistance changes according to the amount of charge that passes through, has become a popular choice for memory arrays and is expected to play a significant role in emerging memory technologies. These memories have a high speed, high density, non-destructive readout, long endurance, low operating voltage, simple structure, low-cost, and also a unique possibility of multilevel data storage. These devices can be constructed with a wide range of materials, such as metal oxides, nanomaterials, and organic materials, polymers with simple crossbar geometry. As a result, wide-range of mechanisms does play a role in electrical memory switching effect. Hence, a brief overview of all such mechanisms has been discussed. In the beginning, zero-dimensional materials, also referred as quantum dots, have become fascinating materials due to their size- and shape-tuneable optoelectronic properties, with a large surface to volume ratio, which can be suitably modified with caping molecules. In the recent past, these are explored as potential candidate for memristor with unique advantages. During this period, a lot of developments are witnessed; the present chapter is aimed to cover all such aspects.

9.1 Introduction

The functionality and performance of modern computing systems significantly depend on its memory subsystems [1, 2]. The electronic memory is the storage of information that can be encoded and randomly accessed and stored temporarily/permanently for influencing the future actions. As a result, the memory devices are

S. Shaikh · M. H. K. Rabinal (✉)
Department of Physics, Karnatak University, Dharwad, India
e-mail: mkrabinal@kud.ac.in

R. Mulla
Energy Safety Research Institute, Swansea University, Swansea, UK

M. N. Kalasad
Department of Physics, Davangere University, Davangere, India

P. Yu, Z. M. Wang (eds.), *Quantum Dot Optoelectronic Devices*,
Lecture Notes in Nanoscale Science and Technology 27,
https://doi.org/10.1007/978-3-030-35813-6_9

crucial components for information and communication technologies of modern world, and almost there is a vertical increasing demand for in-memory data in various sectors like government, defence, healthcare, life science, retail and consumer goods, and many others, and this significance is quite evident from the fact that the global in-memory database marked is expected to touch close to *7 billion US$ in the year 2023* [3]. The present memory devices, basically made up of semiconductors, have wide-ranging applications across the existing and emerging sectors. Many types of memory devices are in use, and hence a clear distinction of these is very much essential. From the embedded design point of view, these memory devices can be broadly classified into two categories, as shown in Fig. 9.1. The first one is random access memory (RAM) and the second is read-only memory (ROM). The major difference between them is ROM holds data without power and RAM cannot. Hence, the former is for permanent storage and the latter is for temporary storage. Further, Fig. 9.1 also shows another class of memory that is referred as hybrids, and due to various developments in technology, the distinction between RAM and ROM is blurred. Some of these memory devices combine the features of both RAM and ROM and are referred to as *hybrid memory devices*. Further, the three main categories have their subset devices. For example, RAM has static RAM (SRAM) and dynamic RAM (DRAM); the primary difference between them is the lifetime of the data storage. The SRAM stores data for a longer period of time as long as there is external electric power, and DRAM, on the other hand, has extremely short period of data storage typically few milliseconds even under the presence of electric power. Similarly, based on the method of writing the new data and the endurance (number of cycles), ROMs are classified as programmable ROM (PROM), erasable-and-programmable ROM (EPROM), and masked ROM (MROM). The common features of all these memory devices are their ability to retain data and programs forever, even without external power. The main difference between PROM and EPROM is in the latter erasing and writing can be done multiple times using a strong ultraviolet light source by the process of photoconductivity. Finally, the hybrids also have few subsets, such as non-volatile RAM (NVRAM), electrically erasable programmable read-only ROM (EEPROM), and flash. As mentioned above, the hybrids have common features of both RAM and ROM memory technologies.

All the above-mentioned memory devices are commonly referred to as *conventional memory technology* and are seriously facing scaling (miniaturization) chal-

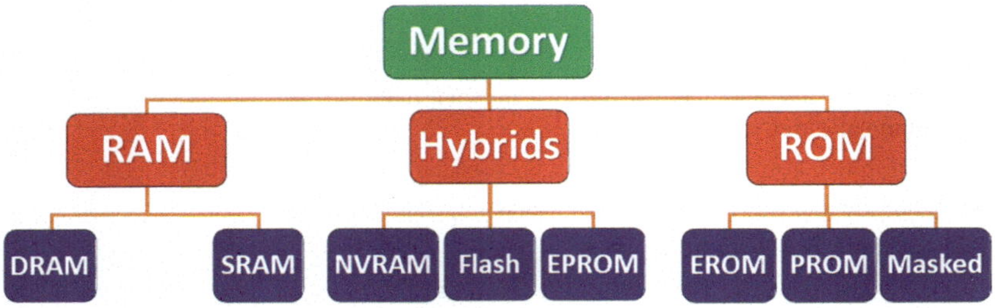

Fig. 9.1 Classification of various types of computer memories

lenges in the coming years to meet the global memory demand [4]. Although these technologies have been significantly improved with the aid of innovative scaling techniques, such as state-of-the-art photolithography, due to inherent limitations of these techniques, such as diffraction limit, etc., it is inevitable to reach cell size to the fundamental physical limit that is close to *10 nm* [5]. As a result, the present technologies face serious challenges on the scaling below *10 nm*. In this regime, the stored charges can be spilled out that leads to serious problem like performance degradation, reliability and noise margin, etc. Looking at these restrictive conditions of charge storage mechanisms, the research and development in industry and academia are seriously seeking non-charge-based memory devices which are easily compatible with existing techniques of integration. The ideal characteristics of emerging memory arrays must have the low operating voltage (<1 V), fast write/read times (<ns), low power consumption (~ femto joule/bit (1 femto = 10^{-15})), long data retention time (>10 years), high write/read cycle (endurance) (>10^{17} cycles), and above all excellent scalability [6]. Probably, it may be quite difficult to meet all such requirements in one type of memory device. As a result, the recent research is seriously focused on a new memory (hybrid) devices, and one of such includes the resistive memory switching, and often these are also referred to as resistive random access memory (RRAM) in short memristor (memory + resistor). Here an active material is suitably sandwiched between two electrodes, whose resistance can be repeatedly tuned from non-conducting state to conducting state and vice versa under the influence of external stimuli, such as electric field, light, temperature, pressure, molecular confirmation, etc. A lot of progress has been made on memory devices using a wide range of nanomaterials, and the present chapter is mainly devoted to these aspects.

9.2 Memristors

A memristor is a two-terminal device having a typical structure of *metal-active material(s)-metal junction* and usually is a non-volatile resistive switch, whose resistance changes in accordance with charge that passes through it. A typical structure of memristor is shown in Fig. 9.2a [7], which is a real crossbar structure, and the device can be made on a wide range of substrates including the flexible ones; one such device is shown in Fig. 9.2b [8]. In 1971, based on the non-linear circuit theory, Leon Chua postulated the existence of such memristor under the title ***Memristor: The Missing Circuit Element*** (as mentioned above, it is a combination of memory + resistor as memristor) [9]. According to this theory, there are only three fundamental circuit elements, i.e. resistor (R), capacitor (C), and inductor (L), corresponding to the four fundamental circuit variables such as current (I), voltage (V), charge (q), and magnetic flux (ϕ).

The fourth missing circuit element that represents the relationship between charge (q) and magnetic flux (ϕ) has been pointed out as the *memristor*; this is schematically shown in Fig. 9.3. Further, Table 9.1 also clearly summarizes the mathe-

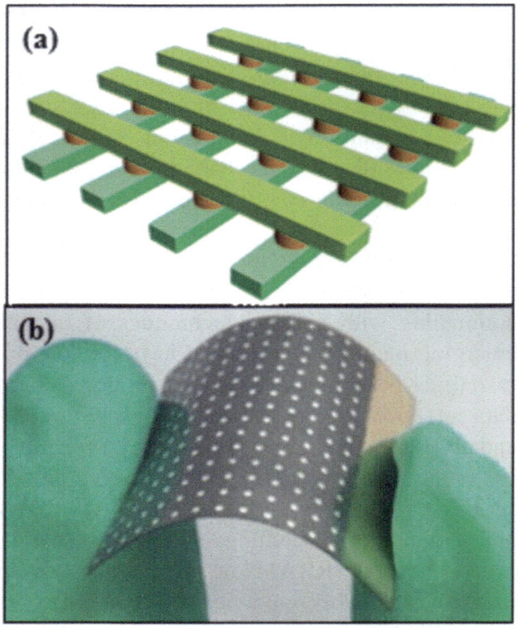

Fig. 9.2 (**a**) Schematic view of crossbar memristors. Reprinted with permission from Copyright (2009) American Chemical Society and [Ref. 7] (**b**) actual memristors fabricated on a flexible substrate [Ref. 8]

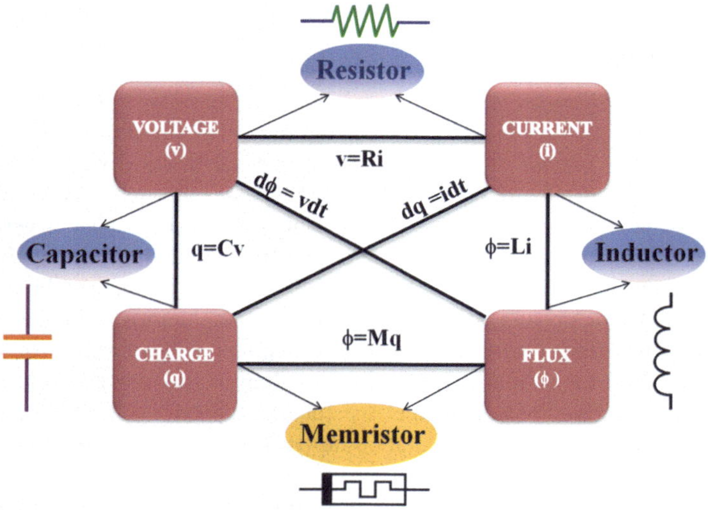

Fig. 9.3 Different types of fundamental circuit elements and their relation with basic parameters, such as current (I), voltage (V), charge (q), and magnetic flux (Φ)

Table 9.1 Relationship between the four fundamental passive elements: resistor, capacitor, inductor, and memristor

Fundamental circuit element	Formula	Constant factor	Physical description	Property
Resistor	$I = KV$	$K = 1/R$	$R = \mathrm{d}v/\mathrm{d}i$	Resistance
Capacitor	$\int I = KV$	$K = C$	$C = \mathrm{d}q/\mathrm{d}v$	Capacitance
Inductor	$I = K \int V$	$K = 1/L$	$L = \mathrm{d}\phi/\mathrm{d}i$	Inductance
Memristor	$\int I = KV$	$K = 1/M$	$M = \mathrm{d}\phi/\mathrm{d}q$	Memristance

matical relationship between the four fundamental circuit elements. Much before Chua's prediction, the researchers were continually experimenting on the current-voltage hysteresis in many types of materials, and some of these milestones are as follows.

The first hysteric resistive switching report was published in case of aluminium oxide (AlO_x) by Hickmott et al. in the year 1962 [10]. Similar behaviour was also observed in the films of NiO_x in 1964 by Gibsson et al. [11] and in films of TiO_2 by Argall et al. in 1968 [12]. For many years, these observations did not attract the attention of the scientific community. Later in the year 1990, Thakoor et al. published a paper on electrically programmable switching in WO_x films [13] and in 1994 by Buot et al. on a paper entitled *Binary Information Storage at Zero Bias in Quantum-Well Diodes* [14].

In this direction, a stronger support emerged from the IBM research lab in the year 2000 by Beck et al. on reproducible resistance switching effect in metal oxide films [15]. Subsequently the first practical memristor was designed by Williams et al. by HP laboratories in the year 2008 [16].

In the later period, the interest in these devices grew quite rapidly as *resistive random access memory* (RRAM) devices. These devices are attractive due to their high performances, such as high resistance (ON/OFF) ratio, non-volatility, high switching speed, low power consumption, good endurance, small size, long life, and low cost. Academia and industries alike realized the potentiality of these as the next-generation non-volatile information storage arrays. This scientific interest is evident from the fact that the number of publications as a function of time on these devices grew quite rapidly, as shown in Fig. 9.4.

The working of typical memristor can be understood by the nature of its current-voltage curves under cyclic variation of applied voltage, for example, when the voltage is varied from zero-to-positive, positive-to-zero, zero-to-negative, and finally negative-to-zero, as a one cycle. Basically there are two types of memristors: one is a unipolar switching and another is a bipolar switching. The current-voltage characteristics of these two types of devices are schematically shown in Fig. 9.5a, b, respectively. In case of unipolar switching (Fig. 9.5a), as voltage is increased from zero-to-positive, the current also increases linearly, obeying Ohm's law, with device exhibiting a high resistance (slope is less), which can be referred to as the high resistance state (HRS). With further continuation of voltage, current also increases superlinearly, and finally at a critical (SET) voltage (V_{set}), there is a rapid and significant change in the current at which the device

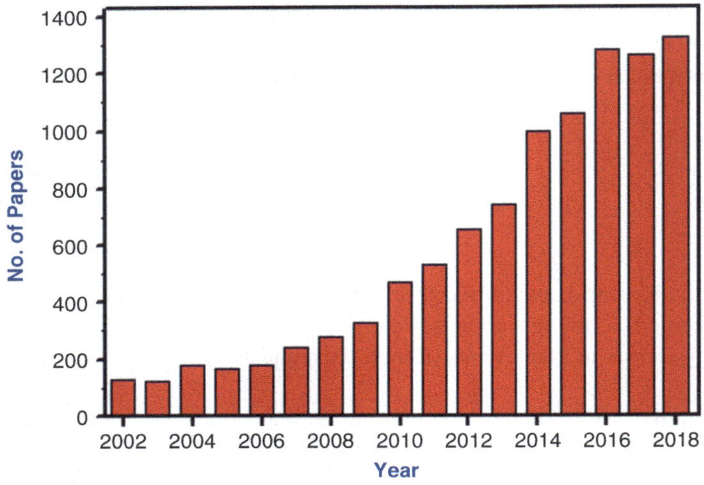

Fig. 9.4 A survey on number of research publications as a function of time on memristors (these numbers are taken through the SciFinder) *Resistive Switching Memory*

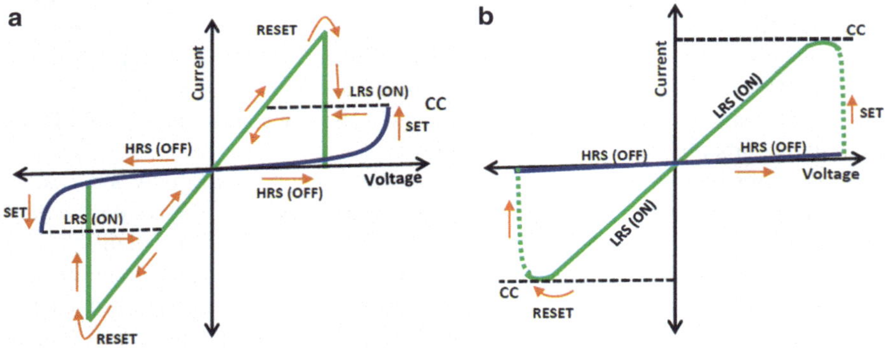

Fig. 9.5 Schematic plots of current-voltage curves in first and third quadrant for (**a**) monostable electrical switching and (**b**) bistable electrical switching. The arrows in these figures clearly indicate the directions of voltage scans

takes a transition from HRS to low resistance state (LRS); in the figure, the current reaches to the set compliance current (CC), a limit that can be imposed through the measuring device. On decrease in voltage, the current remains almost constant due to CC, and on further decrease in voltage towards zero, the current decreases linearly to zero.

The cycle in the first quadrant of current-voltage plot is clearly indicated in Fig. 9.5a. On once again increase in voltage (in first quadrant only), the current increases linearly exactly following the path of LRS, and then there exists a critical voltage, usually much lower than the initial V_{set}, and the device takes a transition from LRS to HRS, as indicated by a sharp transition. On further decrease in voltage, the device follows the HRS and finally reaches the origin, and this path is clearly marked with the set arrows in the above figure. The HRS is usually referred to as 'OFF-STATE' and LRS as 'ON-STATE' of memristor; the former is logical 'zero' state and latter is logical 'one' state. On repeating the same measurements in the third quadrant that is on reversing polarity of voltage, the current-voltage path becomes exactly identical as that in the first quadrant. Hence, this behaviour is referred to as *unipolar electrical switching* in which the SET and RESET occurs with the same voltage polarity. In contrast to this, the electrical switching occurs in the first quadrant and the reverse transition in the third quadrant, which is clearly shown with a set of arrows in Fig. 9.5b. This type of behaviour is referred to as *bipolar electrical switching*. In both the devices, a transition from HRS to LRS is referred to as 'SET' operation and the reverse LRS-to-HRS transition as 'RESET' operation. It is evident from these curves that the unipolar effect is symmetric whereas the bipolar is asymmetric. These operations are repeatable for a large number of cycles. The present RRAM have many interesting characteristics, and the ON/OFF resistance ratio can be as high as 10^3–10^6, with endurance in the range of 10^6–10^{12}, and write energy can be as low as ~0.1 pJ/bit, write/read times just few ns, scalability opportunity down to 2 nm, and data retention for more than 10 years. Further, these devices require a quite low operating current (~ few μA). However, the RRAM technology is expected to face certain challenges, such as process compatibility, device fabrication yield, performance variability, and reliability.

As mentioned above, the ON and OFF states of a memristor with respect to electrical resistance can be achieved under various external stimuli, and based on the mechanism of switching, these devices are further classified as ferroelectric memories, magnetic memories, electrochemical metallization memories, phase change memories, molecular memories, nanocomposite memories, nano-mechanical memories, etc., and these, along with their suitable mechanisms, are schematically represented in Fig. 9.6. Among these, the ferroelectric, magnetic, phase change, and resistance change are considered as focused memories to overcome the limitations posed by the present technologies. These emerging memories have common features: they are non-volatile, two-terminal devices, electrically switched between ON and OFF states triggered by applied electric field. The main device characteristics of mainstream and emerging memory technologies are summarized in Table 9.2.

9.3 Mechanisms of Resistive Switching Effect

There are various mechanisms involved in the two-terminal resistive switches depending upon the nature of active material(s), device configuration, and active material-electrode interactions under the influence of external electric field, magnetic field, mechanical force, etc. Some of these well-noted mechanisms are (1)

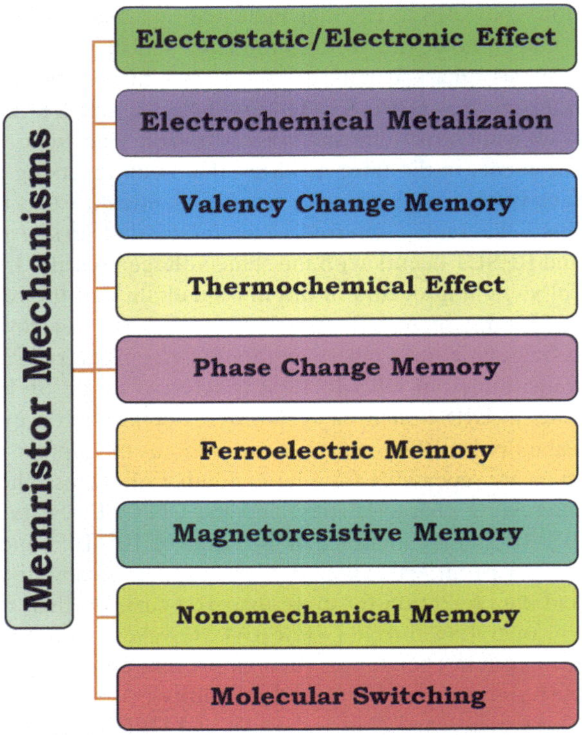

Fig. 9.6 Various mechanisms that are responsible for electrical memory switching in different types of memristors

Table 9.2 The main device characteristics of mainstream and emerging memory technologies

Parameters	Mainstream memories				Emerging memories		
			Flash				
	SRAM	DRAM	NOR	NAND	STT-MRAM	PCRAM	RRAM
Cell area	>100F²	6F²	10F²	<4F²(3D)	6~50F²	4~30F²	4~12F²
Multibit	1	1	2	3	1	2	2
Voltage	<1 V	<1 V	>10 V	>10 V	<1.5 V	<3 V	<3 V
Read time	~1 ns	~10 ns	~50 ns	~10 μs	<10 ns	<10 ns	<10 ns
Write time	~1 ns	~10 ns	10 μs–1 ms	100 μs–1 ms	<10 ns	~50 ns	<10 ns
Retention	N/A	~64 ms	>10 years	>10 years	>10 years	>10 years	>10 years
Endurance	>1E16	>1E16	>1E5	>1E4	>1E15	>1E9	>1E6 ~1E12
Write energy (j/bit)	~fj	~10 fj	~100 pj	~10 fj	~0.1 pj	~10 pj	~0.1 pj

electrostatic/electronic effects, (2) electrochemical metallization, (3) valence change, (4) thermochemical, (5) phase change, (6) ferroelectric tunnelling, (7) magnetoresistive, (8) nano-mechanical, and (9) molecular switching, etc. For clarity, these mechanisms will be discussed here in brief; for more details, the interested readers are advised to refer the earlier literature in this regard [17–21].

9.3.1 Electrostatic/Electronic Effects

There could be several different mechanisms under the electronic/electrostatic effects that control the switching mechanisms in memristors. In the literature, the three prominent electronic effects have been noted, and these are (1) charge injection and trapping, (2) Mott metal-insulator transitions, and (3) ferroelectric polarization reversal. The last one under special condition leads to the realization of ferroelectric tunnel junction, further its combination with ferromagnetic phase forms ferroelectric-ferromagnetic interface gives multiferroic tunnel junctions [22]. Later, the overview of only ferroelectric tunnel junction will be given here. The above-mentioned electronic/electrostatic effects will be discussed herein brief.

9.3.1.1 Charge Injection and Trapping

In this process, the electric charges are injected into the material under consideration with the help of high electric field; the trapping and detrapping of these charges occur that leads to memory switching. It is observed that the injected charges lead to couple of changes in the active material and its interfaces with contacting electrodes, such as changes in defect chemistry, change in band structure, and also change in interface properties, and in turn these effects modify the charge transport properties of these junctions. Some of these processes are clearly depicted in Fig. 9.7 for the case of an n-type semiconductor in contact with a metal through a thin insulating barrier that has electric dipole moment (formation of surface dipole layer). Here, the process (1) represents the charge injection from metal into a semiconductor by overcoming the Schottky barrier, (2) trapping of free carriers by deep trap states, and (3) detrapping of charge carriers. The current-voltage characteristics associated with such mechanisms can be well explained on the basis of Simmons-Verderber model [24]. The effect is studied in *Au/SiOx/Al* junction, and due to a fast switching process, it has been purely accounted as an electronic effect [25]. It is observed that in the formation process, the applied electric field injects the Au ions into the wide band of SiO_x to form highly localized defects, and then the electrons can move through the material by tunnelling between these discrete impurity states. In this way, the stored charge modifies the band structure of insulating material and hence creates the electrostatic effect at the interfaces of Au-SiO and Al-SiO. Under the reverse polarity of voltage, the charge can be detrapped, and hence the device can revert back to its original state. A purely charge trapping modification of inter-

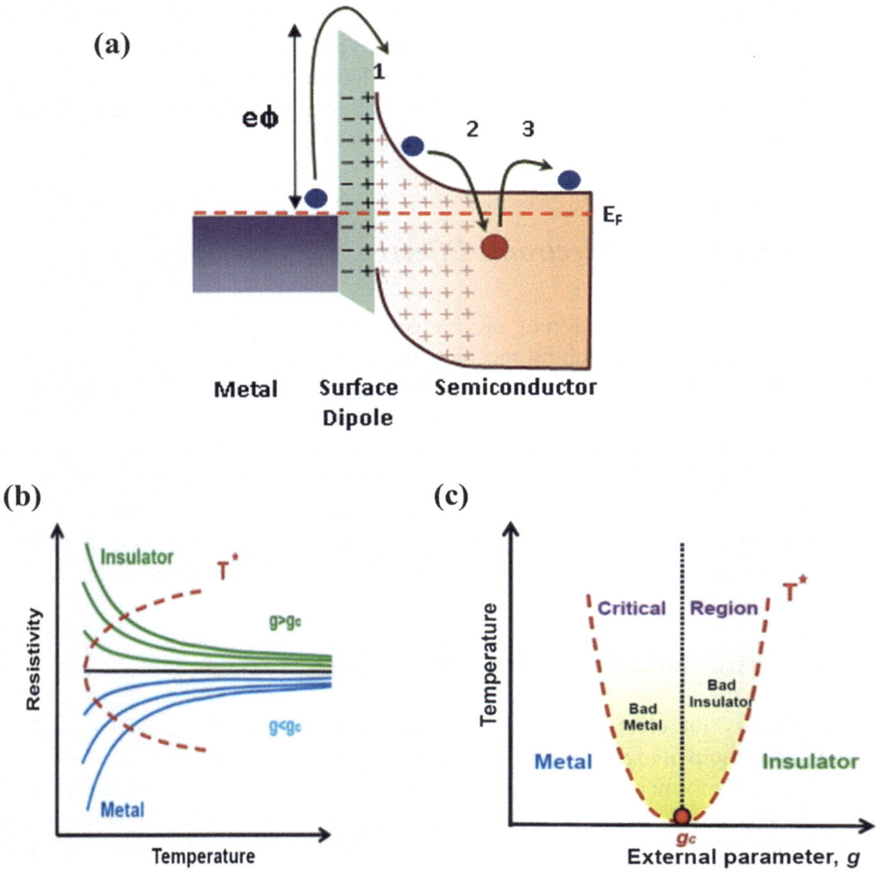

Fig. 9.7 (**a**) The energy band diagram between a metal of work function (eΦ) and an n-type semiconductor with a thin insulating layer having an electric dipole moment. Here process (1) is the thermionic emission, (2) charge trapping, and (3) charge detrapping in a semiconductor. (**b**) Mott's metal insulator transition is represented with density curve as a function of temperature and varying resistivity and (**c**) plot of temperature versus external parameter (g), the curve clearly puts the distinction of Mott's metal insulator transition and its sharpness at critical value of external critical parameter (g_c). (Reprinted with copyright permission from [Ref. 23])

face has been clearly demonstrated in case of $SrRuO_3/SrTi_{0.99}Nb_{0.01}O_3$ junctions [26]. To further support this, an interesting work is carried out by inserting and controlling the thickness of $Sm_{0.7}Ca_{0.3}MnO_3$ oxide between Ti and $La_{0.7}Sr_{0.3}MnO_3$ [27]. Without the oxide layer, the junction lacks the switching; however, the insertion of this layer leads to controlled switching with layer thickness, and the switching time increases with increase in thickness.

9.3.1.2 Mott's Metal-Insulator Transition

The Mott's metal-insulator transition (MIT) is an important theoretical prediction in the field of condensed matter physics [23]. Simply, it states that the change in materials behaviour from insulator to metal can occur due to various factors such as temperature, pressure, composition, magnetic field, electric field, etc. and can happen suddenly at a critical condition. For example, in a semiconductor at a low temperature, it can be explained as follows: at low enough temperature, each atomic site contains certain number of electrons which bind with a nucleus by coulomb interaction and system is electrically neutral. To make electrons free against this coulomb binding, certain amount of energy is required, and if the temperature of system is increased and is high enough than that of energy of each degree of freedom ($1/2k_BT$), as per the Boltzmann distribution, a significant amount of electrons become free by leaving their sites and become available for electrical conduction. Under such condition, the material takes transition from almost like an insulator to metal. Mott's MIT can be explained with the help of a general diagram that is given in Fig. 9.7b. In Figure 9.7b electrical resistivity as a function of temperature is plotted, the two regimes; below and above the horizontal lines are metallic and insulator behaviors, respectively and the set of curves plotted are for different carrier concentrations and the parameter 'g' can be an any external parameter. As the system crosses over metal to insulator, the temperature-dependent electrical resistivity changes its slope from positive to negative. Figure 9.7c shows the plot for temperature versus external parameter (g), and it clearly predicts that a sharp critical transition for MIT can occur only at lowest accessible temperature $T = 0$ K at which there exists a critical parameter g_c.

Hence, the MIT is viewed as quantum critical phenomenon [28]. The MIT has been observed in a wide range of materials such as mercury metal, vapour-liquid, metal-NH_3 solution, transition metal chalcogenides, transition metal oxides, and so on [29]. Mott himself observed this effect in 1949 in case of NiO_x films [30].

The electric charge injection in the material may lead a transition from highly correlated electrons to weakly correlated electrons; as a result, a transition from insulator to metallic can take place. That is quite favourable in case of resistive switching devices. For example, in case of $Au/Pr_{0.7}Ca_{0.3}MnO_3/Pt$ junctions, it is observed that the mixed valence state of ions, i.e. Mn^{4+}/Mn^{3+}, is required at the interface of $Au/Pr_{0.7}Ca_{0.3}MnO_3$ to show electrical switching, and switching is completely absent with only one kind (either type) at the semiconductor-metal interfaces [31]. Based on these observations, it is proposed that Mn^{4+}/Mn^{3+} state modulate the electron carrier density by electrostatic doping process that leads to electrical switching. Such mixed valence state switching by Mott's transition has been also observed in $Ag/CeO_2/L_{0.67}Ca_{0.33}MnO_3$ junctions by Ce^{4+}/Ce^{3+} cation states [32]. Mott's metal-insulator transitions are often accompanied by structural changes in the active material, and such phenomenon leads to a complex process of electrical switching and makes memristor devices less stable.

9.3.1.3 Ferroelectric Polarization Reversal

Suppose a ferroelectric material is inserted in place of an insulator in metal-insulator-metal (MIM) memory device, and then the ferroelectric orientation influences the electron charge transport between two electrodes which leads to electrical bistable switching. However, it must be distinguished from the ferroelectric RAM, in which the dielectric property of active material (as capacitor) is explored to constitute the memory effect [33]. The high field effects such as bulk controlled (space charge limited) or electrode controlled (Schottky emission) do play an important role in these devices [34]. This effect is well studied in $Au/PbTiO_3/La_{0.5}Sr_{0.5}CoO_3$ Schottky junctions [35], and it is noticed that the interface charge and band bending of these can be continuously controlled by an applied electric field. Here the switching voltages are related to coercive field of ferroelectric materials.

The width of Schottky barrier can be either increased or decreased depending on whether the polarization in space charge region is parallel or antiparallel to internal electric field. Hence, at the coercive field, there can be abrupt change in the polarization which causes a switching in electrical resistance. Interestingly, such devices are also constructed by using ferroelectric polymers, such as vinylidene fluoride, trifluoroethylene, etc., with extremely thin films (just two-monolayer thick) [36].

Further, interesting devices were developed based on the above concept; if the ferroelectric material is made quite thin to accomplish the charge tunnelling, then it can be controlled by the polarization in ferroelectric material. Such junctions are called *ferroelectric tunnel junctions* [37]. The external electric field induces the change in polarization that further causes a change in barrier/electrode interfaces, ion position in unit cell, lattice strains, etc. Recently such devices have become quite popular as emerging memory devices. This effect will be further discussed in Sect. 9.3.6. These concepts are further extended to combine the interface between ferroelectric and ferromagnetic materials to control the charge transport in MIM junctions; here the charge transport can be tuned both by electric and magnetic fields. These are normally referred to as *multiferroic tunnel junctions* [38]. However, it is to be noted that the ferroelectric is a cooperative phenomenon; as a result, there exists a critical thickness of film, below which the material may lose the ferroelectric property due to depolarization field and size effects [39]. Systematic experimental investigations are carried out on this aspect, and it is well reported that critical thickness can range between 1 and 5 nm for well-known ferroelectrics [40, 41]. These findings clearly indicate that ferroelectric films of just a few nm thick can be used to constitute ferroelectric tunnel junctions.

9.3.2 Electrochemical Metallization

The effect is observed in memristors wherein an ion conducting solid electrolyte is sandwiched between an active anode and an inert cathode. The electrochemical dissolution and deposition and vice versa is the process that is responsible for resistive

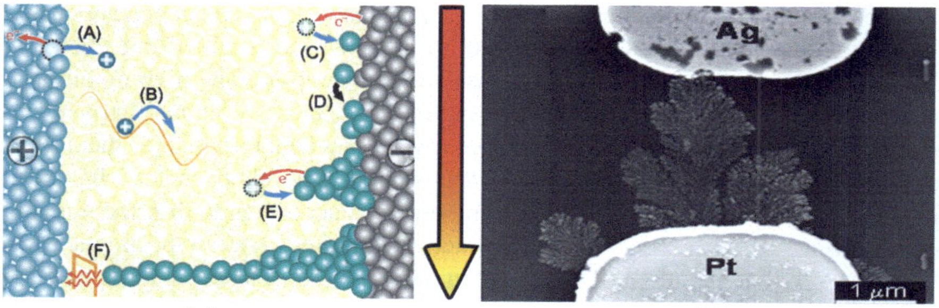

Fig. 9.8 (**a**) Schematic view of electrochemical metallization. Further the figure clearly indicates various subprocesses as explained in the text. Reprinted with copyright permission. (**b**) The experimentally observed dendritic growth of silver between Ag and Pt electrodes during the process of electrochemical metallization. (Reprinted with copyright permission from [Ref. 42])

switching in these devices. Figure 9.8a depicts the various mechanisms that occur during the SET processes in these devices depending on the nature of electrodes and electrolyte materials. During the SET process at sufficient applied voltage, the cations (like Ag^+) get dissolved into solid electrolyte from electrode by charge transfer reaction $M \rightarrow M^{2+} + 2e^-$ (*process A*); under the influence of electric field, these ions migrate towards cathode (*process B*). After reaching the cathode, these ions get reduced to neutral atoms at the inert electrode-electrolyte interface by reaction $M^{2+} + 2e^- \rightarrow M$ (*process C*). These reduced atoms get nucleated on the surface of cathode (*process D*). The filamentary growth occurs by further reduction process throughout the electrolyte medium (*process E*). Finally, when this filamentary growth approaches quite close to the anode surface (just few nm away), electron tunnelling starts between anode and cathode (*process F*). In the ON state, the formation of conducting channels completes; as a result the device goes to a high conducting state. Under the reverse voltage, a reverse mechanism occurs, that is, redissolution of conducting channel and deposition of cation back into anode that leads to RESET state, and process repeats with the voltage cycle. To understand this mechanism more clearly, an interesting experiment has been demonstrated by Gao et al. [42]. Different coplanar structures of *Pt/Ag* electrodes were fabricated on silicon dioxide surface by photolithographic technique, and finally the cell is completed by adding few drops of deionized water as electrolyte which gives $Pt/H_2O/Ag$ junctions. A bistable switching was observed in these devices with applied voltage, and within a few seconds, silver dendrites were clearly observed with the aid of electron microscope, and one such picture is shown in Fig. 9.8b. The usual active anodes are Ag, Cu, or Ni and inert cathodes are Pt, W, Ir, or Au. Many compound materials are explored as active solid electrolytes in these devices; these include the binary and ternary chalcogenide glasses (e.g. basically Cu and Ag doped Ge-sulphides (selenides)) and also the other oxides such as SiO_2, ZrO_2, Al_2O_3, NiO_x, and many more [43]. As mentioned above, the basic requirement on solid electrolyte is that it must exhibit reasonably high ion diffusion under the influence of electric field.

The formation of conducting channels as a dendritic growth between anode and cathode is experimentally well studied in these cells; the result of one such cell *Ag/a-*As_2S_3*-Au* has been reported quite some time back [44]. The high mobility of cations in amorphous matrix is due to random distribution of atoms that leads to more free volume as compared to their crystalline counterparts. A theoretical description of electric field-driven ion diffusion in amorphous matrix has been discussed and is known as Mott-Gurney model [45]; according to this, the hopping of ions occurs due to a series of thermally activated jumps between the atomic sites. It is observed that the initial SET voltage in these devices is always higher than the subsequent SET voltages; the nano-morphological changes that are created for the first time leads to easing out the formation in subsequent cycles. This has been well studied in case of *Cu/*SiO_2*/Pt* cells; the first SET voltages critically depend on the thickness of electrolytic films, whereas the subsequent SET voltage hardly depends on the thickness [46]. This indicates that the first formation process creates the desired structural template that favours easy diffusion of cations. Further, the redox based device is an interesting aspect, the ON state is a dynamic state, it means the appropriate limiting on ON current (compliance current) the ON state resistance can be systematically tuned to many orders of magnitude [47]. The effect has been clearly demonstrated in case of *Ag/GeSe* and *Cu/*SiO_2 junctions [48]. Here the change between ON and OFF states is close to eight orders of magnitude, which is significant and helps to develop multilevel memory devices.

9.3.3 Valence Change

The electrical switching based on this effect has been observed in a wide range of transition metal oxide films [49]. Here the effect has been attributed to the valence change that can be explained as follows: the ionic solids are made up of equal number of cations (+ve) and anions (−ve) that maintain the charge neutrality. Normally, under the influence of electric field, the cations become more mobile as compared to their anion counterparts, because of smaller size due to oxidation process (i.e. loss of electrons). In transition metal oxides, the oxygen defects, particularly oxygen vacancies, are much more mobile than the metal cations. Under the influence of high electric field the SET process occurs due to electroforming, oxygen deficiencies start building up at the cathode and expand towards the anode provided the cathode blocks the ion exchange process. This chemical change is quite balanced by the transition metal cations by accepting the electrons from the cathode. The process fills electrons into 3d band of Ti ions; this electrochemical electron reduction converts the base oxide into metallically conducting phase, e.g. $TiO_{2-n/2}$ for $n > 1.5$ [50]. The formation of highly conducting oxygen-deficient Magneli phase (Ti_4O_7) is experimentally confirmed [51]. As a result, the virtual cathode moves into bulk material and finally shorts with anode by creating conducting path and hence sets the ON state in these devices.

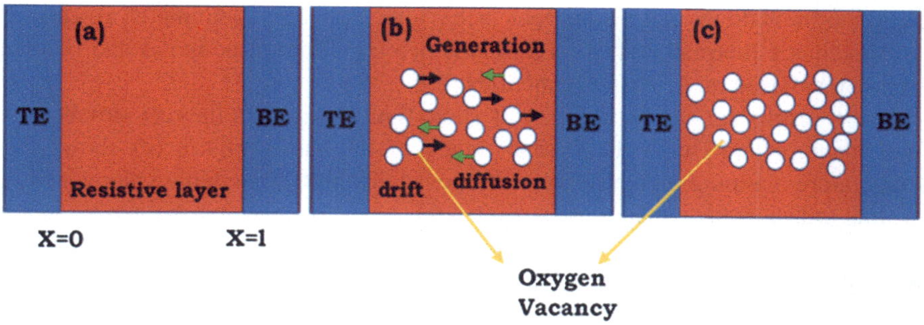

Fig. 9.9 The mechanism of valence change in memory devices: (**a**) initial state, (**b**) generation of and diffusion of oxygen vacancy in an active material, and (**c**) oxygen vacancy creation of electrically conductive channel between two electrodes. [Ref. 50]

Further, at the anode, oxidation reaction takes place to form gases of oxygen also. In the reverse process, the conducting path can be redissolved and hence the ON-to-OFF state and vice versa of such memristors can be repeatedly achieved. The mechanics of valence change leading to ON state is clearly depicted in Fig. 9.9a–c. The valence change mechanism has been extensively studied in wide range of transition metal oxides, such as nanocrystalline Ta_2O_5, Nb_2O_5, VO_2, TiO_2, ZnO_2:Cr, $SrTiO_3$, and WO_3 [52].

It is observed that the valence change mechanism strongly depends on certain standard parameters like device design, electrodes, layer stacks, and local structure around the electrode [53].

9.3.4 Thermochemical

In these cells, the resistive switching occurs due to electric field-induced partial breakdown of active material due to joule heating [54]. In most of the cases, these devices show a unipolar switching due to the set current compliance. This effect was first reported in case of NiO loaded between Pt electrodes in the 1960 [55]; later it is reported that the effect is quite clearly observed in other transition metal oxide devices too [56]. It is believed that switching is triggered by filamentary thermal breakdown of active layer that leads to the formation of conducting channel(s). In the literature, it is argued that the ON state is due to formation of a single filamentary and also by multiple filamentary, but this issue is still unsettled [57].

Experimentally it is observed that the filament channel is composed of electrode material that migrates under electric field and thermal heating; often some organic residues are also observed [58]. Unlike in other types of effect, here the composition of conducting channels is a quite complex. Interestingly, in case of TiO_2 devices, the lower current compliance gives bipolar switching, and the same device exhibits unipolar switching upon increasing the current compliance to a higher value [59]. This clearly indicates that the partial breakdown may give bipolar behaviour and the complete breakdown gives a unipolar behaviour. Recently, Uenuma et al. carried out systematic thermal studies on the joule heating effect in nonpolar and bipolar memristors based on *Pt-NiO-Pt* junctions [60].

These results are shown in Fig. 9.10; here figure (a) shows the real structure of these devices and (b) is the optical image of the same. Figure (c) shows the temperature mapping image in these devices after unipolar switching along with a bar that indicates the temperature scaling.

In this image, a hotspot due to joule heating during the switching is clearly observed. These measurements are shown in figure (d); the figure inset shows the current-voltage characteristics and the switching behaviour. The change of electric power (black dotted line) and the change in temperature ΔT (red solid line) during the switching clearly demonstrate that the temperature change is directly proportional to electric power change that occurs during the operation of these devices.

It must be noted that the temperature change in memristors is a quite common phenomenon that inevitably occurs in various processes of electrical switching as an additional process.

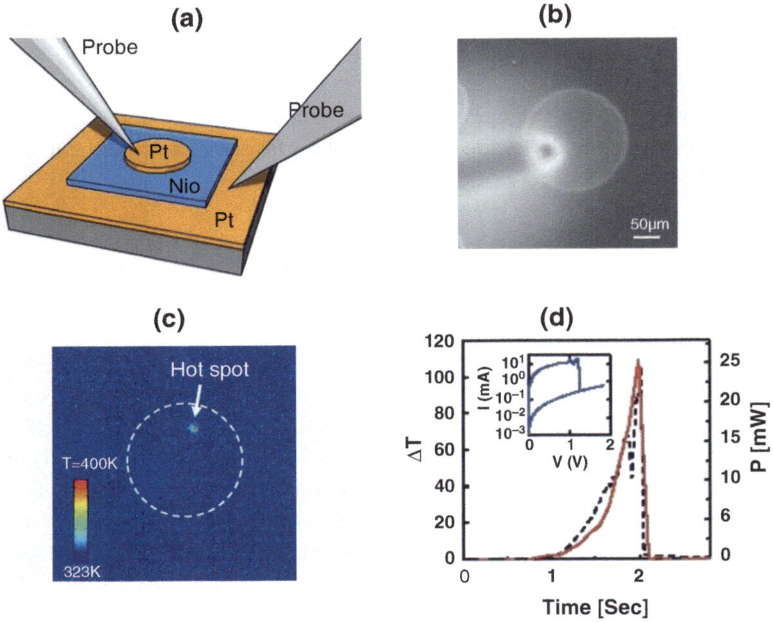

Fig. 9.10 (**a**) Shows the carton of memristor constituted as *Pt/NiO/Pt* junction, (**b**) optical image of top electrode and a sharp probe, (**c**) temperature mapping of image (**b**) and (**d**) variation of temperature difference (ΔT) as a function of time redline, and electrical power as a black line during the reset process of these devices. The inset in the figure is the I-V curve. (Reprinted with copyright permission from [Ref. 60])

9.3.5 Phase Change

The phase change memory effect is generally observed in amorphous semiconductors, particularly in chalcogenide glasses [61]. Here the applied electric pulses generate localized heat, as mentioned in thermochemical effect, due to which these materials undergo a phase change from amorphous state (high resistance) to crystalline state (low resistance). This process has been depicted in Fig. 9.11a. Here the process of write and erase is entirely due to thermal effect. It occurs by the formation of conducting channel in the base material between anode and cathode; this

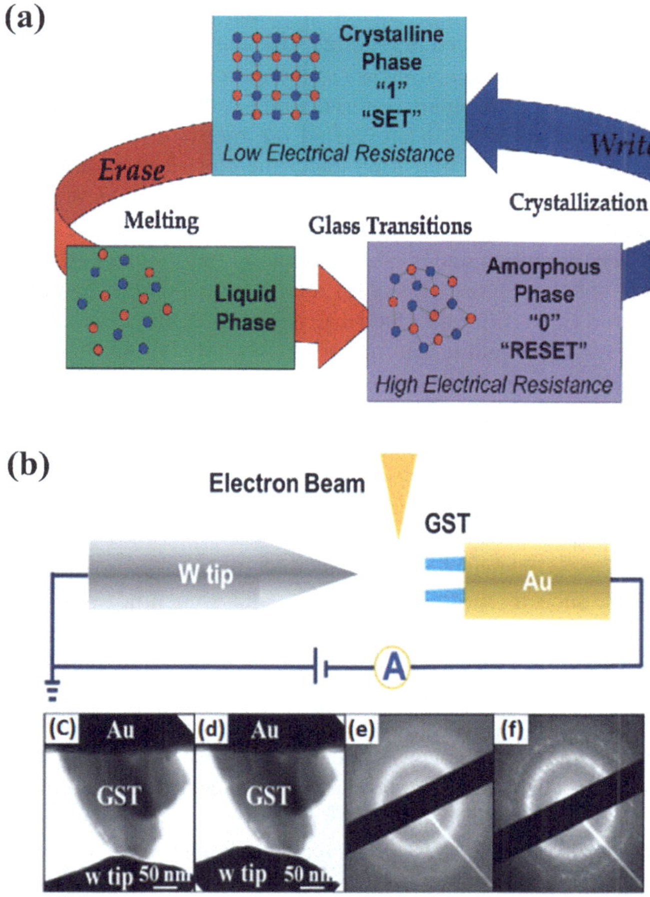

Fig. 9.11 (**a**) Schematic representation of phase change effect in amorphous semiconductor-based memristor. Reprinted with copyright permission, (**b**) An experimental setup W showing a GST film sandwiched between Au electrode and a sharp tip (**c**). (**d**) are two different processes as of GST undergoing crystallization process; first is before and last is after application of voltage. (**e**) and (**f**) are corresponding SAED images. (Reprinted with copyright permission from [Ref. 62])

process is fast and stable. Again a voltage pulse leads to fast melting and redissolution of conducting channel back into amorphous state; as a result there is reversible electrical switching between a high resistance state and a low resistance state with applied voltage pulses. In this mechanism, the memory is based on the phase state of the matter under consideration. All the devices exhibit a forming process; it is the first threshold voltage of switching which is always higher than the subsequent threshold voltages. The effect was first observed by Stanford Ovshinsky in 1968 in amorphous semiconductor of composition $Te_{48}As_{30}Si_{12}Ge_{10}$ and is also referred to as Ovonic diode [63]. Later, the effect was observed in many chalcogenide glasses and is well studied to understand the switching mechanism. Of these systems, the GST ($Ge_2Sb_2Te_5$) is widely a studied system [64]. A ceniographic study in 1970 clearly established that the phase change mechanism in chalcogenide glasses involves the electric field-induced crystalline filament formation [65]. The grown filament always starts at anode and grows in the direction of cathode and is completely dissolves back as amorphous material with application of voltage. Recently an interesting piece of work has been carried out by Shao et al. on GST for the direct observation of structural transition under the electric switching [62]. These results are summarized in Fig. 9.11b–e. Figure (b) shows a cartoon of GST film between gold and sharp tungsten electrode. Figures (b) and (c) are the two independent *Au/GST/W* junctions under observation, (b) before and (c) after application of switching voltage. Figure (d) and (f) are the corresponding selected electron diffraction patterns of in situ and electrically switched films, which shows that these films undergo a reversible amorphous to crystallization phase transitions. Still an interesting piece of work shows that material state can be precisely controlled, such that it can be transformed to one of its four distinct states, one initial amorphous state, two partially crystalline states, and fourth the fully crystalline state. Each of these has distinct electrical properties and hence the memory states. Such possibility helps to create multibit memory in a single cell that can double the memory density [66].

In majority of cases, the materials explored under this effect are multicomponent amorphous solids, such as ternary, quaternary, and sometime even complicated compounds. The continuous scaling towards the higher device densities takes limit close or below 10 nm; this aggressive miniaturization demands precise control of composition of multicomponent materials that is a quite difficult task. In order to overcome such difficulties, recently Salinga et al. explored the phase change memory in monatomic (single component) system [67]. These authors attempted to use the pure antimony as an active material; in amorphous state it exhibits semiconductor behaviour whose electrical conductivity is few orders of magnitude lower as compared to its crystalline counterpart. The main challenge was to achieve its amorphous state; however, the attainment of recent ultrafast techniques of cooling made it possible to create antimony films under the special growth conditions [68]. Still the issue of material stability and data retention timescale needs to be properly addressed.

9.3.6 Ferroelectric Tunnelling

In conventional ferroelectric memory devices, the information is coded through the electric polarization of a thick ferroelectric crystal; the device is like a ferroelectric capacitor in which typically 100 nm ferroelectric film is sandwiched between two metal electrodes as discussed earlier in Sect. 9.3.1.3. A remnant electric polarization is switched between two states, parallel and antiparallel with respect to applied electric field. Due to large thickness, the ferroelectric films of these devices suffer with low readout currents and hence low density [39].

However, the reduction of film thickness just a few nanometres the charge transport can be significantly enhanced due to quantum mechanical tunnelling [69]. Such devices are referred to as ferroelectric tunnelling junctions (FJTs), and the tunnelling probability can be greatly modulated by ferroelectric polarization that produces giant tunnel electro-resistance; hence a reasonably high ON/OFF ratio ~10^4 is observed in recent works [70]. Due to a good readout current and non-destructive read, the FJTs have become the potential candidates for high-density memory arrays. The working of FTJs can be understood with a help of simple diagram that is shown in Fig. 9.12a, wherein a thin ferroelectric film is placed between two different external electrodes (asymmetric configuration). Due to polarization effect, the surface charge appears on film that either attracts or repels the electrons from electrode depending upon the nature of this static charge. This effect persists over a short distance inside the metal that leads to the concept of screening length (δ) in metal and is clearly explained by Thomas-Fermi theory [71]. It predicts that the δ depends on the density of states in metal at the Fermi level; the higher the density, the lower the value of δ. In case of good metals, it is just few nanometres in length, and for a semiconductor, it extends up to tens of nanometres [72]. Later it has been shown that microscopic properties of electrode-ferroelectric interface also strongly influence its magnitude.

In short, this incomplete screening leads to creation of electrostatic potential at the interface between metal and ferroelectric film; it is positive if the polarization points towards the interface and is negative for the opposite polarity. Since the two electrodes are of different density of states and hence there is differing value of δ, as per the figure it is more at the right interface than at the left.

This leads to energy band diagrams that are shown in Fig. 9.12a. The initial rectangular barrier becomes triangular due to interface potential, and hence electrons that flow from left to right face less average potential barrier (Φ_-) as compared to the electrons that flow from right to left with potential barrier (Φ_+), i.e. $\Phi_+ > \Phi_-$. Due to these changes, the tunnel current can be modulated in these junctions. Because of the nature of tunnelling barrier, three different types of charge transport can occur in these devices, which include direct tunnelling, Fowler-Nordheim tunnelling, and also thermal barrier emission [73]. Sometimes the combination of all three can dictate the current-voltage curves of these devices. Recently interesting results based on metal-ferroelectric-semiconductor junctions are reported [74]. Resistance hysteresis loop of one such *Pt/barium titanate/Nb:STO* (Nb: $SrTiO_3$) junction is shown in Fig. 9.12b. The domain evolutions of these FTJs with respect to resistance hysteresis are also shown in the figure as inset.

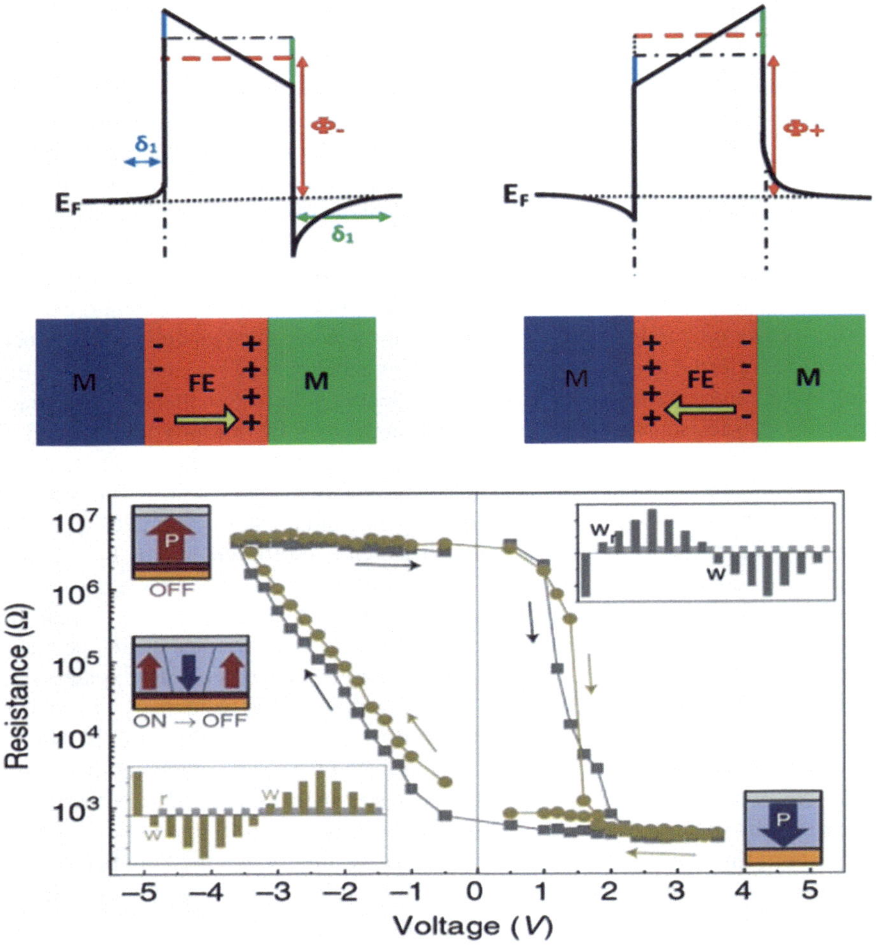

Fig. 9.12 A thin ferroelectric film sandwiched between two different metal electrodes to constitute ferroelectric tunnel junction: (**a**) Energy band bending for two different orientation of electric dipole moments in a ferroelectric film. The parameters δ_1 and δ_2 are screening lengths in metal 1 and 2, and Φ is energy barrier for the flow of charge carriers. (**b**) Resistance hysteresis loops of *Pt/BTO/Nb:STO* ferroelectric junction measured at room temperature. Inset cartoon shows the orientation of ferroelectric polarization under different conditions. The other insets show the pulse train measurements of these junctions. (Reprinted with copyright permission from [Ref. 71, 72])

9.3.7 Magnetoresistive

This memory is based on the magnetoresistance (MR) effect, the ability of materials to change its electrical resistance with a change in applied magnetic field. The effect was first observed by W. Thomson in the year 1856 in pieces of nickel and iron [75].

It has been accounted for the spin-orbit coupling of electrons; the orientation changes in applied magnetic field induce the deformation of electron cloud that leads to more scattering of electron which passes through the lattice. As a result, the electrical resistance of material changes with applied magnetic field; it is observed that magnitude of change depends both on the direction and strength of applied magnetic field. The effect has been also observed in non-magnetic materials like Al, Sb, In, Mo, Cu, Au, Pt, etc. [76]. In conventional materials, the change in magnetoresistance is normally weak; typically it is only of 2% in ferromagnetic materials [77]. In subsequent studies, it is observed that the magnetoresistance can significantly enhance when a thin layer of ferromagnetic (Fe) is laterally combined with non-magnetic (Cr) layer; the new effect is popularly known as giant magnetoresistance (GMR) effect [78]. It is independently discovered by Albert Fert and Peter Grunberg in 1980, and they got the Noble Prize in Physics in 2007. The GMR is due to spin-dependent scattering; this is illustrated with the help of spin-valve diagram shown in Fig. 9.13. Here the two ferromagnetic layers are separated by a non-magnetic layer; if the magnetization of two layers is antiparallel to each other (Fig. 9.13a), then both majority (spin-up) electrons and minority (spin-down) electrons undergo more scattering, and hence device offers a high electrical resistance; on the other hand, if the two magnetic layers are parallel to each other (Fig. 9.13b), the majority of electrons scatter less as compared to minority, hence low electrical resistance.

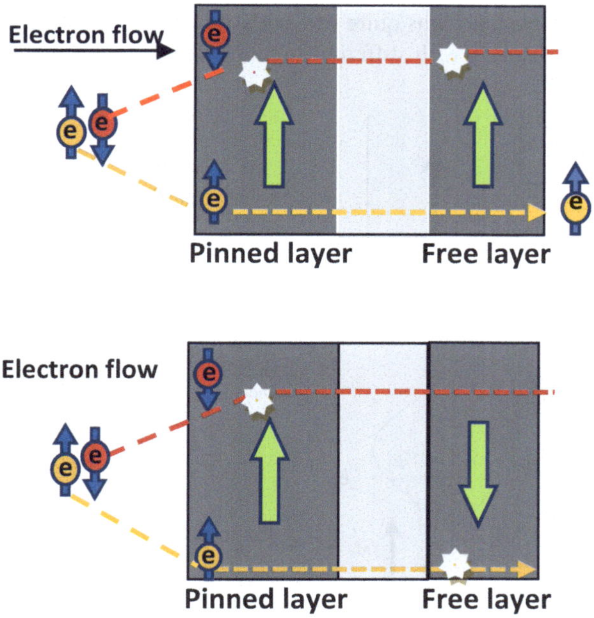

Fig. 9.13 The concept of magnetoresistive effect (spin valve) under two different conditions: one for parallel and another for antiparallel alignments of electrode magnetization. (Reprinted with copyright permission from [Ref. 79])

The GMR effect has been observed in heterostructures such as *Co/Cu*, *Ni/Fe*, *Fe/Au*, etc. [79]. In early years, utilization of spin-valve concept for memory storage was limited due to the fact that it requires field much larger than that can be produced by magnetic medium. However, certain interfaces like *Co/Cu* and *CoFe/Cu* exhibit quite a high GMR. For example, some Heusler alloys show GMR close to 61% [80]. The magnetic tunnel junctions showed a great boost in the development of magnetic memories; in these devices, a thin insulating layer (~ few nm) is sandwiched between two ferromagnetic electrodes. The thickness of insulating layer is quite thin so that the electrons can easily tunnel between the two electrodes.

It is predicted that the orientation of magnetization of two electrodes dictates the amount of tunnel current through these devices under an external electric field. The parallel alignment of electrode magnetization makes these junctions more conducting as compared to the antiparallel alignment. As a result, these junctions can be made more conducting (ON) or less conducting (OFF) by tuning electrode magnetization with external magnetic field. A typical energy band diagram of such junctions is shown in Fig. 9.14; this picture clearly explains how the orientation of electrode magnetization leads to ON (more conducting) and OFF (less conducting) states. In usual device configuration, the bottom electrode magnetization is fixed, also referred to as hard/pinned layer, and only the top electrode magnetization can be free, referred to as soft layer, which is either made parallel or antiparallel with external magnetic field to control ON and OFF states. This effect was first observed in *Fe/GeO/Co* junction at extremely low temperature (~4.2 K) [82]. The ratio between OFF to ON states was quite low with only 14%. Later the effect has been studied in many devices with different combinations of ferromagnetic electrode

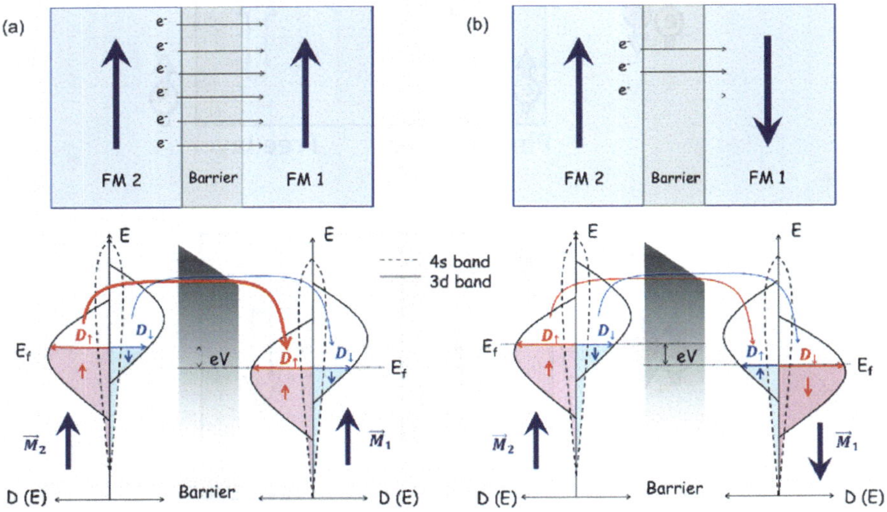

Fig. 9.14 Schematic view of magnetic tunnel junctions under two different conditions, for parallel and antiparallel alignment of electrode magnetization and corresponding energy band diagram for high and low current states which are self-explanatory [Ref. 81]

materials to achieve a better value at room temperature. The best effect is observed with aluminium oxide films, and the change in resistance was quite large (~70%) [83]. The recent work in this direction has been focused on MgO films with various combinations of ferromagnetic electrodes that are made up of both elemental and alloys. Recent work on *CoFeB/MgO/CoFeB* symmetric junction showed a significant enhancement in the ON/OFF ratio which is close to 604% at room temperature that further enhances at low temperature. The magnetic RAM (MRAM) technology gets divided into different generations based on the switching methods employed to write the data [84]. The first generations of MRAM generally include the magnetic-assisted arrays. A main advantage of magnetic writing is it provides unlimited write endurance without any wear-out effects. However, its main difficulty is scaling to smaller size due to several limiting factors associated with cell geometry.

The second generation utilizes the spin property, such spin-transfer torque (STT) in which a spin polarized current for the magnetization of a thin layer of material to program these arrays. The STT is quite compatible with MJT having either an in-plane or perpendicular-to-the-magnetization configurations; such devices have shown a more promising commercial applications [85]. The third-generation MRAM tries to explore many of interesting physical phenomena such as voltage-controlled anisotropy [86], voltage-controlled magnetism [87], spin Hall effect [88], spin-orbit torque switching [81], etc.

One of the major disadvantages of these new-generation memory devices is that they must be developed into three-terminal cell configuration that may be quite incompatible for high-density memory arrays [89].

9.3.8 Nano-Mechanical

The concept of storing and manipulating data by mechanical method was first shown by Babbage (1834) by designing an *analytical engine* [90]; since then the method is in competition with other ways of data storage, such as electronic and magnetic processes which are discussed earlier. It is in continuous evolution to incorporate itself with other electronic devices and is commonly referred to as nano-electromechanical switches (NEMS), which are also referred as nano-mechanical memories, and typically two of their dimensions are smaller than 100 nm. In the conventional data storage systems each bit is accompanied with the field effect transistor, the continues decrease in size of transistor to achieve the high density the dissipated power generates the heat which ultimately limits the size [91]. The proposed nano-mechanical memories are expected to take the problem of power dissipation, which has zero off-state power generation. But some of the challenges are the suitable integration of other electronics, need to have smaller switching times in par with conventional FETs, and the need to operate at expected voltages (~1 V) typically. Further it is expected that the hybrid devices that can combine CMOS and NEMS can also reduce the power consumption. A wide range of materials and device configurations have been explored in the recent past [92]. Some of these results clearly show that nano-mechanical memory elements are quite exceptional because operational frequencies range from megahertz to gigahertz with the expected density reaching up to 100 Gbyte/in^2 [93]. Some theoretical calculations have predicted that the switching times can be close to 1 ns [94], and what is achieved is 10–100 times smaller than the conventional FET timescales [95].

Both single-end and double-end clamped silicon cantilevers and carbon tubes have been utilized to constitute these memory elements. Lefever et al. measured the electromechanical properties of double-walled carbon nanotube, which provide excellent electrical and mechanical properties. To constitute such memories, the carbon tubes were clamped at both ends and were electrostatically deflected with a gate voltage using atomic force microscope [96].

The typical design is shown in Fig. 9.15a; here a heavily doped n-type silicon (with doping 10^{19} cm^{-3}) with silicon oxide on its surface is used as a gate; a multi-walled carbon nanotube of length L (600 nm) and diameter D (10 nm) is clamped between two metallic pads (gold chrome with thickness of 70 nm); the height between the nanotube and gate is denoted as *H*. The real SEM image along with the

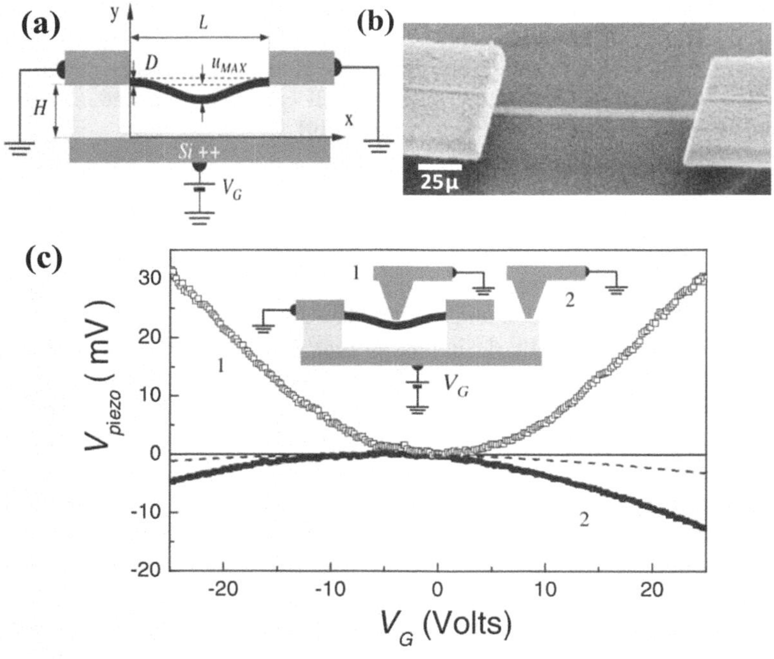

Fig. 9.15 (**a**) Schematic representation of nanoelectromechanical double-side clamped carbon nanotube to constitute nanoelectromechanical three-terminal memory device; the various parameters are explained in the text. (**b**) SEM image of actual device and (**c**) characteristic features of above device; it is plot of feedback voltage applied to piezoelectric element as a function of gate voltage at different positions on the device, (1) at the middle, and (2) on sacrificial oxide of silicon oxide film. (Reprinted with copyright permission from [Ref. 96])

scale of these devices is shown in Fig. 9.15b. The deflection from the initial position of the tube is measured as a function of gate voltage. AFM in tapping mode is used to measure the maximum deflection u_{max} as a function of gate voltage (V_g). These measured results are shown in Fig. 9.15c; instead of deflection *y*(*x*), a piezoelectric voltage that is proportional to deflection is plotted as a function of V_g (both for positive and negative values). Here the curve 1 is midpoint deflection, and curve 2 is at sacrificial oxide of silicon. The behaviour of u_{max} versus *Vg* shows a universal behaviour which depends entirely on the geometrical parameters of above designed structure, such as *L*, *D*, and *H* and Young's modulus of carbon nanotube.

9.3.9 Molecular Switching

The continuous aspiration of miniaturization of electronic devices forced to look for the alternative materials for electronic devices. The use of functional organic molecules, either as an individual or in group, to build the electronic devices has appealed many researchers [97–101]. These molecules can be bonded on varieties of surfaces, such as metals, semiconductors, and insulators, both by weak and strong chemical bonding, to constitute molecular junctions. Just monolayer-thick molecules can be used to constitute *metal-molecules-metal* and *metal-molecules-semiconductor* junctions [102, 103]. The charge transport in such junctions critically depends on many factors, such as intrinsic charge pathway(s) in molecules, electrode-molecule coupling, separation of molecular frontier orbitals, electron spin states, and also molecular conformations. These junctions can be triggered by various stimuli, which include electric field, magnetic field, light, pressure, temperature, chemical, electrochemical, pH of the surrounding liquid phase, etc. Hence, these devices may find a wide range of applications in modern electronics.

Recently there is a growing interest to study the electrical switching in molecular junctions by various external stimuli, particularly by the applied electric field to constitute the memory that is popularly known as *molecular memory*. Such devices are constituted on both a single molecule and on a large number of molecules (film based). A wide range of organic molecules with various electrodes (metals/semiconductors) are explored to study this effect [104]. The actual mechanisms of electrical switching in these devices vary from material to material. However, the effect can be linked to three basic mechanisms: one is molecular conformational-induced switching, second is charging/reduction-oxidation-induced switching, and third is electric field-induced modification of molecular orbitals. The first involves the molecular isomerization due to chemical bonding/cleavage which significantly alters the charge pathway(s), and the second process involves the addition or removal of electron due to applied electric field, and the third leads to effective change in barrier height to accomplish the quantum mechanical tunnelling of charge carriers. These mechanisms are schematically depicted in Fig. 9.16a. From the technical point of view, the challenge is to fabricate the single molecule switch that exhibits a large difference between ON and OFF states with a good endurance.

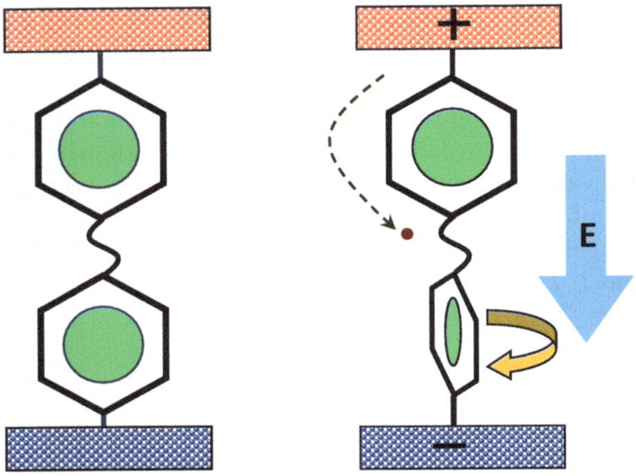

Fig. 9.16 The concept of molecular switching due to charge injection/change in molecular configuration under the influence of applied electric field

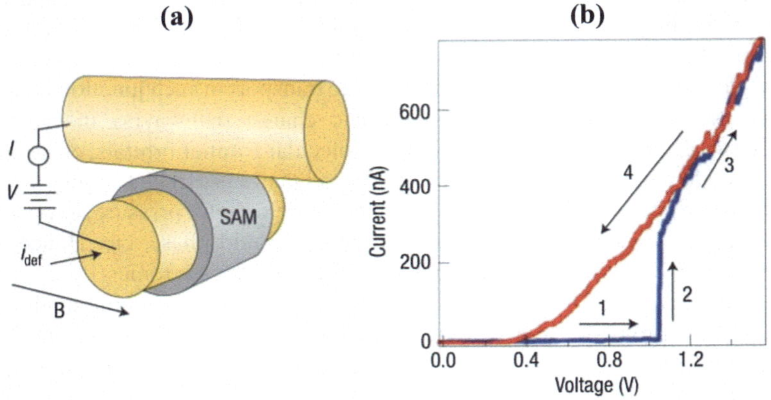

Fig. 9.17 (**a**) Formation of cross-wire molecular junction; self-assembled monolayer (SAM) is sandwiched between two metal electrodes; (**b**) current-voltage characteristics of oligo(phenylene ethylene) (OPE) monolayer molecular junction; the arrows clearly indicate the direction of scan. (Reprinted with copyright permission [Ref. 100])

The single molecule switching is mainly carried out using scanning tunnelling microscopic (STM) studies. Probably the first molecular switching study was demonstrated based on conformational changes of phenylene ethynylene oligomers isolated on self-assembled monolayers of alkanethiolate [105]. These devices showed the conductance change from low resistance state to high resistance state with switching times ranging from few seconds (for patchy monolayers) to few hours (for well-ordered monolayers), and the conductance change has been attributed to confirmation changes rather than electronic/electrostatic effects. The recent studies clearly confirm that the strong coupling of electrodes and the nature of electrode-molecule interfaces are also the deciding factors in the controlling of molecular switching properties [106]. Further, the performances of such devices are found to be environmentally sensitive; the environment induces a profound change in energy landscape of molecules that are bonded to polarizable substrates [107]. Two-state molecular switching behaviour has been clearly demonstrated in number of two-terminal devices [108] (Fig. 9.17).

9.4 Quantum Dots for Memristors

The present chapter is devoted on the recent developments in the field of quantum dots, which include metals, semiconductors, and their composites, for memristor applications; hence only a brief overview on nanomaterials is given here, and the interested readers may refer to the vast literature that is available in the form of books and reviews [109, 110]. Nanomaterials, whose at least one of the dimensions is in few nanometres, have enormous scientific, industrial, and technological potentials. The physical and chemical properties of these materials lie between macroscopic solids and that of atomic/molecular systems. Most importantly, these materials exhibit size-dependent optoelectronic properties, which have purely quantum mechanical origin and are commonly referred to as the *quantum confinement effect (QCE)* [111]. The density of states, number of available energy levels per unit energy range, and intra-energy-level separation can be systematically controlled with size and nanoparticles. The variation in energy discreetness with size in quantum dots is schematically predicted in the below figure (Fig. 9.18).

In metallic nanoparticles, the change in size exhibits surface plasmon effect which is the collective oscillation of free (conduction) electrons in response to applied electro magnetic waves. For conventional metals like Ag, Au, etc this effect gives an absorption band in visible range [112, 113], the details of which can be found elsewhere [114]. The plasmonic materials are found to be the potential candidates for a wide range of modern optoelectronic devices [115]. In case of semiconductor, the QCE induces a significant change in optical band gap (E_g); the gap increases with a decrease in particle size; for example, in case of silicon, E_g increases from bulk value 1.12 eV to 6 eV for 1 nm size quantum dot particles [109]. Apart from QCE, due to a small particle size, the nanomaterials also exhibit few other interesting effects such as high surface-to-volume ratio (i.e. a high fraction of atoms sit at the surface; e.g. almost 96% of atoms occupy surface position as compared to bulk in case of 1-nm-size particle); the surface behaves as elastic film; surface charge can be modified by the choice of suitable organic capping molecules leading to the creation of highly functional materials [116].

Surface chemistry of nanoparticles is an important aspect that controls many of physical/chemical properties [117]. Recent experimental studies clearly confirm that there are various submicroscopic surface parameters that strongly dictate the properties [118]. For example, surface composition and electronic states of surface

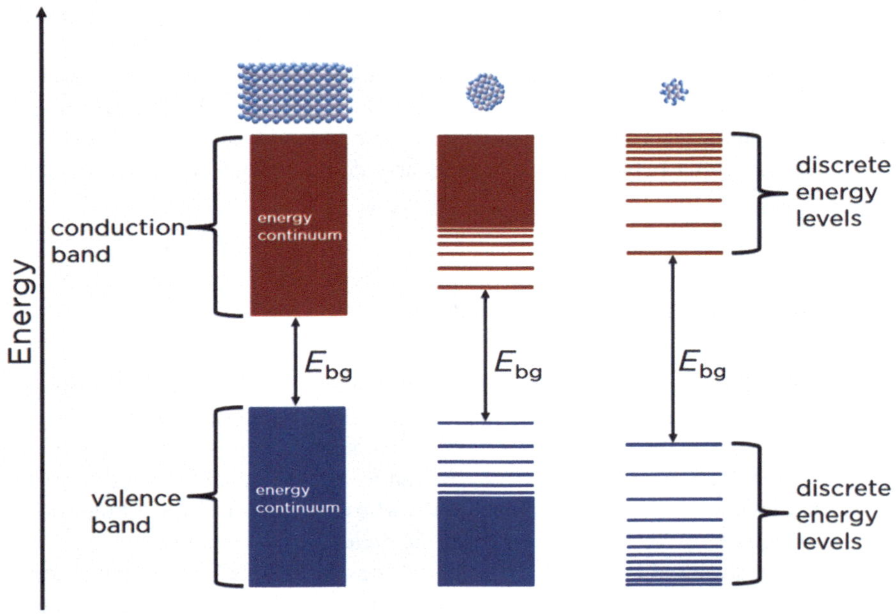

Fig. 9.18 Schematic view of evolution of energy band diagram of a semiconductor with the particle size. (Reprinted with copyright permission [Ref. 111])

atoms decide photoluminescence properties of semiconductor QDs. Further, it is observed that chemical environment around the surface states also significantly influences the photoluminescence properties; in case of polymer-capped CdS nanoparticles, a yellow emission in liquid phase becomes white emission in the solid phase [119]. The effect is accounted for the conformal changes in polymer chains. These effects in silicon nanoparticles are reviewed by Dohnalova et al. [120]. Similar effects are also observed in case of ZnO nanoparticles; there is a good agreement between theory and experiment [32]. Hence, the surface chemistry of nanomaterials can be utilized to design and develop highly specific chemical/biological sensors [121]. Generally, it is believed that the capping molecules are to control nucleation and growth kinetics of nanoparticles and provide temporal stability against the Ostwald ripening and coagulation [122]. However, some of the recent work clearly demonstrates that their role is still more as thought-off. Weiss's group which is activity engaged in such studies has published several interesting articles [123, 124]. An article entitled *Colloidal Quantum Dots: Think Outside the (Particle-in-a-) Box* shows a significant effect of capping molecules on optoelectronic properties of semiconductor nanoparticles [125].

Recently, a significant progress has been made towards fabrication of non-volatile memory devices based on nanoscale materials, such as metallic-, semiconductor-, and carbon-based nanoparticles and their composites [126]. Particularly, quantum dot memristors have emerged as promising candidate for the next-generation memory devices because this is due to their superior functionality, solution processability, easy scalability, easy miniaturization, low power consumption, small size immunity against current leakage, improved retention, and low cost [127–131]. Non-volatile memory based on Si-QDs was first demonstrated by Tiwari et al. [132]; here a floating gate composed of isolated QD islands reduces the problem of charge losses that occurs in conventional flash memories. These devices work with a thin oxide layer with less operating voltage, faster write and erase speeds, better endurance, and a superior non-volatile memory features. Subsequently a significant research effort has been devoted to this technology. Takahashi et al. fabricated single-electron transistors with Si nanocrystal with floating gate and experimentally demonstrated the control of the peak positions of Coulomb blockade oscillations when the positive voltage applied to the gate makes channel electrons tunnel into the floating dots, and the injected electrons raise the potential of quantum dots in single electron transistor, resulting in a shift of peak positions of Coulomb blockade oscillations due to the random distribution of the Si QDs [133]. A wide range of quantum dots and their composites with other materials have been utilized to study the non-volatile memory devices with different configurations. Next few sections will be devoted for the various types of nanomaterials and their composites that are successfully utilized to design the RRAMs.

9.4.1 Metal Nanoparticles

A couple of metal nanoparticles have shown interesting features in the electrical switching. In some cases, metal nanoparticles are incorporated in other compounds to enhance the resistive switching properties of these devices. Recently, Zhang et al. have studied the memory switching effect of functionalized gold nanoparticles [134]. The gold nanoparticles of size 10 nm were functionalized by 3,4-(ethylenedioxy)thiophene molecules and these particles were deposited between two planar electrodes. The device had shown an ON/OFF current ratio of 10^3–10^4. Further, the retention data of the 'ON' and 'OFF' states was observed to be stable for longer duration, and it is clearly tested for about 800 switching cycles (Fig. 9.19).

A bipolar resistive switching has been shown in devices fabricated with a polystyrene film blended with gold nanoparticles [135]. The gold nanoparticles of 1.6–4.4 nm in size capped with conjugated 2-naphthalenethiol were used to form switching devices. There is an almost three orders of change in the resistance between the 'ON' and 'OFF' states. Further, the devices have been tested for several times and found some degradation in the switching properties but are quite stable for couple of hours; the issue was probably related to stability of capped organic molecules. The stability data of these has been shown in the Fig. 9.20, where the current values are tested for about 100 h. Some decrease in resistance ratio has been observed, but still the device maintains enough separation between ON and OFF states. In some reports, the metal nanowires were also used to observe memristive properties; these devices were prepared using Ag nanowires [136].

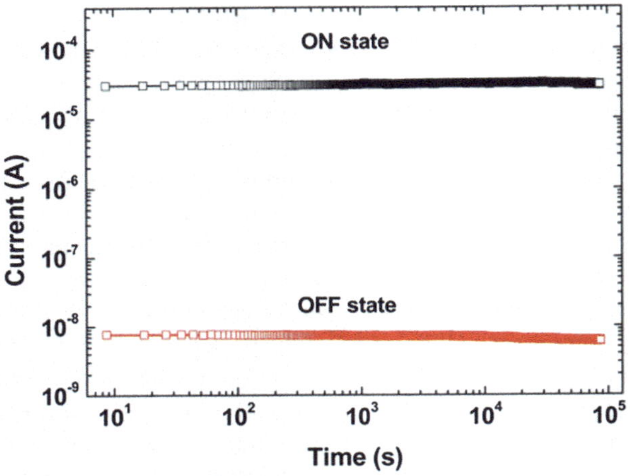

Fig. 9.19 Retention measurements in ON and OFF states (current versus time) for a monolayer device based on gold nanoparticles capped with 3,4-(ethylenedioxy) thiophene. (Reprinted with copyright permission from [Ref. 134])

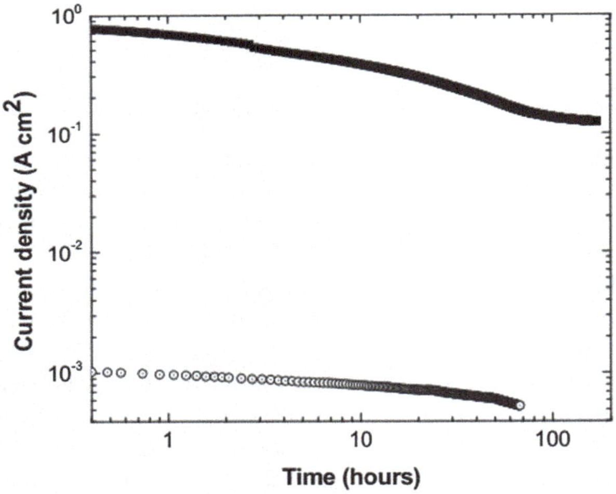

Fig. 9.20 Retention test measurements for polymer: metal nanoparticles, with device configuration *glass/Al/PS+AuNP/Au* junction. (Reprinted with copyright permission from [Ref. 135])

Here $Ag@AgO_x$ core-shell structure was synthesized, and devices of the structure $Ag/Ag@AgO_x/Ag$ were fabricated; an ON/OFF ratio of ~100 was achieved. The resistive switching effects were also proposed in glassy polymer nanocomposites with a series of silver nanowire-polystyrene composites [137].

Similar type of devices using copper phthalocyanine and gold nanoparticle thin films were formed [138]. A very good ON/OFF ratio that is higher than 10^5 was observed with excellent endurance and superior retention properties. Similarly, memory switching effects are observed in devices that were fabricated with organic/polymer layers in between two electrodes [139, 140].

The organic/polymer layers were further added with different metal nanoparticles such as Mg, Au, Ag, Cr, Al, etc. to improve the effect. Figure 9.21a gives current-voltage characteristics of one of the devices wherein gold nanoparticles are trapped inside the polymer.

It can be clearly observed that there is about 10^5 change between the ON and OFF state ratio, where current changes from 10^{-7} to 10^{-2} A.cm^{-2}. Further, the energy level representation of these devices for different metal nanoparticles in the polymer is shown in Fig. 9.21b. However, the study found that there are no systematic changes in the properties with respect to the work functions of the metal nanoparticles. But, it is clear that metal nanoparticles are also quite useful in the memory switching applications if they are used to form hybrid structures.

9.4.2 Metal Oxides

Many metal oxides have been utilized to constitute RRAM devices; here work is mainly focused on some of the binaries such as NiO, ZnO, TiO_2, Cu_2O, ZrO_2, Al_2O_3, TaO_x, etc. [141]. However, there are also studies on multicomponent oxide compounds like $SrZrO_3$, $SrTiO_3$, $Bi_4Ti_3O_{12}$, $PbTiO_3$, $Pr_{0.7}Ca_{0.3}MnO_3$, $La_{0.7}Ca_{0.3}MnO_3$, $PrBa_2Cu_3O_7$, etc. [142–144]. The NiO is a well-known resistive switching material and is studied quite frequently. In 1964, Gibbons and Beadle have found interesting electrical switching properties in thin films of NiO [11]. Two-terminal devices were formed by creating an oxide layer directly on Ni metal sheet and taking contacts from bottom of the metal sheet and a top contact with Ag paste. A typical OFF state resistance was about 10–20 MΩ and ON was about 100–200 Ω with a switching time of 0.1–10 μs. Later, Seo et al. have demonstrated the same reproducible resistance switching in NiO polycrystalline films [145]. Here, films were deposited on desired substrates by dc magnetron reactive sputtering. Further, most importantly, various aspects associated with switching were investigated by preparing these films with different compositions by controlling the ratios of nickel and oxygen feed during synthesis. As expected very interestingly, depending on the ratio between nickel and oxygen, these devices showed different resistance switching properties. Films with a low concentration of metallic defects have shown monostable switching, while those with a high concentration of metallic defects have shown bistable resistive switching; these results are shown in Fig. 9.22.

Lee et al. demonstrated a multilevel tera bit RRAM in NiO-based device fabricated by reactive DC magnetron sputtering [146]. Further, lithium doping has been tried to observe the changes in the device switching properties by having different concentrations of holes. It is proposed that films are compatible and scalable for the multilevel memory switching technology. Resistance switching in *Ni/NiO* core-shell nanowires has been demonstrated by He et al. [147].

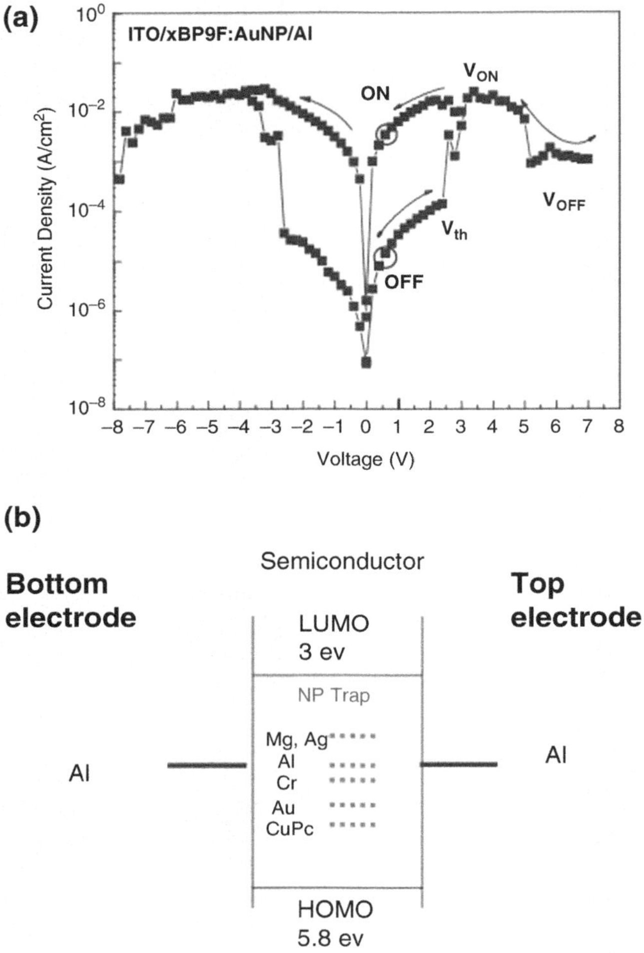

Fig. 9.21 (**a**) Current-voltage characteristics of Au nanoparticle-based memory device. (**b**) Energy band diagram for different choice of metal nanoparticles for *Al/Alq3/NP/Alq₃/Al* structure. (Reprinted with copyright permission [Ref. 140])

The core-shell nanowires were fabricated by electrodepositing Ni in the pores of anodic aluminium oxide template, and then the outer layer is oxidized in an open air. A controlled tuning between memory and threshold switching of these nanowires has been successfully demonstrated. Similarly, Myoung et al. have fabricated nano-filament channels of *Pt-NiO-Pt* structures [56]. The detailed studies have been made to observe the switching time, power, and resistance of the devices, and it is proposed that these structures are ideal for the high-speed, high-density memory applications.

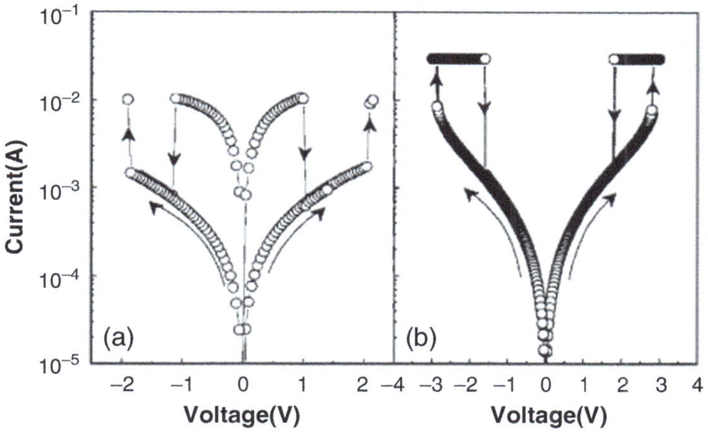

Fig. 9.22 The effect of oxygen content on NiO-based memory device: (**a**) bistable switching with 3% oxygen content and (**b**) monostable switching with 20% oxygen content. (Reprinted with copyright permission [Ref. 145])

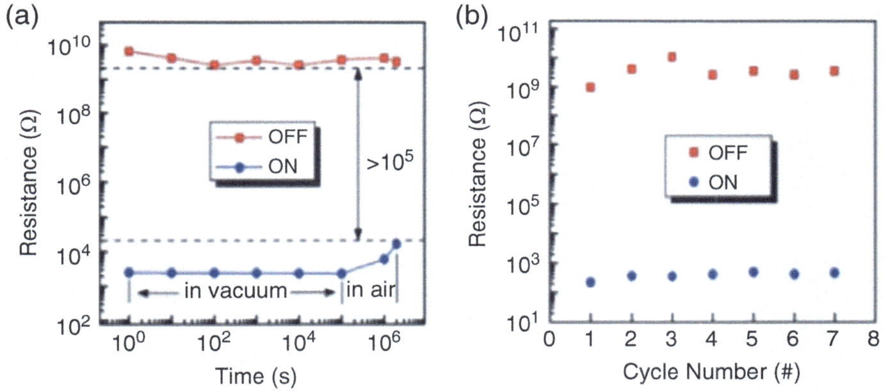

Fig. 9.23 Retention characteristics of memristor based on single crystalline ZnO nanowires (*Cu/ZnO(NW)/Pd* junction): (**a**) resistance as function of time and (**b**) resistance as a function of cycle scans. (Reprinted with copyright permission [Ref. 148])

ZnO is the other compound which is also frequently used to study electrical switching. For illustration, the ZnO nanowires were sandwiched between two dissimilar metal contacts to form *Cu/ZnO/Pd* structures, and a surprising memory switching behaviour is observed [148]. The devices have shown low-threshold voltages and high ON/OFF ratio with a good reproducibility, and these results are shown in Fig. 9.23.

Calculations were done on switching mechanism in ZnO nanowire-based memristors using the density functional theory (DFT) [149]; a novel switching mechanism has been proposed in case of nanowire-based devices.

These authors also fabricated core-shell ZnO nanowires by coating polyacrylic acid-ZnO nanowires and multiple resistive switching observed [150]. This behaviour is accounted for redox reactions of surface states and hence the multiple electrical switching. Memristors were fabricated using solution-processed ZnO nanocrystals [151]. It is shown that the resistance can be modified from 1 to 10^4 Ω. In general, it is proposed that a good electrical hysteresis and memristive effects can be achieved in simple solution processed ZnO nanocrystals. Further, ZnO-based compounds, namely, indium-gallium-zinc oxide (IGZO) memristors, were studied [152]. These were synthesized via solution based method and deposited by spin coating to form *metal-insulator-metal* devices and their bipolar resistive switching behaviour is studied. The ON/OFF ratio higher than 10, is observed with low programming voltage (~ ± 1V) and high endurance. A further improvement in the performance was achieved by annealing these devices at 200 °C. It is proposed that the percentage of oxygen or in other words the oxygen deficiency in these compounds could be responsible for the observed behaviour. Similarly, a bipolar resistive switching is demonstrated in epitaxially grown ZnO nano islands with dimensions of about 30–40 nm [153]. This work illustrates that the fabrication process can be easy to scale up.

There are couples of reports on copper oxides, which are used for the memory applications. A switching is reported in Cu_xO thin films; these films were synthesized by a low-cost solution method [154]. It is shown that resistive switching properties were very sensitive to the concentration of Cu interstitials in these compounds. Further efforts were put to tune the intrinsic point defects in order to change the switching behaviour. The tuning was done by annealing Cu_xO films in the mixture of Ar and O_2 environment and by changing content of latter. It is found that resistivity was mainly dependent on the oxygen content. The measured results of endurance and retention are shown in Fig. 9.24; these results indicate that these materials have potential for memory applications. Similar work is carried out by others as a cost method to fabricate CuO memristive device by successive ionic layer deposition [155]. The TiO_2 nanorod array was synthesized using hydrothermal route for resistive switching [156]. Bipolar resistive switching properties were clearly observed, and the study found that switching voltages, shape of I-V curves, rectifying behaviour, etc. were strongly dependent on the hydrothermal growth conditions. TiO_2-based memristor devices (*Ag/TiO2/Cu*) using electro-hydrodynamic inkjet printing were also fabricated [157]. These exhibit bipolar resistive switching with operating voltage of ± 0.7 V. Similar devices using TiO_2as an active material were prepared by optical lithography and sputtering methods [158]. The Ag@TiO_2 core-shell nanowires were used to fabricate memristors [159]; these were achieved by post-growth shell formation. It is shown that a quite high ON/OFF ratio of order of 10^5 and 10^7 has been observed for both bipolar and unipolar switching.

A transparent resistive switching memory device based on WO_3 has been developed [160]. These films were prepared by cathodic electrodeposition using indium tin oxide (ITO) as electrodes. These devices have shown a good optical transmittance, low operating voltages (+0.25 V to −0.42 V) with uniform operative nature, and long retention time of more than 10^4 s.

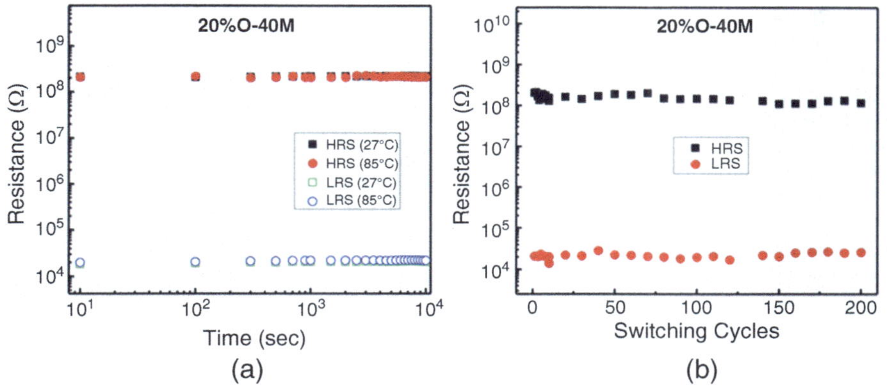

Fig. 9.24 (**a**) Retention characteristics of solution processed Cu_xOmemristor; here the stoichiometry of films is tuned by annealing under oxygen atmosphere hallow circles and hallow squares for low resistance state and solid circles and solid squares for high resistance state recorded at room temperature and at 85 °C. (**b**) Endurance data of same devices for 200 cycles [Ref. 154]

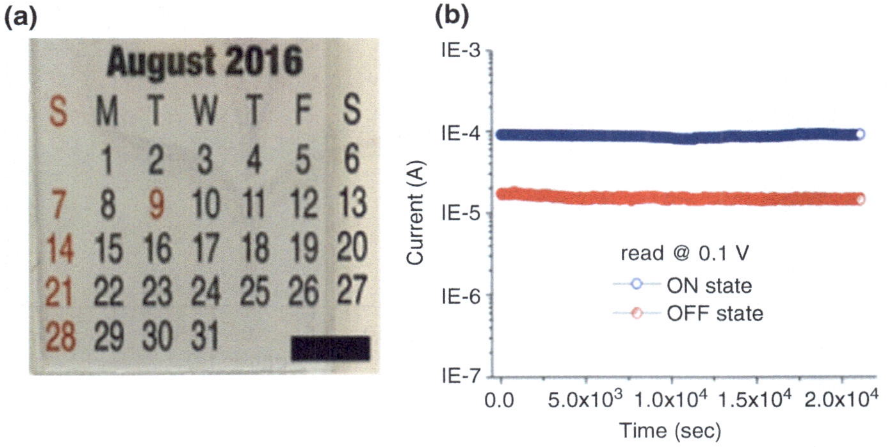

Fig. 9.25 (**a**) Optical image of ITO/WO_3/ITO memristor having complete transparency. (**b**) Retention time of the same device under the read voltage 0.1 V. (Reprinted with copyright permission [Ref. 160])

Figure 9.25a shows the photo of the WO_3/ITO layered designed on a glass substrate, and Fig. 9.25b illustrates the good retention time of these devices. Similarly memristive properties of Al_2O_3 films have been studied [161]. Recently, our group also has studied memristors based on Al_2O_3 films [162]. A simple dip coat patterning method has been implemented to form aluminium oxide-based memristor devices. Figure 9.26a shows the optical image of the as-formed dotted aluminium oxide pattern grown on conducting aluminium substrate. Figure 9.26b clearly illustrates the reproducibility in the charge transport (I-V) curves.

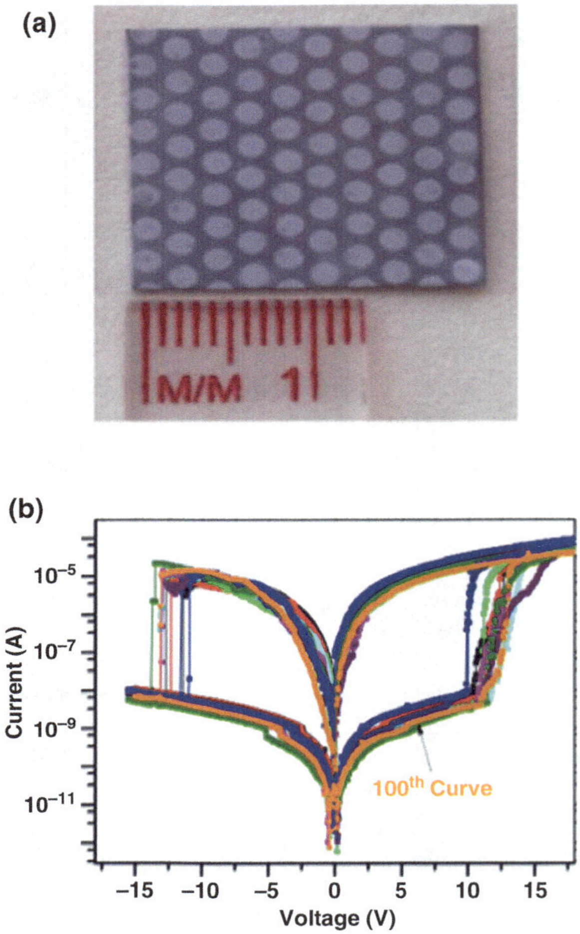

Fig. 9.26 (**a**) The optical image of the as-formed dotted aluminium oxide pattern grown on conducting aluminium substrate. (**b**) Current-voltage characteristics of *Hg/Al2O3/Al* junctions at room temperature. (Reprinted with copyright permission [Ref. 162])

Very recently, an innovative approach has been adopted to form metal oxide bilayers (Al_xO_y/TiO_2) to constitute memristors with a good performance [163]. This approach was successful in making improvements in memory stability via multibit memory operation. Similarly, the resistive switching phenomenon is observed in TaO_x which is studied by several research groups [164–166]. SiO_2 is also another compound frequently investigated for RRAM applications. Devices based on SiO_2 films have shown promising performances, which could operate with very low switching voltage (~250 mV) and low operating currents [167]. Further, Park et al. have also made an effort to make silicon oxide-based devices [168]. Here, direct formation of ordered and dense silicon oxide structures has been achieved on metal and graphene electrodes via self-assembly-induced formation of films. It appears that the approach can be an efficient way to fabricate high-density memory arrays.

9.4.3 Metal Chalcogenides

Majority of semiconductor nanoparticles belong to this class; they are simple to prepare, provide wide variation in bulk optoelectronic properties, and can be easily alloyed with other materials [169]. More importantly, these semiconductors exhibit strong QCE; as a result a wide variation in their optical band gap with particle size has been observed [170]. The peak research on these materials proves that they are also good candidate for RRAM applications. Further, it has been demonstrated that they have a small footprint, large storage capacity, low power consumption, and outstanding rugged characteristics which makes them an ideal storage media for portable devices. Some of the most attractive materials for memory applications are MoS_2, ZnS, CdSe, AgS_2, CdSe, CdS, *CdTe/ZnS*, PbS, CdTe, etc. [171].

The nanocomposites of high and low band gap materials show an excellent bistability. Gosh et al. proposed that the *CdS/PbS* and *CdS/*Ag_2S core-shell nanoparticles grown via ion-exchange reaction exhibit interesting transport and dielectric properties and enhanced charge confinement. The confinement occurred due to the high band gap of CdS, which increases the electrical bistability. The low-band gap material allows facile carrier injection from the electrodes. Here PbS and Ag_2S are low-band gap nanoparticles and they do not show any bistability, whereas CdS with large band gap acts as energy barriers for the carrier's confinement in the core-shell. Hence, high band gap confines carriers to yield a high ON-OFF ratio. The PbS QDs acts as an ambipolar material; both electrons and holes contribute to the conductivity, which leads to a bidirectional shift of threshold voltage to expand the memory window. The thin tunnelling dielectric of nanoparticles is a major drawback for memory application. Instead of traditional designs, the floating gate, discrete nanoparticles embedded into gate dielectric prevent some of these issues and avoid sever charge leakage resulting from the local defects of dielectric. In this context, the double-floating gate memory turns out to be a potential alternative approach to achieve better retention properties by combining two different nanoparticles as double-layer floating gates; these concepts present a new architecture for low-voltage flash memory devices. The solution-processed layer-by-layer assembled graphite oxide (GO) sheets/gold nanoparticles as upper and lower floating gates with PbS quantum dots (QDs) as the semiconductor layer as an alternative to graphene and graphene oxide, a chemically oxidized sheet of graphene, show advantages as a floating gate material due to its numerous charge-trapping sites such as oxygen groups and defects [172]. The use of double-floating gates and PbS QDs in the memory device successfully lowered the programming and operating voltages. Meanwhile, the double-floating gate devices exhibit good memory characteristics, especially in a wide memory window, and promoted retention capability. The introduction of GO covering the Au NP arrays brings extra trapping sites to enhance the memory window as well. More importantly, the energy barrier between GO and Au NPs prevents the trapped charges from leaking back to the channel, which significantly improves the retention capability. The successful solution to enhance the memory window and improve the retention time allows the tunnelling layer to be much thinner, so the programming voltage can be lowered. The non-volatile memory device is constructed with a low-temperature all-solution-processed method. Hence, there is a great potential for the QD-based double-floating gate memory device to achieve wide applications in the field of commercial flash memories; the concept of device formation is shown in Fig. 9.27.

Among most of the 2D monolayer chalcogenides, the molybdenum disulfide (MoS_2) is a rising star owing to its unique band gap, high mobility and quantum confinement effect [173]. It has two phases, one is 2H with hexagonal symmetry and another is 3R with rhombohedral symmetry. Besides these, it has an additional metastable metallic 1T phase with tetragonal symmetry, which can be easily transformed to 2H by annealing at 200 °C is metallic phase with tetragonal symmetry and octahedral coordination. The 2H phase exhibits ohmic behaviour as an active material, whereas 1T phase exfoliated from bulk MoS_2 as nanosheet shows a unique memristive behaviour which is represented in Fig. 9.28.

Hersam et al. have shown that grain boundaries in single-layer MoS_2 generate the memristive phenomena [174]. The 1T phase is 10^7 times more conductive than 2H phase where the hybridization of orbitals leading to overlapping of valence band and conduction band occurs. The cobalt selenium quantum dots (CSQDs-PVP) were synthesized by facile solution-based method and uniformly dispersed in polymer matrix to form the quantum dots/polymer hybrid nanomaterials, CSQDs-PVP, which have a great potential to be used in flexible and high-performance memory devices [175]. Devices were constituted as *Al/CSQDs-PVP/Pt/poly* (ethylene terephthalate) (PET) in that Co_9Se_8 quantum dots act as charge trapping materials in polymer matrix for fabrication of flexible organic memory device. As trap density is higher than free carriers, CSQDs have lower energy levels than PVP; hence, traps charge more effectively leading to space-charge limited conduction. The current-voltage measurement clearly suggests that these devices exhibit a write once-read

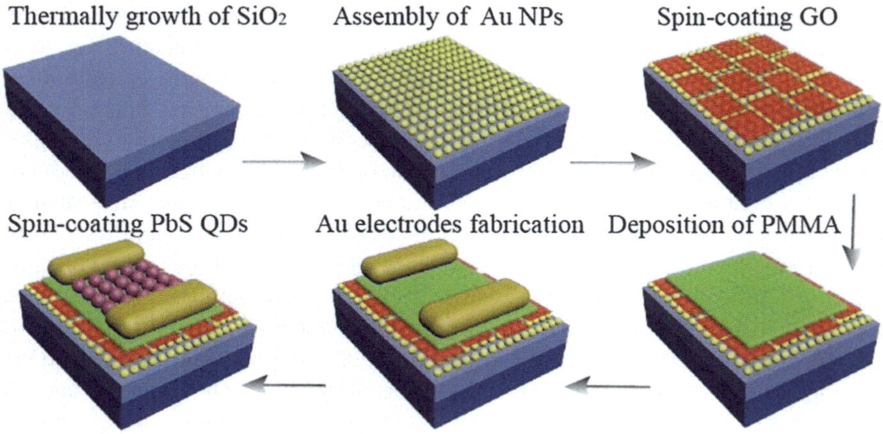

Fig. 9.27 Schematic representation of fabrication of *GO/Au* NP double-floating gate memory devices. (Reprinted with copyright permission from [Ref. 172])

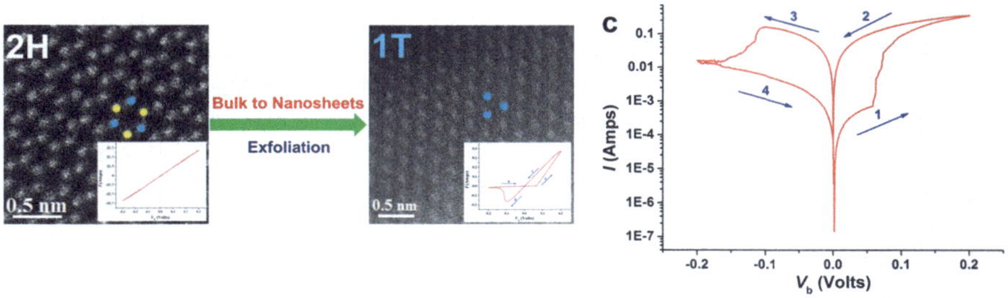

Fig. 9.28 High-resolution TEM images of bulk (2H) to nanosheet (1T) of MoS_2 and current-voltage characteristics of these memory *Ag/MoS$_2$/Ag* devices. (Reprinted with copyright permission from [Ref. 173])

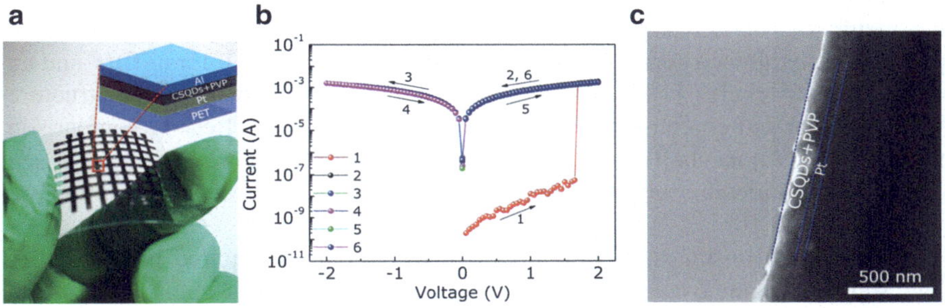

Fig. 9.29 (**a**) A flexible organic-/inorganic-based resistive memory device. (**b**) Current-voltage curve of *Al/Co$_9$Se$_8$ QDs-PVP/Pt/PET* junction. (**c**) SEM cross-sectional image of Co_9Se_8 QDs-PVP hetero-interface. (Reprinted with copyright permission from [Ref. 175])

many times (WROM) type of memory effect. Figure 9.29 shows the result on flexible organic /inorganic memristors. Here a high ON/OFF ratio (~ 10^5) at a read voltage of 0.4 V is observed. Further, these deviation show a negligible degradation.

Silver-based chalcogenide that is Ag_2S is used to form an 'atomic switch' [176]. These devices consist of an inert metallic electrode and Ag layer capped with the solid- state ionic conductor Ag_2S. On positive biasing of Ag electrode with respect to other electrode, a metallic Ag propulsion is grown, and the Ag_2S-Me interface shunts the electrodes, thus creating the non-volatile ON state in these devices. This suggests that a structural-phase transition in the Ag_2S layer also plays a role in the resistive transition. A light-responsive memory, in which the writing, reading, and erasing processes are controlled by light, has become an attractive aspect to form memory devices that may find wide applications [177]. The traditional silicon-based erasable programmable read-only memory could only be erased when exposed to a strong UV light. In order to make the non-volatile memory more sensitive to low-intensity UV light, a suitable charge trapping materials should have a high absorption coefficient of UV light, large quantum efficiency, and exciton generation. In this regard, Han et al. reported a UV-assisted charge trapping/detrapping for multifunctional optoelectronic memory devices; the composite of *CdSe/ZnS* QD monolayer strongly absorbs this light and acts as a good hole trapper. Under forward bias, the conduction channel gets modulated when the charge carriers are transferred from gate electrode. Similarly, in reverse bias, the stored charges are released for programming the operations from one level to another level, which refers to the ON and OFF states [178]. The electrically programmed photonic memory can be completely erased by a low power UV pulse without external gate voltage within a short period of time, which shows the superiority of photonic memory over traditional silicon-based EPROM. The effect is explained on the basis of UV-induced detrapping of charges in *CdSe/ZnS* QDs-organic semiconductor architecture and may lead to a new class of flexible optoelectronic memory devices that are easily scalable at room temperature for large-area manufacturing. Furthermore, various kinds of semiconductor QDs containing Cd and Pb atoms have also been employed in highly stable memristive devices [179]. For example, the memristive device utilizing *CdSe/InP* core-shell nanoparticles embedded in polystyrene exhibits a current-voltage hysteresis with an ON/OFF ratio 10^7 and an endurance 10^5 cycles [180]. Because the *ZnSe/ZnS* (multishell) heterostructure with a similar lattice constant provides a high confinement effect in comparison with the ZnSe (monoshell) structure, hence a highly stable memristive device based on *InP/ZnSe/ ZnS* core has been created, where structure and characterization are shown in Fig. 9.30.

Early studies on the light erasable transistor memories required high-intensity light sources ranging from 1.8 mW/cm^2 to 188 mW/cm^2. There is urgent issue of data erasure under low-intensity light (below 1mW/cm^2), which enables reduction of operating power consumption, integration with highly sensitive photodetector for use in the fields of flexible imaging and biomedical applications, etc. The quantum dot floating gate demonstrates their photo-induced recovery under low intensity for a significantly short erasing time of 1 S and non-destructive readout by engineering the surface chemistry of QDs. In photo-sensitive materials, the photo-generated excitons

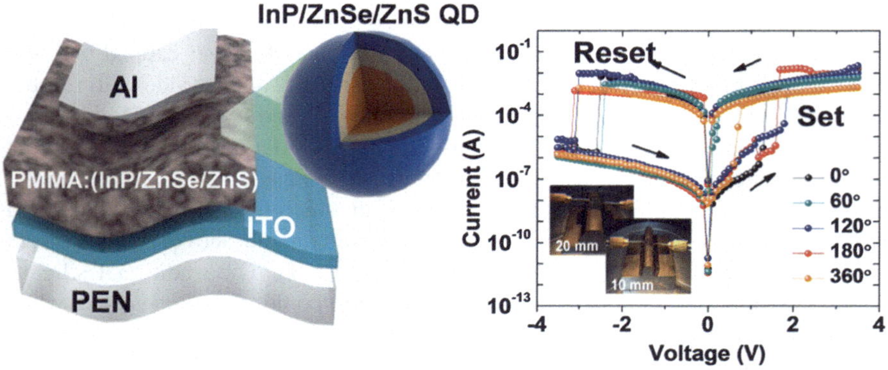

Fig. 9.30 Flexible memristive devices based on *InP/ZnSe/ZnS* core-multishell quantum dot nanocomposite and its memristor characteristics. (Reprinted with copyright permission [Ref. 181])

are separated into hot electrons and hot holes; the hot electrons neutralize with the trap-assisted recombination with trapped holes, whereas hot holes are transferred by multistep hopping across the insulating molecules for favourable diffusion [181]. The dynamics of excited charge carriers including phonon-induced relaxation of charge to its ground state affects many properties of nanoscale devices such as switching speed, carrier mobility, carrier concentration, and luminescence efficiency. The carrier phonon interaction can be enhanced by a strong spatial confinement that occurs in nanocrystal quantum dots. The overcoating of nanocrystals with higher band gap inorganic materials has been shown to improve the photoluminescence quantum yields by passivating surface nanoradiative recombination sites [182]. Core-shell-type quantum dots (QDs) have been the subject of great specific and technological interest owing to their tunable optoelectronic properties. ON/OFF ratio close to four orders of magnitude with the potential window of ±1 V has been observed in certain devices. Unlike conventional memristor, these devices switch under a small applied voltage, for example, high-to-low conducting state with only +0.25 V and low-to-high conducting state at −0.8 V [183]. These devices retain their operational characteristics even after many cycles (more than 1000) with good endurance.

9.4.4 Carbon/Graphene Quantum Dot Composites

Carbon/graphene and its related materials are also promising candidates for the application of resistive switching memories, and these are quite suitable for non-volatile memory technology. Carbon-based switching devices possess low operating voltages (<1 V), low switching times (<10 ns), etc. [184]. However, the parameters like retention, endurance, and power consumption must be improved to make these materials more useful for the application. Switching devices were fabricated using a single and double graphene sheets implanted with organic layers [185]. Resistive switching behaviour has been demonstrated in these by making *ITO/PMMA/graphene/PMMA/Al* and *ITO/PMMA/graphene/PMMA/graphene/PMMA/Al* structures. The results of this work clearly demonstrate that graphene-based devices can be successfully fabricated for memory applications. Similarly, graphene quantum dots (GQDs) with very small size of about ~2 nm were prepared using graphene flakes through grinding assisted co-solvent ultrasonication method [186]. The TEM image of GQDs prepared in this work is shown in Fig. 9.31. These quantum dots were used to fabricate printed resistive switching device to demonstrate their flexible configuration. These worked well for more than 100 switching cycles and have shown robust nature even after 200 bending cycles. Large-area devices of high-quality graphene films using *Melaleuca alternifolia* by plasma-enhanced chemical vapour deposition in the absence of catalyst were fabricated [187].

The memory switching behaviour has been successfully demonstrated by sandwiching pentacene between graphene and aluminium films [188]. Graphene quantum dots were also introduced into memristive devices by Wang et al. [189]. Figure 9.32 gives the TEM images and memristive nature of these devices. Fully

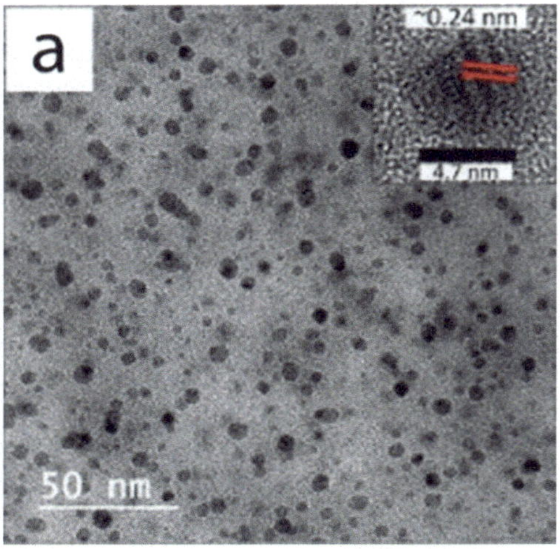

Fig. 9.31 TEM image of graphite quantum dots. (Reprinted with copyright permission from [Ref. 186])

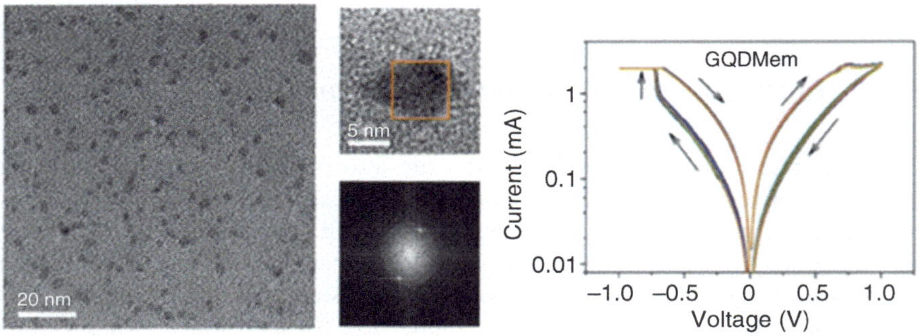

Fig. 9.32 TEM image of graphite quantum dots and its memristor characteristics. (Reprinted with copyright permission from [Ref. 189])

transparent and flexible memory devices have been fabricated by epitaxial growth of thin ferroelectric polymer films on graphene layer [190].

Similarly, graphene oxide quantum dots are also used for memristors. Reduced graphene oxide quantum dots are used for the fabrication of flexible thin film memory devices [191]. Graphene/graphene oxide is also used with other compounds to form a composite kind of structures and to form memristors. For illustration, Yan et al. have found improved performance in resistive switching by inserting graphene oxide quantum dots into $Zr_{0.5}Hf_{0.5}O_2$ films [192]. Further, the graphene is also used as an electrode material to form nanometre-sized *graphene-SiO_x-graphene* switching devices [193]. The improvement in data retention in memristors is a critical issue, particularly in devices having nano-sized conducting filaments. Oxidized carbon quantum dots-graphene oxide nanocomposites are utilized to make memory devices and these are found to have significant improvements in properties [194].

Good retention characteristics were observed with a low compliance current of about 100 μA. Arylamine-based polyazomethine functionalized graphene oxide quantum dots are used to fabricate memristors [195]. It is illustrated that such memory device can significantly increase the memory capacity as compared to the conventional memory arrays. Solution-based synthesis of PbSe quantum dots with embedded graphene oxide has been used to form non-volatile memories with a low operating voltage [196]. These memory devices exhibited a large memory window of ~6 V with a long retention time ~10^4 s. Recently, some theoretical models have also been introduced which explore the advantage of graphene-based compounds in the applications of memory devices [197]. Further, the graphene/graphene oxide has also been utilized to study its memristive nature [198, 199]. A better reproducibility in memory switching has been demonstrated by functionalization of graphene oxide sheets with polyimide which forms a hybrid nanocomposite [200]. This work successfully shows the high uniformity in device-to-device fabrication and also unique memory properties. Further, carbon nanotubes have also been introduced into the memristor applications by Hwang et al. [201]. Carbon nanotubes (CNT) with B and N doping are used for the fabrication of such devices, illustrating the possibility of making solution-processed and mechanically flexible memory structures. A clear demonstration of the fabrication of CNT-based device and its switching properties is illustrated in Fig. 9.33.

9.4.5 Quantum Dot-Organic Composites

Inorganic-organic frameworks (MOFs) have emerged as an important class of materials which are attractive due to their distinct properties and are useful for energy storage and energy conversion [202], chemical/biological sensing [203], catalysis [204], etc. These frameworks have also shown their unique features in memristive applications [205–207]. A couple of recent works on such composites have been reported by several research groups. Recently, memory switching effect of thiol-capped CdS quantum dots which are encapsulated in poly(methyl methacrylate) organic films is presented [208]. A detailed study has been done to understand the role of cell temperature, scan speed, light illumination, film thickness, size of CdS quantum dots, etc. on the switching properties. It is found that the switching events can be tuned by varying the above parameters or experimental conditions. Further, it is proposed that the switching effects are due to the electric field-induced charge transfer from the organic molecule to CdS quantum dots. Yoon et al. have demonstrated switching effects in single crystals of a rubidium-cyclodextrin (Rb-CD)-based MOFs [209].

The *metal/MOF/metal* device structures are used and its electrical properties are shown in Fig. 9.34. Resistive switching is successfully demonstrated in several metal oxide and organic composites such as Al_2O_3, CeO_2, TiO_2, ZrO_2, Y_2O_3, ZnO, etc. with a combination of semiconducting polymers [210].

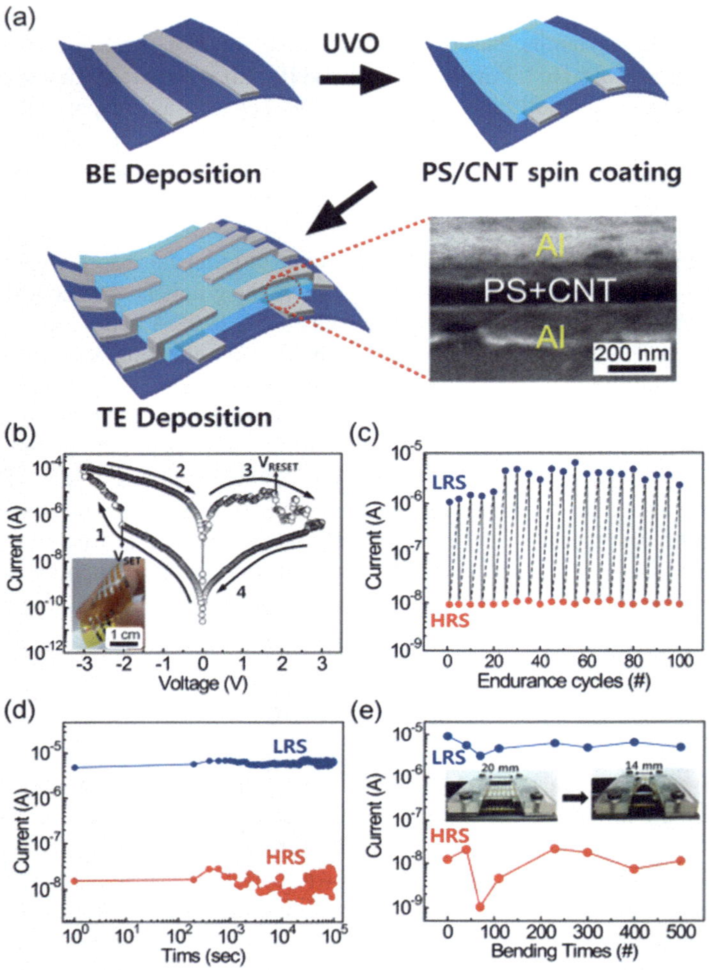

Fig. 9.33 (**a**) Schematics and SEM image of CNT nanocomposite memory devices. (**b**) Current-voltage curves of *Al/(PS+BCNT)/Al* junction. (**c**) Endurance curve. (**d**) Retention curves. (**e**) ON and OFF current characteristics of these flexible devices according to bending times. (Reprinted with copyright permission from [Ref. 201])

Lu et al. synthesized polymer-carbon quantum dot hybrid structures and fabricated the devices with configuration *rGO/PEDOT: PSS/carbon quantum dots/ MEH-PPV/Al* [211]. These devices have shown ON/OFF ratio quite low that is 121 with rectification ratio 103. The switching behaviour in a hybrid structure made up of polyvinyl alcohol and TiO_2 nanotube composites has been also studied [212]. By forming such composite structure, the switching performance has been greatly enhanced with a large resistance ratio $>10^4$ and a good retention 10^4 s. Further, (DFT) calculations have been successfully utilized to understand such enhanced

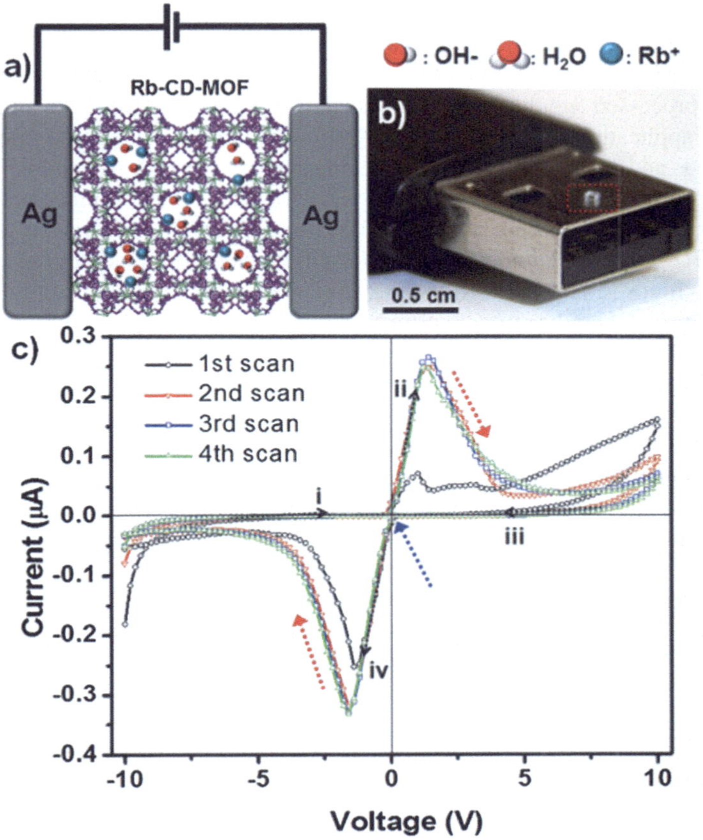

Fig. 9.34 (**a**) Device structure of *metal/organic frame/metal* junction. (**b**) A photograph of circa 1 mm cubic RB-CD-MOF crystal-based real device. (**c**) Its current-voltage characteristics. (Reprinted with copyright permission from [Ref. 209])

properties of these devices. Cheong et al. have prepared nanocomposite films with multilayer structures of organically functionalized nanoparticles [213]. The main highlight here is a very simple and facile synthesis of these materials which does not involve any high-temperature requirements. The devices designed with such nanocomposites having functionalized Fe_3O_4 nanoparticles with the combination of Ag electrodes have shown reversible resistive switching behaviour with an ON/OFF ratio of more than 10^3. Memory devices based on doped carbon quantum dots and poly(vinyl alcohol) have been fabricated [214]. These devices with configuration of *ITO/carbon dots-PVA/Ag* have shown to be operating at low voltage of 1 V and possess high ON/OFF ratio of 10^5. Further, the ON and OFF states of the devices were stable over 3400 s with very less degradation in the performance. Recently, Jeong et al. have prepared memory devices with surface-modified CdSe quantum dots with fluorinated molecules [215]. These devices exhibited a good memory with ON/

OFF ratios of 10^5 and are quite stable for over 10000 s with dynamic switching behaviour. Further, with the voltage-driven programming along with light-assisted erasing mechanisms, the devices have shown electrical switching within 1 s. A solution-processed graphene oxide (GO)-polymer complex has been utilized for memory applications [216]. Functionalization of graphene oxide nanosheets has been tried, and bistable electrical switching and non-volatile rewritable memory effects were observed in the devices. The devices were formed using composite thin films of poly(3-hexylthiophene) (P3HT) having GO sheets with device configuration *Al/GO-polymer complex/ITO*. Similar efforts were also made by several other research groups [182, 217, 218].

Some researchers have made efforts to make use of flexible nature of organic/polymer compounds in fabricating flexible memory devices. For illustration, boron nitride/polymer nanocomposite has been prepared to fabricate flexible resistive memory device by Siddiqui et al. [219], which was designed on a flexible ITO-coated PET substrate. The flexibility of the devices was checked by bending test at different bending levels, and it is found that for over 1500 bending cycles, the devices were quite intact with preserving their original properties which shows their mechanical robustness. This kind of physical stability and reproducibility gives a high hope for flexible memory arrays. In another effort, Yan et al. have made fibre-shaped devices with hybrid organic-inorganic perovskites [220]. This is the first example in which fibre-shaped non-volatile memory device has been fabricated successfully. The devices are based on methylammonium lead halide perovskite, and the synthesis proposed here is quite simple and capable of large-scale production in open air condition which is one of advantageous for future implementation of memristor technology. Further, the switching is observed with low voltages with switch ON at 1.00 V and switch OFF at 1.58 V, with a better stability of about $>10^4$ s.

Some biological molecules are also used in combination with quantum dots to explore their resistive switching properties [221, 222]. Bio-organic molecules such as DNA, RNA, silk fibroin, aloe vera, viruses, etc. have been considered for the implementation of memory applications. We have been working in the field for the last few years, and recently, a new material for non-volatile memory devices based on DNA-PbS nanoparticles has been introduced by our group [223].

A simple electro grafting technique has been developed to deposit large and uniform DNA:PbS films on conducting substrates. The memory switching properties have been studied by fabricating the devices structure *ITO/DNA:PbS/Metal* junction. Electrical bistability is observed in these devices with ON/OFF current ratio over$\sim 10^4$ times. The non-volatile memory device fabricated in this work is quite stable which is illustrated in Fig. 9.35.

The switching properties of these devices are attributed to the strong electrostatic binding between PbS nanoparticles and DNA strands which makes the possible doping of Pb ions/atoms into DNA molecule. Our memristor work is also extended to silk fibroin in combination with semiconductor QDs; the former is an interesting protein molecule and attracted a considerable attention due to its potential use in various electronic devices [224]. Here, silkworm protein has been integrated with CdSe quantum dots to form bilayer memory switching devices. The devices are

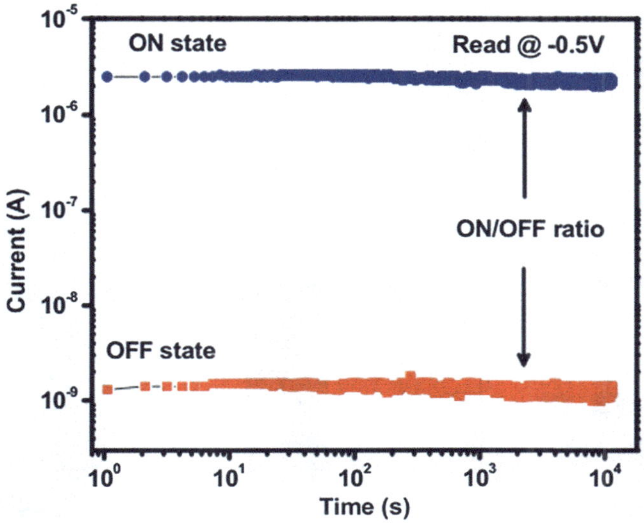

Fig. 9.35 Retention characteristics of ITO/DNA:PbS QD/metal junction memristor. (Reprinted with copyright permission from [Ref. 223])

made by spin coating of individual films on ITO surface to get *ITO/silk fibroin:CdSe QD/metal* structures. All the synthesis techniques adopted for device fabrication is quite simple, and this has been depicted in Fig. 9.36.

The fabricated devices exhibited a clear multilevel electrical switching in both positive and negative voltages, and switching is quite symmetric in nature. The devices possess a good reproducibility and endurance and can be very important for future multistate memory storage applications. These characteristics are well explained on the basis of energy band diagram of these heterointerfaces (Fig. 9.37).

Similarly, Chun et al. have introduced Ni-DNA system as a programmable multistate resistive memory device [225]. It is observed that in the complex host structure of DNA, the conducting Ni ion chain can be made which gives interesting switching effects. The devices were demonstrated successfully for writing, reading, and erasing the information. Lee et al. have worked on RNA, which is another interesting biomolecule [226]. RNA-quantum dot-based memory devices have been designed which exhibit electrical bistability, and electrical switching effects were tested for their reproducibility over time, and it is found that the properties were well preserved for over 13 days. Similarly, silk fibroin is also used along with gold nanoclusters for the memory switching by Xing et al. [227]. To our surprise, virus is also introduced into memory devices by Portney et al. [228]. Virus-templated CdSe/ZnS semiconducting quantum dots were developed for the fabrication of these devices. Further, natural aloe vera has been used for memory switching by Lim et al. [229]. Device has been fabricated with a configuration of *metal/aloe vera/metal* with Al and ITO as electrodes on either side. The memory device has been investigated, and it is observed that it has a large ON/OFF ratio of about 10^3 and a

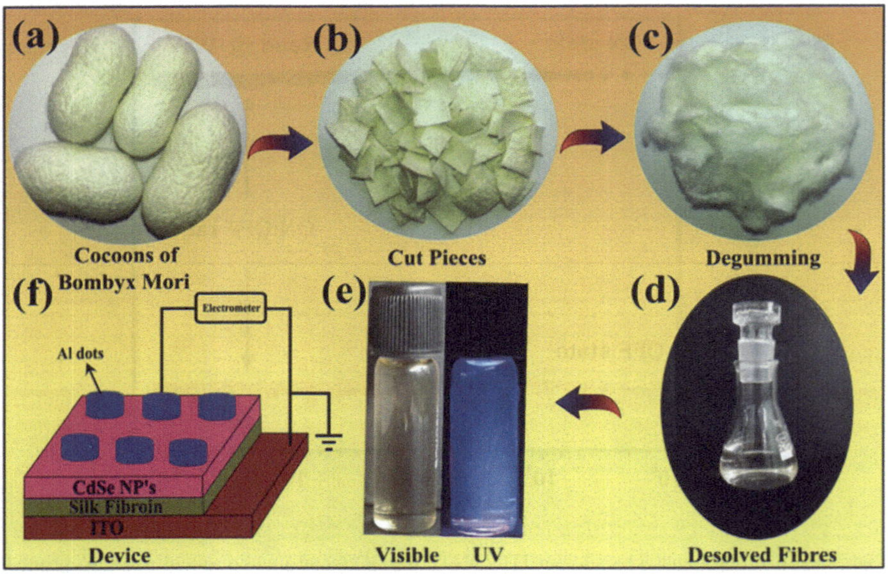

Fig. 9.36 The flowchart describing the (**a**–**f**) preparation of silk fibroin solution from Bombyx Mori silk cocoons for the fabrication of memristors. (Reprinted with copyright permission from [Ref. 224])

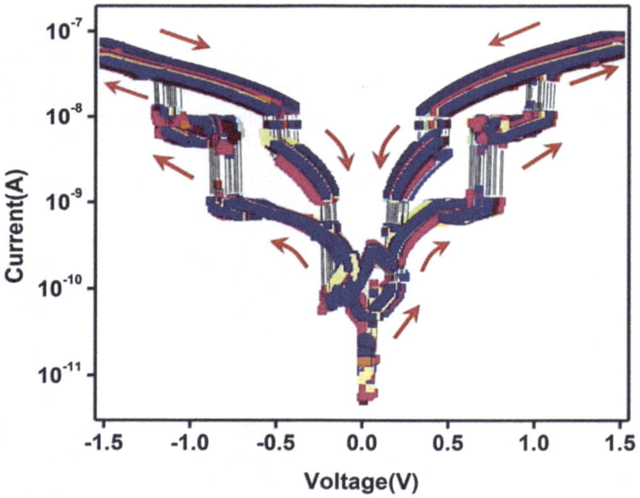

Fig. 9.37 Multilevel memory switching ITO/SF-CdSe QD/Metal junction memristor (data for 50 repeated cycles is shown). (Reprinted with copyright permission from [Ref. 224])

good retention time of more than 10^4 s. So, looking at all these reports and results obtained by the organic/biomolecule in combination with quantum dot-based memory devices, it is possible that they can establish a promising future in memory technology while giving a green touch to the future technology.

9.5 Future Scope and Conclusions

During the past decades, semiconductor device technology has continuously improved through active developments and innovative achievements. With the rapid development of electronic industries, information technology-related devices have become an essential part of our daily life where the memory technology has very important role. In this chapter, we discussed the basics of resistive switching mechanism and several effects involved in it. Subsequently, applications of quantum dots/nanomaterials for the memory switching devices have been presented for different classes of materials such as metal nanoparticles, metal oxides, metal chalcogenides, carbon/graphene composites, quantum dots/organic composites, etc. Along with that, synthesis of these nanostructured materials and fabrication of devices have been discussed in detail.

Recent studies have clearly shown the promising nature of these nanostructures in the field of resistive memory switching. Several novel nanomaterials have made considerable progress, and they can be new building blocks for future applications. In many works, the quantum dots have been introduced in a bulk host compound for enhancing their performances. By introducing these quantum dots, the fundamental parameters such as endurance, switching time, retention, power consumption, operation voltage, etc. have been successfully improved. However, the challenges at the materials level still need to be addressed the successful use of nanomaterials into the technology and for large-scale applications. More and detailed studies are important to realize the possibilities of commercializing the quantum dot-based memory devices.

In conclusion, the quantum dots are excellent choice for resistive switching applications which offer a number of advantages. It is true that the barriers to device scaling have always been very high, but a rigorous research work can overcome these difficulties. It is also very important to develop a proper technology that provides a clear method on how to implement the nanostructured material memory devices. Looking into the future, challenges still need to overcome to bring these nanostructures or quantum dot-based memory switching devices to practical use.

References

1. Yu, S., & Chen, P.-Y. (2016). Emerging memory technology recent trends and prospects. *IEEE Solid-State Circuits Magazine, 8*, 43.
2. Zidan, M. A., Strachan, J. P., & Lu, W. D. (2018). The future of electronics based on memristive systems. *Nature Electronics, 1*(1), 22–29.

3. https://www.marketwatch.com/press-release/in-memory-database-market-2019-size-share-trends-growth-global-analysis-with-focus-on-opportunities-and-forecast-by-2023-2019-02-08.
4. Chen, A. (2016). A review of emerging non-volatile memory (NVM) technologies and applications. *Solid-State Electronics, 125*, 25–38.
5. Wong, H. S., & Salahuddin, S. (2015). Memory leads the way to better computing. *Nature Nanotechnology, 10*(3), 191–194.
6. Jian-Gang, Z. (2008). Magnetoresistive random access memory: The path to competitiveness and scalability. *Proceedings of the IEEE, 96*(11), 1786–1798.
7. Jo, S. H., Kim, K.-H., & Lu, W. (2009). High-density crossbar arrays based on a Si memristive system. *ACS Nano Letters, 9*(2), 870–874.
8. Zhao, X., Fan, Z., Xu, H., Wang, Z., Xu, J., Ma, J., & Liu, Y. (2018). Reversible alternation between bipolar and unipolar resistive switching in Ag/MoS2/Au structure for multilevel flexible memory. *Journal of Material Chemistry C, 6*, 7195–7200. https://doi.org/10.1039/C8TC01844H.
9. Chua, L. O. (1971). Memristor-the missing circuit element. *IEEE Transactions on Circuit Theory, 18*, 507–519.
10. Hickmott, T. W. (1962). Low-frequency negative resistance in thin anodic oxide films. *Journal of Applied Physics, 33*, 2669–2682.
11. Gibbons, J. F., & Beadle, W. E. (1964). Switching properties of thin Nio films. *Solid-State Electronics, 7*, 785–790.
12. Argall, F. (1968). Switching phenomena in titanium oxide thin films. *Solid-State Electronics, 11*, 535–541.
13. Thakoor, S., Moopenn, A., Daud, T., & Thakoor, A. P. (1990). Solid-state thin-film memistor for electronic neural networks. *Journal of Applied Physics, 67*, 3132.
14. Buot, F. A., & Rajagopal, A. K. (1994). Binary information storage at zero bias in quantum-well diodes. *Journal of Applied Physics, 76*, 5552.
15. Beck, A., Bednorz, J. G., Gerber, C., Rossel, C., & Widmer, D. (2000). Reproducible switching effect in thin oxide films for memory applications. *Applied Physics Letters, 77*, 139.
16. Strukov, D. B., Snider, G. S., Stewart, D. R., & Williams, R. S. (2008). The missing memristor found. *Nature, 453*, 80–83.
17. Lim, E., & Ismail, R. (2015). Conduction mechanism of valence change resistive switching memory: A survey. *Electronics, 4*(3), 586–613.
18. Jeong, D. S., Thomas, R., Katiyar, R. S., Scott, J. F., Kohlstedt, H., Petraru, A., & Hwang, C. S. (2012). Emerging memories: resistive switching mechanisms and current status. *Reports on Progress in Physics, 75*(7), 076502.
19. Mickel, P. R., Lohn, A. J., & Marinella, M. J. (2014). Memristive switching: physical mechanisms and applications. *Modern Physics Letters B, 28*(10), 1430003.
20. Waser, R., & Aono, M. (2007). Nanoionics-based resistive switching memories. *Natuture Materials, 6*, 833–840.
21. Menzel, S., Böttger, U., Wimmer, M., & Salinga, M. (2015). Physics of the switching kinetics in resistive memories. *Advanced Functional Materials, 25*(40), 6306–6325.
22. Zhuravlev, M. Y., Jaswal, S. S., Tsymbal, E. Y., & Sabirianov, R. F. (2005). Ferroelectric switch for spin injection. *Applied Physics Letters, 87*(22).
23. Dobrosavljevi, V. (2011). *Introduction to metal-insulator transitions*. Oxford University Press. https://doi.org/10.1093/acprof:oso/9780199592593.003.0001.
24. Simmons, J. G., & Verderber, R. R. (1967). New conduction and reversible memory phenomena in thin insulating films. *Proceedings of the Royal Society of London. Series A. Mathematical and Physical Sciences, 301*(1464), 77–102.
25. Thurstans, R. E., Wild, P. C., & Oxley, D. P. (1974). Enhanced forming in *Al/SiOx/Au* structures under pulsed bias. *Thin Solid Films, 20*, 281–286.
26. Fujii, T., & Kawasaki, M. (2005). Hysteretic current–voltage characteristics and resistance switching at an epitaxial oxide Schottky junction SrRuO3/SrTi0.99Nb0.01O3. *Applied Physics Letters, 86*(1).

27. Sawa, A., Fujii, T., Kawasaki, M., & Tokura, Y. (2006). Interface resistance switching at a few nanometer thick perovskite manganite active layers. *Applied Physics Letters, 88*(23), 232112.
28. Hoyos, J. A., Kotabage, C., & Vojta, T. (2007). Effects of dissipation on a quantum critical point with disorder. *Physical Review Letters, 99*(23), 230601.
29. Ahn, C. H., Bhattacharya, A., Ventra, M. D., et al. (2006). Electrostatic modification of Novel materials. *Review of Modern Physics, 78*(4), 1185–1212.
30. Mott, N. F. (1949). The basis of the electron theory of metals, with special reference to the transition metals. *Proceedings of the Physical Society A, 62.*
31. Kim, D. S., Kim, Y. H., Lee, C. E., & Kim, Y. T. (2006). Colossal electroresistance mechanism in Au/Pr0.7Ca0.3MnO3/Ptsandwich structure: Evidence for a Mott transition. *Physical Review B, 74*(17).
32. Fors, R., Khartsev, S. I., & Grishin, A. M. (2005). Giant resistance switching in metal-insulator-manganite junctions: Evidence for Mott transition. *Physical Review B, 71*(4).
33. Chen, A. (2015). Electronic effect resistive switching memories. In *Emerging nanoelectronic devices* (pp. 164–180).
34. Tsymbal, E. Y., & Kohlstedt, H. (2006). Applied physics. Tunneling across a ferroelectric. *Science, 313*(5784), 181–183.
35. Blom, P. W. M., Wolf, R. M., Cillessen, J. F. M., & Krijn, M. P. C. M. (1994). Ferroelectric Schottky diode. *Physical Review Letters, 73*(15), 2107–2110.
36. Qu, H., Yao, W., Garcia, T., & Zhang, J. (2003). Nanoscale polarization manipulation and conductance switching in ultrathin films of a ferroelectric copolymer. *Applied Physics Letters, 82*, 4322–4324.
37. Kohlstedt, H., Pertsev, N. A., Rodríguez Contreras, J., & Waser, R. (2005). Theoretical current-voltage characteristics of ferroelectric tunnel junctions. *Physical Review B, 72*, 125341.
38. Velev, J. P., Duan, C.-G., Burton, J. D., Smogunov, A., Niranjan, M. K., Tosatti, E., Jaswal, S. S., & Tsymbal, E. Y. (2009). Magnetic tunnel junctions with ferroelectric barriers: prediction of four resistance States from first principles. *Nano Lett, 9*(1), 427–432.
39. Scott, J. F. (2007). Applications of modern ferroelectrics source. *Science, 315*, 954–959.
40. Batra, I. P., & Silverman, B. D. (1972). Thermodynamic stability of thin ferroelectric films. *Solid State Communications, 11*, 291–294.
41. Hou, P., Wang, J., & Zhong, X. (2017). Investigation of multilevel data storage in silicon-based polycrystalline ferroelectric tunnel junction. *Scientific Reports, 7*(1), 4525.
42. Guo, X., Schindler, C., Menzel, S., & Waser, R. (2007). Understanding the switching-off mechanism in Ag+ migration based resistively switching model systems. *Applied Physics Letters, 91*(13).
43. Menzel, S., Waters, M., Marchewka, A., Böttger, U., Dittmann, R., & Waser, R. (2011). Origin of the ultra-nonlinear switching kinetics in oxide-based resistive switches. *Advanced Functional Materials, 21*(23), 4487–4492.
44. Buroff, A., Nebauer, E., Suptiz, P., & Willert, I. (1977). Migration of silver and gold in Amorphous As_2S_3. *Physica status Solidi.* a(40), K195.
45. Clymer, D. A., & Matin, M. A. (2005). Application of Mott-Gurney law to model the current-voltage relationship of PPV/CN-PPV with a thin-metal anode buffer. In *Photonic Devices and Algorithms for Computing VII.*
46. Liu, C.-Y., Sung, P.-W., Lai, C.-H., & Wang, H.-Y. (2011). Resistive Switching Characteristics of *Cu/SiO$_2$/Pt* Structure. *Materials Science Forum, 687*, 167–173.
47. Waser, R., Dittmann, R., Staikov, G., & Szot, K. (2009). Redox-based resistive switching memories - Nanoionic mechanisms, prospects, and challenges. *Advanced Materials, 21*(25-26), 2632–2663.
48. Schindler, C., Weides, M., Kozicki, M. N., & Waser, R. (2008). Low current resistive switching in Cu–SiO_2 cells. *Applied Physics Letters, 92*(12).
49. Ee, W. L., & Ismail, R. (2015). Conduction mechanism of valence change resistive switching memory: A survey. *Electronics, 4*, 586–613.

50. Sun, P., Li, L., Lu, N., Lv, H., Liu, M., & Liu, S. (2014). Physical model for electroforming process in valence change resistive random access memory. *Journal of Computational Electronics, 14*(1), 146–150.
51. You, S., Liu, B., Gao, Y., Wang, Y., Tang, C. Y., Huang, Y., & Ren, N. (2016). Monolithic porous Magnéli-phase Ti_4O_7 for electro-oxidation treatment of industrial wastewater. *Electrochimica Acta, 214*, 326–335.
52. Akinaga, H., & Shima, H. (2010). Resistive random access memory (ReRAM) based on metal oxides. *Proceedings of the IEEE, 98*(12), 2237–2251.
53. Munjal, S., & Khare, N. (2017). Valence change bipolar resistive switching accompanied with magnetization switching in $CoFe_2O4$ thin film. *Scientific Reports, 7*(1), 12427.
54. Lee, D., Woo, J., Cha, E., Park, S., Lee, S., Park, J., & Hwang, H. (2013). Defect engineering using bilayer structure in filament-type RRAM. *IEEE Electron Device Letters, 34*(10), 1250–1252.
55. Phark, S.-H., & Chae, S. C. (2015). Initial defect configuration in NiO film for reliable unipolar resistance switching of Pt/NiO/Pt structure. *Journal of Physics D: Applied Physics, 48*, 155102.
56. Lee, M.-J., Han, S., Jeon, S. H., Park, B. H., Kang, B. S., Ahn, S.-E., Kim, K. H., Lee, C. B., Kim, C. J., Yoo, I.-K., Seo, D. H., Li, X.-S., Park, J.-B., Lee, J.-H., & Park, Y. (2009). Electrical manipulation of nanofilaments in transition-metal oxides for resistance-based memory. *Nano Letters, 9*(4), 1476–1481.
57. Kondo, H., Arita, M., Fujii, T., Kaji, H., Moniwa, M., Yamaguchi, T., Fujiwara, I., Yoshimaru, M., & Takahashi, Y. (2011). The observation of "Conduction Spot" on NiO resistance random access memory. *Japanese Journal of Applied Physics, 50*(8), 081101.
58. Larentis, S., Nardi, F., Balatti, S., Member, S., Gilmer, D. C., & Ielmini, D. (2012). Resistive switching by voltage-driven ion migration in bipolar RRAM—Part II: modeling. *IEEE Transactions on Electron Devices, 59*(9), 2468–2475.
59. Jeong, D. S., Schroeder, H., & Waser, R. (2007). Coexistence of bipolar and unipolar resistive switching behaviors in a Pt/TiO_2/Pt stack. *Electrochemical and Solid-State Letters, 10*(8), G51–G53.
60. Uenuma, M., Ishikawa, Y., & Uraoka, Y. (2015). Joule heating effect in nonpolar and bipolar resistive random access memory. *Applied Physics Letters, 107*(7).
61. Caldwell, M. A., Jeyasingh, R. G. D., Wong, H.-S. P., & Milliron, D. J. (2012). Nanoscale phase change memory materials. *Nanoscale, 4*(15), 4382–4392.
62. Shao, R., Zheng, K., Chen, Y., Zhang, B., Deng, Q., Jiao, L., Liao, Z., Zhang, Z., Zou, J., & Han, X. (2016). Direct observation of structural transitions in the phase change material $Ge_2Sb_2Te_5$. *Journal of Materials Chemistry C, 4*(39), 9303–9309.
63. Ovshinsky, S. R. (1968). Reversible electrical switching phenomena in disordered structures. *Physical Review Letters, 21*(20), 1450–1453.
64. Lyeo, H.-K., Cahill, D. G., Lee, B.-S., Abelson, J. R., Kwon, M.-H., Kim, K.-B., Bishop, S. G., & Cheong, B.-K. (2006). Thermal conductivity of phase-change material $Ge_2Sb_2Te_5$. *Applied Physics Letters, 89*(15).
65. Stocker, H. J. (1970). Phenomenology of switching and memoryeffects in semiconducting chalcogenide glasses. *Journal of Non-Crystalline Solids, 2*, 371–381.
66. Matsunaga, T., & Yamada, N. (2004). Structural investigation of $GeSb_2Te_4$: A high-speed phase-change material. *Physical Review B, 69*(10), 104111.
67. Salinga, M., Kersting, B., Ronneberger, I., Jonnalagadda, V. P., Vu, X. T., Le Gallo, M., Giannopoulos, I., Cojocaru-Mirédin, O., Mazzarello, R., & Sebastian, A. (2018). Monatomic phase change memory. *Nature Materials, 17*(8), 681–685.
68. Raoux, S., Cheng, H.-Y., Jordan-Sweet, J. L., Muñoz, B., & Hitzbleck, M. (2009). Influence of interfaces and doping on the crystallization temperature of Ge–Sb. *Applied Physics Letters, 94*(18).
69. Tsymbal, E. Y., Gruverman, A., Garcia, V., Bibes, M., & Barthélémy, A. (2012). Ferroelectric and multiferroic tunnel junctions. *MRS Bulletin, 37*, 138–143.

70. Stengel, M., Vanderbilt, D., & Spaldin, N. A. (2009). Enhancement of ferroelectricity at metal-oxide interfaces. *Nature Materials, 8*(5), 392–397.
71. Garcia, V., & Bibes, M. (2014). Ferroelectric tunnel junctions for information storage and processing. *Nature Communications, 5*, 4289.
72. Wen, Z., Li, C., Wu, D., Li, A., & Ming, N. (2013). Ferroelectric-field-effect-enhanced electroresistance in metal/ferroelectric/semiconductor tunnel junctions. *Nature Materials, 12*(7), 617–621.
73. Pantel, D., & Alexe, M. (2010). Electroresistance effects in ferroelectric tunnel barriers. *Physical Review B, 82*(13), 134105.
74. Boyn, S., Girod, S., Garcia, V., Fusil, S., Xavier, S., Deranlot, C., Yamada, H., Carrétéro, C., Jacquet, E., Bibes, M., Barthélémy, A., & Grollier, J. (2014). High-performance ferroelectric memory based on fully patterned tunnel junctions. *Applied Physics Letters, 104*, 052909.
75. Thomson, W. (1856). On the electro-dynamic qualities of metals:—Effects of magnetization on the electric conductivity of nickel and of iron. *Proceedings of the Royal Society of London. Series A. Mathematical and Physical Sciences, 8*, 546–550.
76. Nickel, J. (1995). Magnetoresistance, SAL, dual stripe, spin valve. In *GMR magnetoresistance overview* (pp. 95–60).
77. Tsang, C., Chen, M.-M., Yogi, T., & Ju, K. (1990). Gigabit density recording using dual-element MR/inductive heads on thin-film disks. *IEEE Transactions on Magnetics, 26*, 1689–1693.
78. Baibich, M. N., Broto, J. M., Fert, A., Nguyen Van Dau, F., & Petroff, F. (1988). Giant magnetoresistance of (001) Fe/(001)Cr magnetic superlattices. *Physics Review Letters, 61*(21), 2472–2475.
79. Bhatti, S., Sbiaa, R., Hirohata, A., Ohno, H., Fukami, S., & Piramanayagam, S. N. (2017). Spintronics based random access memory: a review. *Materials Today, 20*, 530–548.
80. Wang, D., Anderson, J., & Daughton, J. M. (1997). Thermally stable, low saturation field, low hysteresis, high GMR *CoFe/Cu* multilayers. *IEEE Transactions on Magnetics, 33*, 3520–3522.
81. Kowalska, E. (2018). *Current-induced dynamics in hybrid geometry MgO-based spin-torque nano-oscillators* (Dissertation, Technische Universitat Dresden).
82. Julliere, M. (1975). Tunnelling between ferromagnetic films. *Physics Letters, 54A*, 225–226.
83. Wang, D., Nordman, C., Daughton, J. M., Qian, Z., & Fink, J. (2004). 70% TMR at room temperature for SDT sandwich junctions with CoFeB as free and reference layers. *IEEE Transactions on Magnetics, 40*(4), 2269–2271.
84. Djayaprawira, D. D., Tsunekawa, K., Nagai, M., Maehara, H., Yamagata, S., & Watanabe, N. (2005). 230% room-temperature magnetoresistance in *CoFeB/MgO/CoFeB* magnetic tunnel junctions. *Applied Physics Letters, 86*(9).
85. Akerman, J. (2005). Toward a universal memory. *Science, 308*, 508–510.
86. Slaughter, J. M., Rizzo, N. D., Janesky, J., Whig, R., Mancoff, F. B., Houssameddine, D., Sun, J. J., Aggarwal, S., Nagel, K., Deshpande, S., Alam, S. M., Andre, T., & LoPresti, P. (2012). High density ST-MRAM technology. *IEEE.*
87. Amiri, P. K., Wang, K. L., & Galatsis, K. (2013). *Voltage-controlled magnetic anisotropy (VCMA) switch and magneto-electric memory (MERAM).* U.S. patent application Ser., No. 14/082,118 filed.
88. Valenzuela, S. O., & Tinkham, M. (2006). Direct electronic measurement of the spin Hall effect. *Nature, 442*(7099), 176–179.
89. Chung, S. W., Kishi, T., Park, J. W., Yoshikawa, M., Park, K. S., Nagase, T., Sunouchi, K., Kanaya, H., Kim, G. C., Lee, M. S., Noma, K., Yamamoto, A., Rho, K. M., Tsuchida, K., Chung, S. J., Yi, J. Y., Kim, H. S., Chun, Y. S., Oyamatsu, H., & Hong, S. J. (2016). 4Gbit density STT-MRAM using perpendicular MTJrealized with compact cell structure. *IEEE.*
90. Babbage's Analytical Engine, 1834–1871. (Trial model). Science Museum. Retrieved 23 August 2017.
91. Badzey, R. L., Zolfagharkhani, G., Gaidarzhy, A., & Mohanty, P. (2004). A controllable nanomechanical memory element. *Applied Physics Letters, 85*(16), 3587–3589.

92. Czaplewski, D. A., Patrizi, G. A., Kraus, G. M., Wendt, J. R., Nordquist, C. D., Wolfley, S. L., Baker, M. S., & de Boer, M. P. (2009). A nanomechanical switch for integration with CMOS logic. *Journal of Micromechanics and Microengineering, 19*(8), 085003.
93. Pantazi, A., et al. (2008). Probe-based ultrahigh-density storage technology. *IBM Journal of Research and Development, 52*(4/5), 493–511.
94. Rueckes, T., Kim, K., Joselevich, E., Tseng, G. Y., Cheung, C.-L., & Lieber, C. M. (2000). Carbon nanotube-based nonvolatile random access memory for molecular computing. *Science, 289*, 94–97.
95. Feng, X. L., Matheny, M. H., Zorman, C. A., Mehregany, M., & Roukes, M. L. (2010). Low voltage nanoelectromechanical switches based on silicon carbide nanowires. *Nano Letters, 10*(8), 2891–2896.
96. Lefèvre, R., Goffman, M. F., Derycke, V., Miko, C., Forro, L., Bourgoin, J. P., & Hesto, P. (2005). Scaling law in carbon nanotube electromechanical devices. *Physics Review Letters, 95*(18), 185504.
97. Hiremath, R. K., Rabinal, M. H. K., & Mulimani, B. G. (2010). Dipole tuning of charge transport in molecular junctions. *Physical Chemistry Chemical Physics, 12*(11), 2564–2568.
98. Avouris, P. (2002). Molecular electronics with carbon nanotubes. *Accounts of Chemical Research, 35*(12), 1026–1034.
99. Kim, W. Y., Choi, Y. C., Min, S. K., Cho, Y., & Kim, K. S. (2009). Application of quantum chemistry to nanotechnology: Electron and spin transport in molecular devices. *Chem Soc Rev, 38*(8), 2319–2333.
100. Tao, N. J. (2006). Electron transport in molecular junctions. *Nature Nanotechnology, 1*, 173–181.
101. Sun, L., Diaz-Fernandez, Y. A., Gschneidtner, T. A., Westerlund, F., Lara-Avilab, S., & Moth-Poulsen, K. (2014). Single-molecule electronics: From chemical design to functional devices. *Chemical Society Review, 43*(21), 7378–7411.
102. Hsu, J. W. P., Loo, Y. L., Lang, D. V., & Rogers, J. A. (2003). Nature of electrical contacts in a metal–molecule–semiconductor system. *Journal of Vacuum Science & Technology B: Microelectronics and Nanometer Structures, 21*(4), 1928–1935.
103. Wold, D. J., & Frisbie, C. D. (2000). Formation of metal−molecule−metal tunnel junctions: Microcontacts to alkanethiol monolayers with a conducting AFM tip. *Journal of the American Chemical Society, 122*(12), 2970–2971.
104. Zhang, J. L., Zhong, J. Q., Lin, J. D., Hu, W. P., Wu, K., Xu, G. Q., Weeb, A. T. S., & Chen, W. (2015). Towards single molecule switches. *Chemical Society Review, 44*(10), 2998–3022.
105. Xue, Y., & Datta, S. (1999). Negative differential resistance in the scanning-tunneling spectroscopy of organic molecules. *Physical Review B, 59*, 7852–7855.
106. Nitzan, A. (2001). A relationship between electron-transfer rates and molecular conduction. *The Journal of Physical Chemistry A, 105*(12), 2677–2679.
107. Moore, A. M., Yeganeh, S., Yao, Y., Claridge, S. A., Tour, J. M., Ratner, M. A., & Weiss, P. S. (2010). Polarizabilities of adsorbed and assembled molecules: measuring the conductance through buried contacts. *ACS Nano, 4*(12), 7630–7636.
108. Vahl, A., Carstensen, J., Kaps, S., Lupan, O., Strunskus, T., Adelung, R., & Faupel, F. (2019). Concept and modelling of memsensors as two terminal devices with enhanced capabilities in neuromorphic engineering. *Nature Scientific Reports, 9*, 4361.
109. Kulkarni, S. (2007). *Nanotechnology, principles and practices* (1st ed.). New Delhi: Capital Publishing Company. Retrieved from https://www.springer.com/gp/book/9783319091709.
110. Han, S.-T., Hu, L., Wang, X., Zhou, Y., Zeng, Y.-J., Ruan, S., Pan, C., & Peng, Z. (2017). Black phosphorus quantum dots with tunable memory properties and multilevel resistive switching characteristics. *Advanced Science, 4*(8), 1600435.
111. Marta, R. Quantum dots: Harnessing the nanoscopic rainbow, chem13 news magazine. Retrieved from https://uwaterloo.ca/chem13-news-magazine/november017/feature/quantum-dots-harnessing-nanoscopic-rainbow. University of Waterloo, Waterloo, Ontario.
112. Seung-Yun, T., Seng, T. & Simon, M. (2012). *Non-volatile memory devices and application*. ISBN: 1-58883-250-3.

113. Ganeshan, D., Xie, F., Sun, Q., Li, Y., & Wei, M. (2018). Plasmonic effects of silver nanoparticles embedded in the counter electrode on the enhanced performance of dye-sensitized solar cells. *Langmuir, 34*(19), 5367–5373.
114. Chen, H., Shao, L., Lia, Q., & Wang, J. (2013). Gold nanorods and their plasmonic properties. *Chemical Society Reviews, 42*(7), 2679–2724.
115. Shahbazyan, T. V. (2014, 22 January). (challenges)Plasmonics theory and advances in computational chemistry and physics. In *Plasmonic: Theory & applications*.
116. Zhong, Y., Malagari, S. D., Hamilton, T., & Wasserman, D. (2015). Review of mid-infrared plasmonic materials. *Journal of Nanophotonics, 9*, 093791.
117. Green, M. (2010). The nature of quantum dot capping ligands. *Journal of Materials Chemistry, 20*, 5797–5809.
118. Diwan, B. D. & Murugan, S. (2013). Role of size on effective band gap in silicon nano-solid. *Proceedings of ICANMEET*.
119. Campisi, S., Schiavoni, M., Chan-Thaw, C. E., & Villa, A. (2016). Untangling the role of the capping agent in nanocatalysis: Recent advances and perspectives. *Catalysts, 6*(12), 185–206.
120. Vempati, S., Ertas, Y., & Uyar, T. (2013). Sensitive surface states and their passivation mechanism in CdS quantum dots. *The Journal of Physical Chemistry C, 117*(41), 21609–21618.
121. Schoenhalz, A. L., Arantes, J. T., Fazzio, A., & Dalpian, G. M. (2010). Surface and quantum confinement effects in ZnO nanocrystals. *The Journal of Physical Chemistry C, 114*(43), 18293–18297.
122. Rowland, C. E., Brown, C. W., III, Delehanty, J. B., & Medintz, I. L. (2016). Nanomaterial-based sensors for the detection of biological threat agents. *Materials Today, 19*(8), 464–477.
123. Thanh, N. T. K., Maclean, N., & Mahiddine, S. (2014). Mechanisms of nucleation and growth of nanoparticles in solution. *Chemical Reviews, 114*, 7610–7630.
124. Rossmanith, R., Weiss, C. K., Geserick, J., Hüsing, N., Hörmann, U., Kaiser, U., & Landfester, K. (2008). Porous anatase nanoparticles with high specific surface area prepared by miniemulsion technique. *Chemistry of Materials, 20*(18), 5768–5780.
125. Vogel, N., Fischer, J., Mohammadi, R., Retsch, M., Butt, H.-J., Landfester, K., Weiss, C. K., & Kreiter, M. (2011). Plasmon hybridization in stacked double crescents arrays fabricated by colloidal lithography. *Nano Letters, 11*(2), 446–454.
126. Knowles, K. E., Frederick, M. T., Tice, D. B., Morris-Cohen, A. J., & Weiss, E. A. (2011). Colloidal quantum dots: Think outside the (particle-in-a-)box. *The Journal of Physical Chemistry Letters, 3*(1), 18–26.
127. Ko, Y., Baek, H., Kim, Y., Yoon, M., & Cho, J. (2013). Hydrophobic nanoparticle-based nanocomposite films using in situ ligand exchange layer-by-layer assembly and their nonvolatile memory applications. *ACS Nano, 7*(1), 143–153.
128. Balocco, C., Song, A. M., & Missous, M. (2004). Room-temperature operations of memory devices based on self-assembled InAs quantum dot structures. *Applied Physics Letters, 85*(24), 5911–5913.
129. Geller, M., Marent, A., Nowozin, T., Bimberg, D., & Akçay, N. (2008). A write time of 6ns for quantum dot–based memory structures. *Applied Physics Letters, 92*(9), 092108.
130. Lee, J.-S., Kim, Y.-M., Kwon, J.-H., Shin, H., Sohn, B.-H., & Lee, J. (2009). Tunable memory characteristics of nanostructured, nonvolatile charge trap memory devices based on a binary mixture of metal nanoparticles as a charge trapping layer. *Advanced Materials, 21*(2), 178–183.
131. Jung, S. M., Kim, H.-J., Kim, B.-J., Kim, Y.-S., Yoon, T.-S., & Lee, H. H. (2010). Electrical charging of Au nanoparticles embedded by streptavidin-biotin biomolecular binding in organic memory device. *Applied Physics Letters, 97*(15), 153302.
132. Tiwari, S., Rana, F., Hanafi, H., Hartstein, A., Crabbe, E. F., & Chan, K. (1996). A silicon nanocrystals based memory. *Applied Physics Letters, 68*, 1377–1379.
133. Takahashi, N., Ishikuro, H., & Hiramoto, T. (2000). Control of Coulomb blockade oscillations in silicon single electron transistors using silicon nanocrystal floating gates. *Applied Physics Letters, 76*(2), 209–211.

134. Zhang, T., Guerin, D., Alibart, F., Vuillaume, D., Lmimouni, K., Lenfant, S., Yassin, A., Ocafrain, M., Blanchard, P., & Roncali, J. (2017). Negative differential resistance, memory, and reconfigurable logic functions based on monolayer devices derived from gold nanoparticles functionalized with electropolymerizable TEDOT units. *The Journal of Physical Chemistry C, 121*, 10131–10139.
135. Ouyang, J., & Yang, Y. (2010). Polymer: Metal nanoparticle devices with electrode-sensitive bipolar resistive switchings and their application as nonvolatile memory devices. *Applied Physics Letters, 96*, 063506.
136. Haiwei, D., Wan, T., Qu, B., Cao, F., Lin, Q., Chen, N., Lin, X., & Chu, D. (2017). Engineering silver nanowire networks: From transparent electrodes to resistive switching devices. *ACS Applied Materials & Interfaces, 9*, 20762–20770.
137. White, S. I., Vora, P. M., Kikkawa, J. M., & Winey, K. I. (2010). Resistive switching in bulk silver nanowire-polystyrene composites. *Advanced Functional Materials, 21*, 233–240.
138. Padma, N., Betty, C. A., Samanta, S., & Nigam, A. (2017). Tunable switching characteristics of low operating voltage organic bistable memory devices based on gold nanoparticles and copper phthalocyanine thin films. *The Journal of Physical Chemistry C, 121*, 5768–5778.
139. Hwang, B., & Lee, J.-S. (2018). Recent advances in memory devices with hybrid materials. *Advanced Electronic Materials, 5*, 1800519.
140. Bozano, L. D., Kean, B. W., Beinhoff, M., Center, K. R., Rice, P. M., & Scott, J. C. (2005). Organic materials and thin-film structures for cross-point memory cells based on trapping in metallic nanoparticles. *Advanced Functional Materials, 15*, 1933–1939.
141. Sawa, A. (2008). Resistive switching in transition metal oxides. *Materials Today, 11*, 28–39.
142. Liu, X. J., Li, X. M., Yu, W. D., Wang, Q., Yang, R., Cao, X., & Chen, L. D. (2009). Bipolar resistance switching property of Al-Ag/$La_{0.3}Ca_{0.7}MnO_3$/Pt sandwitches. *Journal of Ceramic Society of Japan, 117*(6), 732.
143. Fujimoto, M., Koyama, H., Kobayashi, S., Tamai, Y., Awaya, N., Nishi, Y., & Suzuki, T. (2005). Resistivity and resistive switching properties of Pr 0.7 Ca $_{0.3}$ MnO_3 thin films. *Journal of Applied Physics, 97*, 10H709.
144. Liu, C.-Y., & Tseng, T.-Y. (2007). Resistance switching properties of sol–gel derived SrZrO3 based memory thin films. *Journal of Physics D: Applied Physics, 40*, 2157.
145. Seo, S., Lee, M. J., Seo, D. H., Jeoung, E. J., Suh, D.-S., Joung, Y. S., & Yoo, I. K. (2004). Reproducible resistance switching in polycrystalline NiO films. *Applied Physics Letters, 85*, 5655–5657.
146. Lee, M.-J., Ahn, S.-E., Lee, C. B., Kim, C.-J., Jeon, S., Chung, U.-I., Yoo, I.-K., Park, G.-S., Han, S., Hwang, I. R., & Park, B.-H. (2011). A simple device unit consisting of All NiO storage and switch elements for multilevel terabit nonvolatile random access memory. *ACS Applied Materials & Interfaces, 3*, 4475–4479.
147. He, L., Liao, Z.-M., Wu, H.-C., Tian, X.-X., Xu, D.-S., Cross, G. L. W., Duesberg, G. S., Shvets, I. V., & Yu, D.-P. (2011). Memory and threshold resistance switching in Ni/NiO core–shell nanowires. *Nano Letters, 11*, 4601–4606.
148. Yang, Y., Zhang, X., Gao, M., Zeng, F., Zhou, W., Xie, S., & Pan, F. (2011). Nonvolatile resistive switching in single crystalline ZnO nanowires. *Nanoscale, 3*, 1917.
149. Raffone, F., Risplendi, F., & Cicero, G. (2016). A new theoretical insight into ZnO NWs memristive behavior. *Nano Letters, 16*, 2543–2547.
150. Porro, S., Risplendi, F., Cicero, G., Bejtka, K., Milano, G., Rivolo, P., Jasmin, A., Chiolerio, A., Pirriab, C. F., & Ricciardi, C. (2017). Multiple resistive switching in core–shell ZnO nanowires exhibiting tunable surface states. *Journal of Materials Chemistry, C5*, 10517–10523.
151. Wang, J., Sun, B., Gao, F., & Greenham, N. C. (2010). Memristive devices based on solution-processed ZnO nanocrystals. *Physica Status Solidi (a), 207*, 484–487.
152. Rosa, J., Kiazadeh, A., Santos, L., Deuermeier, J., Martins, R., Gomes, H. L., & Fortunato, E. (2017). Memristors Using Solution-Based IGZO Nanoparticles. *ACS Omega, 2*, 8366–8372.
153. Qi, J., Olmedo, M., Ren, J., Zhan, N., Zhao, J., Zheng, J.-G., & Liu, J. (2012). Resistive switching in single epitaxial ZnO nanoislands. *ACS Nano, 6*, 1051–1058.

154. Rehman, S., Hur, J.-H., & Kim, D.-k. (2018). Resistive switching in solution-processed copper oxide (CuxO) by stoichiometry tuning. *The Journal of Physical Chemistry C, 122*, 11076–11085.
155. Pawara, P. S., Tikkea, R. S., Patila, V. B., Mullania, N. B., Waifalkarb, P. P., Khotc, K. V., Telib, A. M., Sheikha, A. D., & Dongale, T. D. (2017). A low-cost copper oxide thin film memristive device based on successive ionic layer adsorption and reaction method. *Materials Science in Semiconductor Processing, 71*, 102–108.
156. Khot, A. C., Desai, N. D., Khot, K. V., Salunkhe, M. M., Chougule, M. A., Bhave, T. M., Kamat, R. K., Musselman, K. P., & Dongale, T. D. (2018). Bipolar resistive switching and memristive properties of hydrothermally synthesized TiO2 nanorod array: Effect of growth temperature. *Materials & Design, 151*, 37–47.
157. Duraisamy, N., Muhammad, N. M., Kim, H.-C., Jo, J.-D., & Choi, K.-H. (2012). Fabrication of TiO2 thin film memristor device using electrohydrodynamic inkjet printing. *Thin Solid Films, 520*, 5070–5074.
158. Qingjiang, L., Khiat, A., Salaoru, I., Papavassiliou, C., Hui, X., & Prodromakis, T. (2014). Memory Impedance in TiO_2 based metal-insulator-metal devices. *Scientific Reports, 4*, 4522.
159. Manning, H. G., Biswas, S., Holmes, J. D., & Boland, J. J. (2017). Nonpolar resistive switching in Ag@TiO2 core–shell nanowires. *ACS Applied Materials & Interfaces, 9*, 38959–38966.
160. Qian, K., Cai, G., Nguyen, V. C., Chen, T., & Lee, P. S. (2016). Direct observation of conducting filaments in tungsten oxide based transparent resistive switching memory. *ACS Applied Materials & Interfaces, 8*, 27885–27891.
161. Hubbard, W. A., Kerelsky, A., Jasmin, G., White, E. R., Lodico, J., Mecklenburg, M., & Regan, B. C. (2015). Nanofilament formation and regeneration during Cu/Al_2O_3 resistive memory switching. *Nano Letters, 15*, 3983.
162. Sharma, B., & Rabinal, M. K. (2016). A simple dip coat patterning of aluminum oxide to constitute a bistable memristor. *Materials Research Express, 3*, 126302.
163. Stathopoulos, S., Khiat, A., Trapatseli, M., Cortese, S., Serb, A., Valov, I., & Prodromakis, T. (2017). Multibit memory operation of metal-oxide bi-layer memristors. *Scientific Reports, 7*, 17532.
164. Rahaman, S. Z., Maikap, S., Tien, T.-C., Lee, H.-Y., Chen, W.-S., Chen, F. T., Kao, M.-J., & Tsai, M.-J. (2012). Excellent resistive memory characteristics and switching mechanism using Ti nanolayer at Cu/TaO_X interface. *Nanoscale Research Letters, 7*, 345.
165. Dearnaley, G., Stoneham, A. M., & Morgan, D. V. (1970). Electrical phenomena in amorphous oxide films. *Reports on Progress in Physics, 33*, 1129.
166. Mardare, A. I., Mardare, C. C., Kollender, J. P., Huber, S., & Hassel, A. W. (2018). Basic properties mapping of anodic oxides in the hafnium–niobium–tantalum ternary system. *Science and Technology of Advanced Materials, 19*, 554.
167. Nandakumar, S. R., Minvielle, M., Nagar, S., Dubourdieu, C., & Rajendran, B. (2016). A 250 mV $Cu/SiO_2/W$ memristor with half-integer quantum conductance states. *Nano Letters, 16*, 1602–1608.
168. Yoon, C., Lee, J. H., Lee, S., Jeon, J. H., Jang, J. T., Kim, D. H., Kim, Y. H., & Park, B. H. (2017). Synaptic plasticity selectively activated by polarization-dependent energy-efficient ion migration in an ultrathin ferroelectric tunnel junction. *ACS Nano Letters, 17*, 1949–1955.
169. Hwang, C. S. (2015). Prospective of semiconductor memory devices: From memory system to materials. *Advanced Electronic Materials, 1*(6).
170. Onyia, A. I., Ikeri, H. I., & Nwobodo, A. N. (2008). Theoretical study of the quantum confinement Effects on quantum dots using particle in a box model. *Journal of Physical Chemistry C, 112*, 11290–11294.
171. Zhuge, F., Li, K., Bing, F., Zhang, H., Li, J., Chen, H., Liang, L., Gao, J., Cao, H., Liu, Z., & Luo, H. (2015). Mechanism for resistive switching in chalcogenide-based electrochemical metallization memory cells. *AIP Advances, 5*, 057125.
172. Zhang, H., Zhang, Y., Yu, Y., Song, X., Zhang, H., Cao, M., Che, Y., Dai, H., Yang, J., & Yao, J. (2017). Ambipolar quantum-dot-based low-voltage nonvolatile memory with double floating Gates. *ACS Photonics, 4*(9), 2220–2227.

173. Cheng, P., Sun, K., & Hu, Y. H. (2016). Memristive behavior and ideal memristor of 1T phase MoS2 nanosheets. *ACS Nano Letters, 16*, 572–576.
174. Jariwala, D., Sangwan, V. K., Lauhon, L. J., Marks, T. J., & Hersam, M. C. (2014). Emerging device applications for semiconducting two-dimensional transition metal dichalcogenides. *ACS Nano, 8*(2), 1102–1120.
175. Zhang, P., Xu, B., Gao, C., Chen, G., & Gao, M. (2016). Facile synthesis of Co9Se8 quantum dots as charge traps for flexible organic resistive switching memory device. *ACS Applied Materials & Interfaces, 8*(44), 30336–30343.
176. Lutz, C., Hasegawa, T., & Chikyow, T. (2016). Ag_2S atomic switch-based 'tug of war' for decision making. *Nanoscale, 8*(29), 14031–14036.
177. Gubicza, A., Csontos, M., Halbritterab, A., & Mihályab, G. (2015). Non-exponential resistive switching in Ag_2S memristors: a key to nanometer-scale non-volatile memory devices. *Nanoscale, 7*(10), 4394–4399.
178. Han, S.-T., Zhou, Y., Zhou, L., Yan, Y., Huang, L.-B., Wuab, W., & Roy, V. A. L. (2015). cdse/zns core–shell quantum dots charge trapping layer for flexible photonic memory. *Journal of Materials Chemistry C, 3*(13), 3173–3180.
179. Kilina, S. V., Kilin, D. S., Prezhdo, V. V., & Prezhdo, O. V. (2011). Theoretical study of electron–phonon relaxation in PbSe and CdSe quantum dots: Evidence for phonon memory. *The Journal of Physical Chemistry C, 115*(44), 21641–21651.
180. Yun, D. Y., Song, W. S., Kim, T. W., Kim, S. W., & Kim, S. W. (2012). Electrical stabilities and carrier transport mechanisms of flexible organic bistable devices based on CdSe-InP core-shellnanoparticle/polystyrene nanocomposites. *Applied Physics Letters, 101*(10), 103305.
181. Kim, D. H., Wu, C., Park, D. H., Kim, W. K., Seo, H. W., Kim, S. W., & Kim, T. W. (2018). Flexible memristive devices based on InP/ZnSe/ZnS core-multishell quantum dot nanocomposites. *ACS Applied Materials & Interfaces, 10*(17), 14843–14849.
182. Mohanta, K., Rivas, J., & Pai, R. K. (2012). Reverse switching phenomena in hybrid organic–inorganic thin film composite material. *The Journal of Physical Chemistry C, 117*(1), 124–130.
183. Ghosh, B., Sahu, S., & Pal, A. J. (2008). Core-shell nanoparticles: An approach to enhance electrical bistability. *The Journal of Physical Chemistry C, 112*, 11290–11294.
184. Hui, F., Grustan-Gutierrez, E., Long, S., Liu, Q., Ott, A. K., Ferrari, A. C., & Lanza, M. (2017). Graphene and related materials for resistive random access memories. *Advanced Electronic Materials, 3*, 1600195.
185. Park, M., Park, S., & Yoo, K.-H. (2016). Multilevel nonvolatile memristive and memcapacitive switching in stacked graphene sheets. *ACS Applied Materials & Interfaces, 8*, 14046–14052.
186. Ali, J., Siddiqui, G.-u.-d., Yang, Y. J., Lee, K. T., Umb, K., & Choi, K. H. (2016). Direct synthesis of graphene quantum dots from multilayer graphene flakes through grinding assisted co-solvent ultrasonication for all-printed resistive switching arrays. *RSC Advances, 6*, 5068–5078.
187. Jacob, M. V., Rawat, R. S., Ouyang, B., Bazaka, K., Kumar, D. S., Taguchi, D., Iwamoto, M., Neupane, R., & Varghese, O. K. (2015). Catalyst-free plasma enhanced growth of graphene from sustainable sources. *Nano Letters, 15*, 5702–5708.
188. Kim, H.-J., Jung, S. M., Kim, Y.-H., Kim, B.-J., Ha, S., Kim, Y.-S., Yoon, T.-S., & Lee, H. H. (2011). Characterization of gold nanoparticle pentacene memory device with polymer dielectric layer. *Thin Solid Films, 519*, 6140–6143.
189. Wang, C., He, W., Tong, Y., Zhang, Y., Huang, K., Song, L., Zhong, S., Ganeshkumar, R., & Zhao, R. (2017). Memristive devices with highly repeatable analog states boosted by graphene quantum dots. *Small, 13*, 1603435.
190. Kim, K. L., Lee, W., Hwang, S. K., Joo, S. H., Cho, S. M., Song, G., Cho, S. H., Jeong, B., Hwang, I., Ahn, J.-H., Yu, Y.-J., Shin, T. J., Kwak, S. K., Kang, S. J., & Park, C. (2015). Epitaxial growth of thin ferroelectric polymer films on graphene layer for fully transparent and flexible nonvolatile memory. *Nano Letters, 16*, 334–340.

191. Kim, Y. H., Lee, E. Y., Lee, H. H., & Seo, T. S. (2017). Characteristics of reduced graphene oxide quantum dots for a flexible memory thin film transistor. *ACS Applied Materials & Interfaces, 9*, 16375–16380.
192. Yan, X., Zhang, L., Yang, Y., Zhou, Z., Zhao, J., Zhang, Y., Liud, Q., & Chen, J. (2017). Highly improved performance in Zr0.5Hf0.5O2 films inserted with graphene oxide quantum dots layer for resistive switching non-volatile memory. *Journal of Materials Chemistry C, 5*, 11046.
193. Pósa, L., Abbassi, E. M., Makk, P., Sánta, B., Nef, C., Csontos, M., Calame, M., & Halbritter, A. (2017). Multiple physical time scales and dead time rule in few-nanometers sized graphene–SiOx-graphene memristors. *Nano Letters, 17*, 6783–6789.
194. Qi, M., Liang, B., Xu, H., Wang, Z., Kang, Z., Zhao, X., Liu, W., Jiangang, M., & Liua, Y. (2018). Oxidized carbon quantum dot–graphene oxide nanocomposites for improving data retention of resistive switching memory. *Journal of Materials Chemistry C, 6*, 2026–2033.
195. Fan, F., Zhang, B., Cao, Y., Yang, X., Gu, J., & Chen, Y. (2017). Conjugated polymer covalently modified graphene oxide quantum dots for ternary electronic memory devices. *Nanoscale, 9*, 10610–10618.
196. Che, Y., Zhang, Y., Cao, X., Song, X., Cao, M., Dai, H., Yang, J., Zhangab, G., & Yao, J. (2016). Low operating voltage ambipolar graphene oxide-floating-gate memory devices based on quantum dots. *Journal of Materials Chemistry C, 4*, 1420–1424.
197. Wu, C., Li, F., Zhang, Y., Guo, T., & Chen, T. (2011). Highly reproducible memory effect of organic multilevel resistive-switch device utilizing graphene oxide sheets/polyimide hybrid nanocomposite. *Applied Physics Letters, 99*, 042108.
198. Pan, X., & Skafidas, E. (2016). Resonant tunneling based graphene quantum dot memristors. *Nanoscale, 8*, 20074–20079.
199. Zhuge, F., Hu, B.-L., He, C., Li, R.-W., Zhou, X., & Liu, Z. (2011). Mechanism of nonvolatile resistive switching in graphene oxide thin films. *Carbon, 49*, 3796–3802.
200. Jeong, H. Y., Kim, J. Y., Kim, J. W., Hwang, J. O., Kim, J. E., Lee, J. Y., Yoon, T. H., Cho, B. J., Kim, S. O., Ruoff, R. S., & Choi, S. Y. (2010). Graphene oxide thin films for flexible nonvolatile memory applications. *Nano Letters, 10*, 4381–4386.
201. Hwang, S. K., Lee, J. M., Kim, S., Park, J. S., Park, H. I., Ahn, C. W., Lee, K. J., Lee, T., & Kim, S. O. (2012). Flexible multilevel resistive memory with controlled charge trap B- and N-doped carbon nanotubes. *Nano Letters, 12*, 2217–2221.
202. Li, H., Eddaoudi, M., O'Keeffe, M., & Yaghi, O. M. (1999). Design and synthesis of an exceptionally stable and highly porous metal-organic framework. *Nature, 402*, 276–279.
203. Kreno, L. E., Leong, K., Farha, O. K., Allendorf, M., Van Duyne, R. P., & Hupp, J. T. (2011). Metal–organic framework materials as chemical sensors. *Chemical Reviews, 112*, 1105–1125.
204. Seo, J. S., Whang, D., Lee, H., Jun, S. I., Oh, J., Jeon, Y. J., & Kim, K. (2000). A homochiral metal–organic porous material for enantioselective separation and catalysis. *Nature, 404*, 982–986.
205. Huang, R., Cai, Y., Liu, Y., Bai, W., Kuang, Y., & Wang, Y. (2014). Resistive switching in organic memory devices for flexible applications. *IEEE, 838*.
206. Gao, S., Yi, X., Shang, J., Liu, G., & Li, R.-W. (2019). Organic and hybrid resistive switching materials and devices. *Chemical Society Reviews, 48*, 1531. https://doi.org/10.1039/c8cs00614h.
207. Qia, Y., Suna, B., Fuc, G., Lib, T., Zhua, S., Liang, Z., Maoa, S., Kanb, X., Leib, M., & Chen, Y. (2019). A nonvolatile organic resistive switching memory based on lotus leaves. *Chemical Physics, 516*, 168–174.
208. Biswas, B., Chowdhury, A., Sanyal, M. K., Majumder, M., & Mallik, B. (2013). Electric field induced tunable bistable conductance switching and the memory effect of thiol capped CdS quantum dots embedded in poly(methyl methacrylate) thin films. *Journal of Materials Chemistry C, 1*, 1211–1222.
209. Yoon, S. M., Warren, S. C., & Grzybowski, B. A. (2014). Storage of Electrical Information in Metal-Organic-Framework Memristors. *Angewandte Chemie, 126*, 4526–4530.

210. Verbakel, F., Meskers, S. C. J., de Leeuw, D. M., & Janssen, R. A. J. (2008). Resistive switching in organic memories with a spin-coated metal oxide nanoparticle layer. *The Journal of Physical Chemistry C, 112*, 5254–5257.
211. Lu, H., Chen, Y., Chang, Q., Cheng, S., Ding, Y., Chen, J., Xiu, F., Wang, X., Ban, C., Liu, Z., Liu, J., & Huang, W. (2018). Polymer–carbon dot hybrid structure for a self-rectifying memory device by energy level offset and doping. *RSC Advances, 8*, 13917–13920.
212. Pham, N. K., Vu, N. H., Van Pham, V., Ta, H. K. T., Cao, T. M., Thoaid, N., & Tran, V. C. (2018). Comprehensive resistive switching behavior of hybrid polyvinyl alcohol and TiO2 nanotube nanocomposites identified by combining experimental and density functional theory studies. *Journal of Materials Chemistry C, 6*, 1971–1979.
213. Cheong, S., Kim, Y., Kwon, T., Kim, B. J., & Cho, J. (2013). Inorganic nanoparticle multilayers using photo-crosslinking layer-by-layer assembly and their applications in nonvolatile memory devices. *Nanoscale, 5*, 12356.
214. Meng, L., Lan, M., Liang, G., Xie, L., Wang, H., Ge, J., Liu, W., Wang, Y., & Wang, P. (2015). Nonvolatile memory devices based on carbon nano-dot doped poly(vinyl alcohol) composites with low operation voltage and high ON/OFF ratio. *RSC Advances, 5*, 26886–26890.
215. Jeong, Y. J., Yun, D. J., Noh, S. H., Park, C. E., & Jang, J. (2018). Surface modification of CdSe quantum-dot floating gates for advancing light-erasable organic field-effect transistor memories. *ACS Nano, 12*, 7701–7709.
216. Li, G. L., Liu, G., Li, M., Wan, D., Neoh, K. G., & Kang, E. T. (2010). Organo- and water-dispersible graphene oxide−polymer nanosheets for organic electronic memory and gold nanocomposites. *The Journal of Physical Chemistry C, 114*, 12742–12748.
217. Alibart, F., Pleutin, S., Bichler, O., Gamrat, C., Serrano-Gotarredona, T., Linares-Barranco, B., & Vuillaume, D. (2011). A memristive nanoparticle/organic hybrid synapstor for neuroinspired computing. *Advanced Functional Materials, 22*, 609–616.
218. Shao, Y., Fang, Y., Li, T., Wang, Q., Dong, Q., Deng, Y., Yuan, Y., Wei, H., Wang, M., Gruverman, A., Shielda, J., & Huang, J. (2016). Grain boundary dominated ion migration in polycrystalline organic–inorganic halide perovskite films. *Energy & Environmental Science, 9*, 1752–1759.
219. Siddiqui, G. U., Rehman, M. M., Yang, Y.-J., & Choi, K. H. (2017). A two-dimensional hexagonal boron nitride/polymer nanocomposite for flexible resistive switching devices. *Journal of Materials Chemistry C, 5*, 862–871.
220. Yan, K., Chen, B., Hu, H., Chen, S., Xue, G., Xiao, X., Zhou, J., & Zou, D. (2016). First fiber-shaped non-volatile memory device based on hybrid organic-inorganic perovskite. *Advanced Electronic Materials, 2*, 1600160.
221. Hota, M. K., Bera, M. K., Kundu, B., Kundu, S. C., & Maiti, C. K. (2012). A natural silk fibroin protein-based transparent bio-memristor. *Advanced Functional Materials, 22*(21), 4493–4499.
222. Lv, Z., Zhou, Y., Han, S.-T., & Roy, V. A. L. (2018). From biomaterial-based data storage to bio-inspired artificial synapse. *Materials today, 21*, 537–552.
223. Murgunde, B. K., Rabinal, M. K., & Kalasad, M. N. (2018). Biologically active nanocomposite of DNA-PbS nanoparticles: A new material for non-volatile memory devices. *Applied Surface Science, 427*, 344–353.
224. Murgunde, B. K., & Rabinal, M. K. (2017). Solution processed bilayer junction of silk fibroin and semiconductor quantum dots as multilevel memristor devices. *Organic Electronics, 48*, 276–284.
225. Chu, H.-L., Chiu, S.-C., Sung, C.-F., Tseng, W., Chang, Y.-C., Jian, W.-B., Chen, Y.-C., Yuan, C.-J., Li, H.-Y., Gu, F. X., Di Ventra, M., & Chang, C.-C. (2014). Programmable redox state of the nickel ion chain in DNA. *Nano Letters, 14*, 1026–1031.
226. Lee, T., Yagati, A. K., Pi, F., Sharma, A., Choi, J.-W., & Guo, P. (2015). Construction of RNA–quantum dot chimera for nanoscale resistive biomemory application. *ACS Nano, 9*, 6675–6682.

227. Xing, Y., Shi, C., Zhao, J., Qiu, W., Lin, N., Wang, J., Yan, X. B., Yu, W. D., & Liu, X. Y. (2017). Mesoscopic-functionalization of silk fibroin with gold nanoclusters mediated by keratin and bioinspired silk synapse. *Small, 13*, 1702390.
228. Portney, N. G., Martinez-Morales, A. A., & Ozkan, M. (2008). Nanoscale memory characterization of virus-templated semiconducting quantum dots. *ACS Nano, 2*, 191–196.
229. Lim, Z. X., & Cheong, K. Y. (2015). Effects of drying temperature and ethanol concentration on bipolar switching characteristics of natural Aloe vera-based memory devices. *Physical Chemistry Chemical Physics, 17*, 26833–26853.

Index

P. Yu, Z. M. Wang (eds.), *Quantum Dot Optoelectronic Devices*,
Lecture Notes in Nanoscale Science and Technology 27,
https://doi.org/10.1007/978-3-030-35813-6

V

W

X

The manufacturer's authorised representative in the EU is Springer Nature Customer Service Centre GmbH, Europaplatz 3, 69115 Heidelberg, Germany. If you have any concerns regarding our products, please contact ProductSafety@springernature.com

Printed and bound by CPI Group (UK) Ltd, Croydon, CR0 4YY
05/10/2026
02228120-0013